Cornerstones

More information about this series at http://www.springer.com/series/7161

John P. D'Angelo

Hermitian Analysis

From Fourier Series to Cauchy-Riemann
Geometry

Second Edition

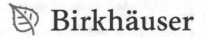 Birkhäuser

John P. D'Angelo
Department of Mathematics
University of Illinois, Urbana-Champaign
Urbana, IL, USA

ISSN 2197-182X   ISSN 2197-1838   (electronic)
Cornerstones
ISBN 978-3-030-16516-1   ISBN 978-3-030-16514-7   (eBook)
https://doi.org/10.1007/978-3-030-16514-7

Library of Congress Control Number: 2019935545

Mathematics Subject Classification (2010): 46C05, 42A16, 42A38, 51F20

1st edition: © Springer Science+Business Media New York 2013
2nd edition: © Springer Nature Switzerland AG 2019

This book is published under the imprint Birkhäuser, www.birkhauser-science.com by the registered company Springer Nature Switzerland AG
The registered company address is: Gewerbestrasse 11, 6330 Cham, Switzerland

Preface to the second edition

The second edition of the book differs from the first edition in several ways. Most important is the inclusion of a new chapter. This new chapter considers the CR geometry of the unit sphere. It is primarily based on recent joint work of the author and Ming Xiao, and the author thanks Ming for his contribution to this research. This work concerns groups associated with holomorphic mappings; some material along these lines appeared in Chapter 4 of the first edition, and that material remains in Chapter 4, little changed. Other material from Chapter 4 has been rewritten, and some of it has been moved to the new Chapter 5. This chapter develops some ideas from CR geometry and, by including the work on groups associated with sphere maps, heads in a new direction. As usual in mathematics, new results open more doors than they close. For example, we use these groups to provide a criterion for a holomorphic mapping to be a polynomial. Chapter 5 concludes with a test when a formal power series in several variables is a rational function.

The author has added a section at the end of Chapter 4 called *Unifying remarks*. This section makes a modest effort toward showing how some of the topics in the book fit together. The author wishes to thank an anonymous reviewer for pointing out that some of the exercises, scattered throughout the book, are in fact intimately connected. The reviewer also suggested that making these connections would enable more to be said about the proof of a higher-dimensional analogue of the Riesz-Fejer theorem used in Chapter 5. Following these suggestions has improved the book. Several small improvements in Chapers 2 and 3 also help unify topics.

A third difference is that typos and other small errors have been corrected. Of these, the author wishes to note one error that has been fixed. Chapter 4 included two proofs of Wirtinger's inequality; the second proof was sketched and contained a subtle mistake. We correct that error in this edition. In several places, we have improved the way formulas look on the page and we rewrote a few sentences for clarity. We have also included some material from basic differential geometry, such as a discussion of pushforwards and pullbacks. The material on differential forms remains intuitive rather than formal, but perhaps has been improved. Several references have been added to the bibliography. Many new exercises appear.

A few words about prerequisites are appropriate. The first three chapters assume that the reader has taken three semesters of calculus, is comfortable with linear algebra, is acquainted with complex numbers and ODE, and knows some basic real analysis. Chapters 4 and 5 require more. The author hopes that the first three chapters, some elementary complex analysis, and some elementary abstract algebra (what a group is) will provide the necessary background. Much of the material in the last two chapters is oriented toward current research, but understanding the results and their proofs does not require much jargon. Perhaps, the lack of jargon explains why this material appeals so much to the author. The question "what are the rational mappings sending the unit sphere in $\mathbf{C}^n$ to the unit sphere in $\mathbf{C}^N$ for some N" is easy to state. It is possible to write an entire book answering this question. Chapters 4 and 5 begin to do so. The author therefore hopes researchers in several complex variables and CR geometry find these chapters to be useful.

The author modestly hopes that the changes made justify the second edition. He wishes to add Ming Xiao, Dusty Grundmeier, Fritz Haslinger, Martino Fassina, Xiaolong Han, Bernhard Lamel, and several anonymous reviewers to the list of people acknowledged in the original preface. The author spent a wonderful week at the Erwin Schrödinger Institute (ESI) in December 2018. Although this edition was nearly completed by then, both preparing a talk and having discussions with other participants helped the author polish some aspects of the book. He therefore wishes to acknowledge support from ESI for that week. In addition, he wishes to acknowledge support from the American Institute of Mathematics (AIM) for several earlier meetings. Finally, he acknowledges support from NSF Grant DMS 13-61001.

Preface to the first edition

This book aims both to synthesize much of undergraduate mathematics and to introduce research topics in geometric aspects of complex analysis in several variables. The topics all relate to orthogonality, real analysis, elementary complex variables, and linear algebra. I call the blend *Hermitian analysis*. The book developed from my teaching experiences over the years and specifically from Math 428, a capstone Honors course taught in Spring 2013 at the University of Illinois. Many of the students in Math 428 had taken honors courses in analysis, linear algebra, and beginning abstract algebra. They knew differential forms and Stokes' theorem. Other students were strong in engineering, with less rigorous mathematical training, but with a strong sense of how these ideas get used in applications.

Rather than repeating and reorganizing various parts of mathematics, the course began with Fourier series, a new topic for many of the students. Developing some of this remarkable subject and related parts of analysis allows the synthesis of calculus, elementary real and complex analysis, and algebra. Proper mappings, unitary groups, complex vector fields, and differential forms eventually join this motley crew. Orthogonality and Hermitian analysis unify these topics. In the process, ideas arising on the unit circle in $\mathbf{C}$ evolve into more subtle ideas on the unit sphere in complex Euclidean space $\mathbf{C}^n$.

The book includes numerous examples and more than two-hundred seventy exercises. These exercises sometimes appear, with a purpose, in the middle of a section. The reader should stop reading and start computing. Theorems, lemmas, propositions, etc. are numbered by chapter. Thus Lemma 2.4 means the fourth lemma in Chapter 2. The only slight exception concerns figures. Figure 8 means the eighth figure in whatever chapter it appears.

Chapter 1 begins by considering the conditionally convergent series $\sum_{n=1}^{\infty} \frac{\sin(nx)}{n}$. We verify its convergence using summation by parts, which we discuss in some detail. We then review constant coefficient ordinary differential equations, the exponentiation of matrices, and the wave equation for a vibrating string. These topics motivate our development of Fourier series. We prove the Riesz-Fejer theorem characterizing non-negative trigonometric polynomials. We develop topics such as approximate identities and summability methods, enabling us to conclude the discussion on the series $\sum_{n=1}^{\infty} \frac{\sin(nx)}{n}$. The chapter closes with two proofs of Hilbert's inequality.

Chapter 2 discusses the basics of Hilbert space theory, motivated by orthonormal expansions, and includes the spectral theorem for compact Hermitian operators. We return to Fourier series after these Hilbert space techniques have become available. We also consider Sturm-Liouville theory in order to provide additional examples of orthonormal systems. The exercises include problems on Legendre polynomials, Hermite polynomials, and several other collections of special functions. The chapter ends with a section on spherical harmonics, whose purpose is to indicate one possible direction for Fourier analysis in higher dimensions. As a whole, this chapter links classical and modern analysis. It considerably expands the material on Hilbert spaces from my Carus monograph *Inequalities From Complex Analysis*. Various items here help the reader to think in a magical Hermitian way. Here are two specific examples:

- There exist linear transformations A, B on a real vector space satisfying the relationship $A^{-1} + B^{-1} = (A + B)^{-1}$ if and only if the vector space admits a complex structure.
- It is well known that a linear map on a complex space preserves inner products if and only it preserves norms. This fact epitomizes the polarization technique which regards a complex variable or vector z and its conjugate $\bar{z}$ as independent objects.

Chapter 3 considers the Fourier transform on the real line, partly to glimpse higher mountains and partly to give a precise meaning to distributions. We also briefly discuss Sobolev spaces and pseudo-differential operators. This chapter includes several standard inequalities (Young, Hölder, Minkowski) from real analysis and Heisenberg's inequality from physics. Extending these ideas to higher dimensions would be natural, but since many books treat this material well, we head in a different direction. This chapter is therefore shorter than the other chapters and it contains fewer interruptions.

Chapter 4, the heart of the book, considers geometric issues revolving around the unit sphere in complex Euclidean space. We begin with Hurwitz's proof (using Fourier series) of the isoperimetric inequality for smooth curves. We prove Wirtinger's inequality in two ways. We continue with an inequality on the areas of complex analytic images of the unit disk, which we also prove in two ways. One of these involves differential forms. This chapter therefore includes several sections on vector fields and differential forms, including the complex case. Other geometric considerations in higher dimensions include topics from my own research: finite unitary groups, group-invariant mappings, and proper mappings between balls. We use the notion of *orthogonal homogenization* to prove a sharp inequality on the volume of the images of the unit ball under certain polynomial mappings. This material naturally leads to the Cauchy-Riemann (CR) Geometry of the unit sphere. The chapter closes with a brief discussion of positivity conditions for Hermitian polynomials, connecting the work on proper mappings to an analogue of the Riesz-Fejer theorem in higher dimensions. Considerations of orthogonality and Hermitian geometry weave all these topics into a coherent whole.

The prerequisites for reading the book include three semesters of calculus, linear algebra, and basic real analysis. The reader needs some acquaintance with complex numbers but does not require all of the material in the standard course. The appendix summarizes the prerequisites. We occasionally employ the notation of Lebesgue integration, but knowing measure theory is not a prerequisite for reading this book. The large number of exercises, many developed specifically for this book, should be regarded as crucial. They link the abstract and the concrete.

Books in the Cornerstone series are aimed at aspiring young mathematicians ranging from advanced undergraduates to second-year graduate students. This audience will find the first three chapters accessible. The many examples, exercises, and motivational discussions make these chapters also accessible to students in Physics and Engineering. While Chapter 4 is more difficult, the mathematics there flows naturally from the earlier material. These topics require the synthesis of diverse parts of mathematics. The unity and beauty of the mathematics rewards the reader while leading directly to current research. The author hopes someday to write a definitive account describing where in complex analysis and CR Geometry these ideas lead.

I thank the Department of Mathematics at Illinois for allowing me to teach various Honors courses and in particular the one for which I used these notes. I acknowledge various people for their insights into some of the mathematics here, provided in conversations, published writing, or e-mail correspondences. Such people include Phil Baldwin, Steve Bradlow, David Catlin, Geir Dullerud, Ed Dunne, Charlie Epstein, Burak Erdogan, Jerry Folland, Jen Halfpap, Zhenghui Huo, Robert Kaufman, Rick Laugesen, Jeff McNeal, Tom Nevins, Mike Stone, Emil Straube, Jeremy Tyson, Bob Vanderbei, and others, including several unnamed reviewers.

I also thank Charlie Epstein and Steve Krantz for encouraging me to write a book for the Birkhäuser Cornerstone Series. I much appreciate the efforts of Kate Ghezzi, Associate Editor of Birkhäuser Science, who guided the evolution of my first draft into this book. I thank Carol Baxter of the Mathematical Association of America for granting me permission to incorporate some of the material from Chapter 2 of my Carus monograph *Inequalities From Complex Analysis*. I acknowledge Jimmy Shan for helping prepare pictures and solving many of the exercises.

I thank my wife Annette and our four children for their love.

I acknowledge support from NSF grant DMS-1066177 and from the Kenneth D. Schmidt Professorial Scholar award from the University of Illinois.

Contents

CHAPTER 1

Introduction to Fourier series

1. Introduction

We start the book by considering the series $\sum_{n=1}^{\infty} \frac{\sin(nx)}{n}$, a nice example of a Fourier series. This series converges for all real numbers x, but the issue of convergence is delicate. We introduce summation by parts as a tool for handling some conditionally convergent series of this sort. After verifying convergence, but before finding the limit, we pause to introduce and discuss several elementary differential equations. This material also leads to Fourier series. We include the exponentiation of matrices here. The reader will observe these diverse topics begin being woven into a coherent whole.

After these motivational matters, we introduce the fundamental issues concerning the Fourier series of Riemann integrable functions. We define trigonometric polynomials, Fourier series, approximate identities, Cesàro and Abel summability, and related topics enabling us to get some understanding of the convergence of Fourier series. We show how to use Fourier series to establish some interesting inequalities.

In Chapter 2 we develop the theory of Hilbert spaces, which greatly clarifies the subject of Fourier series. We prove additional results about Fourier series there, after we know enough about Hilbert spaces. The manner in which concrete and abstract mathematics inform each other is truly inspiring.

2. A famous series

Consider the infinite series $\sum_{n=1}^{\infty} \frac{\sin(nx)}{n}$. This sum provides an example of a Fourier series, a term we will define precisely a bit later. Our first agenda item is to show that this series converges for all real x. After developing a bit more theory, we determine the sum for each x; the result defines the famous *sawtooth function*.

Let $\{a_n\}$ be a sequence of (real or) complex numbers. We say that $\sum_{n=1}^{\infty} a_n$ *converges* to L if

$$\lim_{N \to \infty} \sum_{n=1}^{N} a_n = L.$$

We say that $\sum_{n=1}^{\infty} a_n$ *converges absolutely* if $\sum_{n=1}^{\infty} |a_n|$ converges. In case $\sum a_n$ converges, but does not converge absolutely, we say that $\sum a_n$ *converges conditionally* or is *conditionally convergent*. Note that *absolute convergence* implies *convergence*, but that the converse fails. See for example Corollary 1.3. See the exercises for subtleties arising when considering conditionally convergent series.

The expression $A_N = \sum_{n=1}^{N} a_n$ is called the N-th partial sum. In this section we will consider two sequences $\{a_n\}$ and $\{b_n\}$. We write their partial sums, using capital letters, as A_N and B_N. We regard the sequence of partial sums as an

© Springer Nature Switzerland AG 2019
J. P. D'Angelo, *Hermitian Analysis*, Cornerstones,
https://doi.org/10.1007/978-3-030-16514-7_1

analogue of the integral of the sequence of terms. Note that we can recover the terms from the partial sums because $a_n = A_n - A_{n-1}$, and we regard the sequence $\{a_n\}$ of terms as an analogue of the derivative of the sequence of partial sums. The next result is extremely useful in analyzing conditionally convergent series. One can remember it by analogy with the integration by parts formula

$$\int aB' = aB - \int a'B. \tag{1}$$

PROPOSITION 1.1 (Summation by parts). *For $1 \leq j \leq N$ consider complex numbers a_j and b_j. Then*

$$\sum_{j=1}^{N} a_j b_j = a_N B_N - \sum_{j=1}^{N-1} (a_{j+1} - a_j) B_j. \tag{2}$$

PROOF. We prove the formula by induction on N. When $N = 1$ the result is clear, because $a_1 b_1 = a_1 B_1$ and the sum on the right-hand side of (2) is empty.

Assume the result for some N. Then we have

$$\sum_{j=1}^{N+1} a_j b_j = a_{N+1} b_{N+1} + \sum_{j=1}^{N} a_j b_j = a_{N+1} b_{N+1} + a_N B_N - \sum_{j=1}^{N-1} (a_{j+1} - a_j) B_j$$

$$= a_{N+1} b_{N+1} + a_N B_N - \sum_{j=1}^{N} (a_{j+1} - a_j) B_j + (a_{N+1} - a_N) B_N$$

$$= a_{N+1} b_{N+1} + a_{N+1} B_N - \sum_{j=1}^{N} (a_{j+1} - a_j) B_j = a_{N+1} B_{N+1} - \sum_{j=1}^{N} (a_{j+1} - a_j) B_j.$$

The induction is complete. □

COROLLARY 1.1. *Suppose $a_n \to 0$ and that $\sum |a_{n+1} - a_n|$ converges. Assume also that the sequence $\{B_N\}$ of partial sums is bounded. Then $\sum_{j=1}^{\infty} a_j b_j$ converges.*

PROOF. We must show that the limit, as N tends to infinity, of the left-hand side of (2) exists. The limit of the first term on the right-hand side of (2) exists and is 0. The limit of the right-hand side of (2) is the infinite series

$$-\sum_{j=1}^{\infty} (a_{j+1} - a_j) B_j. \tag{3}$$

We claim that the series (3) is absolutely convergent. By hypothesis, there is a constant C with $|B_j| \leq C$ for all j. Hence, for each j we have

$$|(a_{j+1} - a_j) B_j| \leq C |a_{j+1} - a_j|. \tag{4}$$

The series $\sum |a_{j+1} - a_j|$ converges. By (4) and the comparison test (Proposition 6.1 from the Appendix), $\sum_{j=1}^{\infty} |(a_{j+1} - a_j) B_j|$ converges as well. Thus the claim holds and the conclusion follows by letting N tend to ∞ in (2). □

COROLLARY 1.2. *Suppose a_n decreases monotonically to 0. Then $\sum (-1)^n a_n$ converges.*

PROOF. Put $b_n = (-1)^n$. Then $|B_N| \leq 1$ for all N. Since $a_{n+1} \leq a_n$ for all n,

$$\sum_{j=1}^{N} |a_{j+1} - a_j| = \sum_{j=1}^{N} (a_j - a_{j+1}) = a_1 - a_{N+1}.$$

Since a_{N+1} tends to 0, we have a convergent telescoping series. Thus Corollary 1.1 applies. □

COROLLARY 1.3. $\sum_{n=1}^{\infty} \frac{(-1)^n}{n}$ converges.

PROOF. Put $a_n = \frac{1}{n}$ and $b_n = (-1)^n$. Corollary 1.2 applies. □

PROPOSITION 1.2. $\sum_{n=1}^{\infty} \frac{\sin(nx)}{n}$ converges for all real x.

PROOF. Let $a_n = \frac{1}{n}$ and let $b_n = \sin(nx)$. First suppose x is an integer multiple of π. Then $b_n = 0$ for all n and the series converges to 0. Otherwise, suppose x is not a multiple of π; hence $e^{ix} \neq 1$. We then claim that B_N is bounded. In the next section, for complex z we define $\sin(z)$ by $\frac{e^{iz} - e^{-iz}}{2i}$ and we justify this definition. Using it we have

$$b_n = \sin(nx) = \frac{e^{inx} - e^{-inx}}{2i}.$$

Since we are assuming $e^{ix} \neq 1$, the sum $\sum_{n=1}^{N} e^{inx}$ is a finite geometric series which we can compute explicitly. We get

$$\sum_{n=1}^{N} e^{inx} = e^{ix} \frac{1 - e^{iNx}}{1 - e^{ix}}. \tag{5}$$

The right-hand side of (5) has absolute value at most $\frac{2}{|1-e^{ix}|}$ and hence the left-hand side of (5) is bounded independently of N. The same holds for $\sum_{n=1}^{N} e^{-inx}$. Thus B_N is bounded. The proposition now follows by Corollary 1.1. □

REMARK 1.1. The partial sums B_N depend on x. We will see later why the limit function fails to be continuous in x.

REMARK 1.2. The definition of convergence of a series involves the partial sums. Other summability methods will arise soon. For now we note that conditionally convergent series are quite subtle. In Exercise 1.1 you are asked to verify Riemann's remark that the sum of a conditionally convergent series depends in a striking way on the order in which the terms are added. Such a reordered series is called a *rearrangement* of the given series.

EXERCISE 1.1. (Subtle) (Riemann's remark on rearrangement). Let $\sum a_n$ be a conditionally convergent series of real numbers. Given any real number L (or ∞), prove that there is a rearrangement of $\sum a_n$ that converges to L (or diverges). (Harder) Determine and prove a corresponding statement if the a_n are allowed to be complex. (Hint: For some choices of complex numbers a_n, not all complex L are possible as limits of rearranged sums. If, for example, all the a_n are purely imaginary, then the rearranged sum must be purely imaginary. Figure out all possible alternatives.)

EXERCISE 1.2. Show that $\sum_{n=2}^{\infty} \frac{\sin(nx)}{\log(n)}$ and, (for $\alpha > 0$), $\sum_{n=1}^{\infty} \frac{\sin(nx)}{n^{\alpha}}$ converge.

EXERCISE 1.3. Suppose that $\sum c_j$ converges and that $\lim_n nc_n = 0$. Determine

$$\sum_{n=1}^{\infty} n(c_{n+1} - c_n).$$

EXERCISE 1.4. Find a sequence of complex numbers such that $\sum a_n$ converges but $\sum (a_n)^3$ diverges.

EXERCISE 1.5. This exercise has two parts.

(1) Assume that Cauchy sequences (see Section 1 of the Appendix) of real numbers converge. Prove the following statement: if $\{a_n\}$ is a sequence of complex numbers and $\sum_{n=1}^{\infty} |a_n|$ converges, then $\sum_{n=1}^{\infty} a_n$ converges.
(2) Next, do not assume that Cauchy sequences of real numbers converge; instead assume that whenever $\sum_{n=1}^{\infty} |a_n|$ converges, then $\sum_{n=1}^{\infty} a_n$ converges. Prove that Cauchy sequences of real (or complex) numbers converge.

EXERCISE 1.6. (Difficult) For $0 < x < 2\pi$, show that $\sum_{n=0}^{\infty} \frac{\cos(nx)}{\log(n+2)}$ converges to a non-negative function. Suggestion: Sum by parts twice and find an explicit formula for $\sum_{n=1}^{N} \sum_{k=1}^{n} \cos(kx)$. If needed, look ahead to formula (49).

EXERCISE 1.7. Consider the sequence defined by $a_N = (\sum_{j=1}^{N} \frac{1}{j}) - \log N$. Show that this sequence is decreasing and bounded below, and therefore converges. The limit is called Euler's constant and equals approximately .5772.

EXERCISE 1.8. Verify that $\sum_{n=1}^{\infty} \frac{(-1)^{n+1}}{n}$ converges to $\log(2)$.

EXERCISE 1.9. a. Consider the series $(1 + \frac{1}{3} - \frac{1}{2}) + (\frac{1}{5} + \frac{1}{7} - \frac{1}{4}) + \ldots$ Show that this series, a rearrangement of the series in the previous exercise, converges to $\frac{3}{2} \log(2) = \log(\sqrt{8})$. Suggestion: Write the partial sums in terms of the partial sums of the harmonic series $\sum \frac{1}{n}$ and use Exercise 1.7.

b. Use the method in part a. to show that

$$\sum_{k=1}^{\infty} \left(\frac{1}{6k-5} + \frac{1}{6k-3} + \frac{1}{6k-1} - \frac{1}{2k} \right) = \log(6) - \frac{1}{2} \log(3) = \log(\sqrt{12}).$$

3. Trigonometric polynomials

We let S^1 denote the unit circle in $\mathbf{C}$. Our study of Fourier series involves functions defined on the unit circle, although we sometimes work with functions defined on $\mathbf{R}$, on the interval $[-\pi, \pi]$, or on the interval $[0, 2\pi]$. In order that such functions be defined on the unit circle, we must assume that they are periodic with period 2π, that is, $f(x+2\pi) = f(x)$. The most obvious such functions are sine and cosine. We will often work instead with complex exponentials.

We therefore begin by defining, for $z \in \mathbf{C}$,

$$e^z = \sum_{n=0}^{\infty} \frac{z^n}{n!} = \lim_{N \to \infty} \sum_{n=0}^{N} \frac{z^n}{n!}. \tag{6}$$

Using the ratio test, we see that the series in (6) converges absolutely for each $z \in \mathbf{C}$. It converges uniformly on each closed ball in $\mathbf{C}$. Hence the series defines a *complex analytic* function of z whose derivative (which is also e^z) is found by differentiating term-by-term. See the appendix for the definition of *holomorphic* or complex analytic function.

Note that $e^0 = 1$. Also it follows from (6) that, for all complex numbers z and w, $e^{z+w} = e^z e^w$. (See Exercise 1.17.) From these facts we can also see for $\lambda \in \mathbf{C}$ that $\frac{d}{dz}e^{\lambda z} = \lambda e^{\lambda z}$. Since complex conjugation is continuous, we also have $\overline{e^z} = e^{\overline{z}}$ for all z. (Continuity is used in passing from the partial sum in (6) to the infinite series.) Hence when t is real we see that e^{-it} is the conjugate of e^{it}. Therefore

$$|e^{it}|^2 = e^{it}e^{-it} = e^0 = 1$$

and hence e^{it} lies on the unit circle. All trigonometric identities follow from the definition of the exponential function. The link to trigonometry (trig) comes from the definitions of sine and cosine for a complex variable z:

$$\cos(z) = \frac{e^{iz} + e^{-iz}}{2} \tag{7}$$

$$\sin(z) = \frac{e^{iz} - e^{-iz}}{2i}. \tag{8}$$

We obtain $e^{iz} = \cos(z) + i\sin(z)$ for all z. In particular, when t is real, $\cos(t)$ is the real part of e^{it} and $\sin(t)$ is the imaginary part of e^{it}. Consider the bijection from $[0, 2\pi)$ to the unit circle given by $t \mapsto e^{it}$. We can *define* the radian measure of an angle as follows. Given the point e^{it} on the unit circle, we form two line segments: the segment from 0 to 1, and the segment from 0 to e^{it}. Then t is the angle between these segments, measured in radians. Hence $\cos(t)$ and $\sin(t)$ have their usual meanings when t is real.

Although we started the book with the series $\sum \frac{\sin(nx)}{n}$, we prefer using complex exponentials instead of cosines and sines to express our ideas and formulas.

DEFINITION 1.1. A complex-valued function on the circle is called a *trigonometric polynomial* or *trig polynomial* if there are complex constants c_j such that

$$f(\theta) = \sum_{j=-N}^{N} c_j e^{ij\theta}.$$

It is of degree N if $c_N \neq 0$ or $c_{-N} \neq 0$. The complex numbers c_j are called the (Fourier) coefficients of f.

LEMMA 1.1. *A trig polynomial f is the zero function if and only if all its coefficients vanish.*

PROOF. If all the coefficients vanish, then f is the zero function. The converse is less trivial. We can recover the coefficients c_j of a trig polynomial by integration:

$$c_j = \frac{1}{2\pi} \int_0^{2\pi} f(\theta)e^{-ij\theta}d\theta. \tag{9}$$

If $f(\theta) = 0$ for all θ, then each of these integrals vanishes and the converse assertion follows. $\square$

See Theorem 1.12 for an important generalization. Lemma 1.1 has a geometric interpretation, which we will develop and generalize in Chapter 2. The functions $x \to e^{inx}$ for $-N \leq n \leq N$ form an orthonormal basis for the $(2N+1)$-dimensional vector space of trig polynomials of degree at most N. The lemma states that f is the zero vector if and only if all its components with respect to this basis are 0.

We need the following result about real-valued trig polynomials. See Lemma 1.7 for a generalization.

LEMMA 1.2. *A trig polynomial is real-valued if and only if $c_j = \overline{c_{-j}}$ for all j.*

PROOF. The trig polynomial is real valued if and only if $f = \overline{f}$, which becomes

$$\sum_{j=-N}^{N} c_j e^{ij\theta} = \sum_{j=-N}^{N} \overline{c_j} e^{-ij\theta}. \tag{10}$$

Replacing j by $-j$ in the second sum in (10) shows that f is real-valued if and only if

$$\sum_{j=-N}^{N} c_j e^{ij\theta} = \sum_{j=-N}^{N} \overline{c_j} e^{-ij\theta} = \sum_{j=N}^{-N} \overline{c_{-j}} e^{ij\theta} = \sum_{j=-N}^{N} \overline{c_{-j}} e^{ij\theta}. \tag{11}$$

The difference of the two far sides of (11) is the zero function; hence the conclusion follows from Lemma 1.1.   □

We sometimes call this condition on the coefficients the *palindrome property*; it characterizes *real-valued* trig polynomials. Our next result, which is considerably more difficult, characterizes *non-negative* trig polynomials.

THEOREM 1.1 (Riesz-Fejer 1916). *Let f be a trig polynomial with $f(\theta) \geq 0$ for all 0. Then there is a complex polynomial $p(z)$ such that $f(\theta) = |p(e^{i\theta})|^2$.*

PROOF. Assume f is of degree d and write

$$f(\theta) = \sum_{j=-d}^{d} c_j e^{ij\theta},$$

where $c_{-j} = \overline{c_j}$ since f is real-valued. Note also that $c_{-d} \neq 0$. Define a polynomial q in one complex variable by

$$q(z) = z^d \sum_{j=-d}^{d} c_j z^j. \tag{12}$$

Let $\xi_1, ..., \xi_{2d}$ be the roots of q, repeated if necessary. We claim that the reality of f, or equivalently the palindrome property, implies that if ξ is a root of q, then $(\overline{\xi})^{-1}$ also is a root of q. This point is called the *reflection* of ξ in the circle. See Figure 1. Because 0 is not a root, the claim follows from the formula

$$q(z) = z^{2d} \, \overline{q((\overline{z})^{-1})}. \tag{13}$$

To check (13), we use (12) and the palindrome property. Also, we replace $-j$ by j in the sum to get

$$z^{2d} \, \overline{q((\overline{z})^{-1})} = z^d \, \overline{\sum c_j (\frac{1}{\overline{z}})^j} = z^d \sum \overline{c_j} z^{-j} = z^d \sum \overline{c_{-j}} z^j = z^d \sum c_j z^j = q(z).$$

Thus (13) holds. It also follows that each root on the circle must occur with even multiplicity. Thus the roots of q occur in pairs, symmetric with respect to the unit circle.

By the Fundamental Theorem of Algebra, we may factor the polynomial q into linear factors. For z on the circle we can replace the factor $z - (\bar{\xi})^{-1}$ with

$$\frac{1}{\bar{z}} - \frac{1}{\bar{\xi}} = \frac{\bar{\xi} - \bar{z}}{\bar{z}\bar{\xi}}.$$

Let $p(z) = C \prod_{j=1}^{d}(z - \xi_j)$. Here we use all the roots in the unit disk and half of those where $|\xi_j| = 1$. Note that $|q| = |p|^2$ on the circle. Since $f \geq 0$ on the circle, we obtain

$$f(\theta) = |f(\theta)| = |q(e^{i\theta})| = |p(e^{i\theta})|^2.$$

$\square$

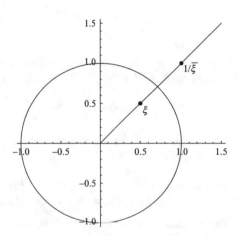

FIGURE 1. Reflection in the circle

EXERCISE 1.10. Put $f(\theta) = 1 + a\cos(\theta)$. Note that $f \geq 0$ if and only if $|a| \leq 1$. In this case find p such that $|p(e^{i\theta})|^2 = f(\theta)$ on the circle.

EXERCISE 1.11. Put $f(\theta) = 1 + a\cos(\theta) + b\cos(2\theta)$. Find the condition on a, b for $f \geq 0$. Carefully graph the set of (a, b) for which $f \geq 0$. Find p such that $f = |p|^2$ on the circle. Suggestion: To determine the condition on a, b, rewrite f as a polynomial in x on the interval $[-1, 1]$.

EXERCISE 1.12. Find a polynomial p such that $|p(e^{i\theta})|^2 = 4 - 4\sin^2(\theta)$. (The roots of p lie on the unit circle, illustrating part of the proof of Theorem 1.1.)

In anticipation of later work, we introduce Hermitian symmetry and rephrase the Riesz-Fejer theorem in this language.

DEFINITION 1.2. Let $R(z, \overline{w})$ be a polynomial in the two complex variables z and $\overline{w}$. R is called *Hermitian symmetric* if $R(z, \overline{w}) = \overline{R(w, \overline{z})}$ for all z and w.

The next lemma characterizes Hermitian symmetric polynomials.

LEMMA 1.3. *The following statements about a polynomial in two complex variables are equivalent:*

(1) *R is Hermitian symmetric.*

(2) *For all z, $R(z, \overline{z})$ is real.*

(3) $R(z, \overline{w}) = \sum_{a,b} c_{ab} z^a \overline{w}^b$ *where* $c_{ab} = \overline{c_{ba}}$ *for all* a, b.

PROOF. Left to the reader. □

The next result, together with various generalizations, justifies considering z and its conjugate $\overline{z}$ to be independent variables. When a polynomial identity in z and $\overline{z}$ holds for all z, it holds when we vary z and $\overline{z}$ separately.

LEMMA 1.4 (Polarization). *Let R be a Hermitian symmetric polynomial. If $R(z, \overline{z}) = 0$ for all z, then $R(z, \overline{w}) = 0$ for all z and w.*

PROOF. Write $z = |z|e^{i\theta}$. Plugging into the third item from Lemma 1.3, we are given

$$\sum_{a,b} c_{ab} |z|^{a+b} e^{i(a-b)\theta} = 0$$

for all $|z|$ and for all θ. Put $k = a - b$, which can be positive, negative, or 0. By Lemma 1.1, the coefficient of each $e^{ik\theta}$ is 0. Thus, for all k and z,

$$|z|^k \sum_b c_{(b+k)b} |z|^{2b} = 0. \tag{14}$$

After dividing by $|z|^k$, for each k (14) defines a polynomial in $|z|^2$ that is identically 0. Hence each coefficient $c_{(b+k)b}$ vanishes and R is the zero polynomial. □

EXAMPLE 1.1. Note that $|z + i|^2 = |z|^2 - iz + i\overline{z} + 1$. Polarization implies that $(z + i)(w - i) = zw - iz + iw + 1$ for all z and w. We could also replace w with $\overline{w}$.

REMARK 1.3. We can restate the Riesz-Fejer Theorem in terms of Hermitian symmetric polynomials: If r is Hermitian symmetric and non-negative on the circle, then $r(z, \overline{z}) = |p(z)|^2$ there. (Note that there are many Hermitian symmetric polynomials agreeing with a given trig polynomial on the circle.) The higher dimensional analogue of the Riesz-Fejer theorem uses Hermitian symmetric polynomials. See Theorem 4.13 and [D1].

EXERCISE 1.13. Prove Lemma 1.3.

EXERCISE 1.14. Verify the second sentence of Remark 1.3.

EXERCISE 1.15. Explain why the factor z^{2d} appears in (13).

EXERCISE 1.16. Assume $a \in \mathbf{R}$, $b \in \mathbf{C}$, and $c > 0$. Find the minimum of the Hermitian polynomial R:

$$R(t, \overline{t}) = a + bt + \overline{bt} + c|t|^2.$$

Compare with the proof of the Cauchy-Schwarz inequality, Theorem 2.1.

EXERCISE 1.17. Prove that $e^{z+w} = e^z e^w$.

EXERCISE 1.18. Simplify the expression $\sum_{j=1}^k \sin((2j-1)x)$.

EXERCISE 1.19. Prove the following statement from plane geometry. Let ξ be a point in the complex plane other than the origin, and let w lie on the unit circle. Then every circle, perpendicular to the unit circle, and containing both ξ and w, also contains $(\overline{\xi})^{-1}$.

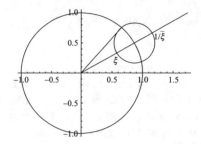

FIGURE 2. Reflection and perpendicularity

4. Constant coefficient differential equations

Our work thus far has begun to prepare us for the study of Fourier series. Fourier series also arise in solving both ordinary and partial differential equations. In order to develop this connection, we recall some things from that realm, thereby providing us with additional motivation.

The differential equation $y' = \lambda y$ has the obvious solution $y(t) = y(0)e^{\lambda t}$, for λ a real or complex constant, and t a real variable. How do we know that this solution is the only one? In fact we have a simple lemma.

LEMMA 1.5. *Suppose y is a differentiable function of one real variable t and $y' = \lambda y$. Then $y(t) = y(0)e^{\lambda t}$.*

PROOF. Let y be differentiable with $y' = \lambda y$. Put $f(t) = e^{-\lambda t}y(t)$. The product rule for derivatives gives $f'(t) = e^{-\lambda t}(-\lambda y(t) + y'(t)) = 0$. The mean-value theorem from calculus guarantees that the only solution to $f' = 0$ is a constant c. Hence $e^{-\lambda t}y(t)$ is a constant c, which must be $y(0)$, and $y(t) = y(0)e^{\lambda t}$. □

This result generalizes to constant coefficient equations of higher order; see Theorem 1.2. Such equations reduce to first order systems. Here is the simple idea. Given a k-times differentiable function y of one variable, we form the vector-valued function $Y : \mathbf{R} \to \mathbf{R}^{k+1}$ as follows:

$$Y(t) = \begin{pmatrix} y(t) \\ y'(t) \\ \dots \\ y^{(k)}(t) \end{pmatrix}. \tag{15}$$

The initial vector $Y(0)$ in (15) tells us the values for $y(0), y'(0), \dots, y^{(k)}(0)$.

Consider the differential equation $y^{(m)} = c_0 y + c_1 y' + \dots + c_{m-1} y^{(m-1)}$. Here y is assumed to be an m-times differentiable function of one variable t, and each c_j is a constant. Put $k = m - 1$ in (15). Define an m-by-m matrix A as follows:

$$A = \begin{pmatrix} 0 & 1 & 0 & \dots & 0 \\ 0 & 0 & 1 & \dots & 0 \\ \dots & \dots & \dots & 1 & 0 \\ 0 & 0 & \dots & 0 & 1 \\ c_0 & c_1 & c_2 & \dots & c_{m-1} \end{pmatrix} \tag{16}$$

Consider the matrix product AY and the equation $Y' = AY$. The matrix A has been constructed such that the first $m - 1$ rows of this equation tell us that

$\frac{d}{dt} y^{(j)} = y^{(j+1)}$, and the last row tells us that $y^{(m)} = \sum c_j y^{(j)}$. The equation

$$y^{(m)} = c_0 y + c_1 y' + \ldots + c_{m-1} y^{(m-1)}$$

is therefore equivalent to the first-order matrix equation $Y' = AY$. In analogy with Lemma 1.5 we solve this system by exponentiating the matrix A.

Let M_ν denote a sequence of matrices of complex numbers. We say that M_ν converges to M if each entry of M_ν converges to the corresponding entry of M.

Let M be a square matrix, say n-by-n, of real or complex numbers. We define e^M, the exponential of M, by the series

$$e^M = \sum_{k=0}^{\infty} \frac{M^k}{k!} = \lim_{N \to \infty} \sum_{k=0}^{N} \frac{M^k}{k!}. \tag{17}$$

It is not difficult to show that this series converges and also, when $MK = KM$, that $e^{M+K} = e^M e^K$. Note also that $e^O = I$, where I denotes the identity matrix. As a consequence of these facts, for each M the matrix e^M is invertible and e^{-M} is its inverse. It is also easy to show that $M e^M = e^M M$, that is, M and its exponential e^M commute. We also note that e^{At} is differentiable and $\frac{d}{dt} e^{At} = A e^{At}$.

EXERCISE 1.20. Prove that the series in (17) converges for each square matrix of complex numbers. Suggestion. Use the Weierstrass M-test to show that each entry converges.

EXERCISE 1.21. If B is invertible, prove for each positive integer k that

$$(BMB^{-1})^k = BM^k B^{-1}.$$

EXERCISE 1.22. If B is invertible, prove that $B e^M B^{-1} = e^{BMB^{-1}}$.

EXERCISE 1.23. Find a simple expression for $\det(e^M)$ in terms of a trace.

A simple generalization of Lemma 1.5 enables us to solve constant coefficient ordinary differential equations (ODE)s of higher order m. As mentioned above, the initial vector $Y(0)$ provides m pieces of information.

THEOREM 1.2. *Suppose* $y : \mathbf{R} \to \mathbf{R}$ *is m times differentiable and there are constants* c_j *such that*

$$y^{(m)} = \sum_{j=0}^{m-1} c_j y^{(j)}. \tag{18}$$

Define Y *as in (15) and* A *as in (16) above. Then* $Y(t) = e^{At} Y(0)$, *and* $y(t)$ *is the first component of* $e^{At} Y(0)$.

PROOF. Suppose Y is a solution. Differentiating $e^{-At} Y(t)$ gives

$$\frac{d}{dt} \left(e^{-At} Y(t) \right) = e^{-At} (Y'(t) - A Y(t)). \tag{19}$$

Since y satisfies (18), the expression in (19) is the zero element of $\mathbf{R}^m$. Hence $e^{-At} Y(t)$ is a constant element of $\mathbf{R}^m$ and the result follows.

$\square$

In order to apply this result we need a good way to exponentiate matrices (linear mappings). Let $A : \mathbf{C}^n \to \mathbf{C}^n$ be a linear transformation. Recall that λ is called an *eigenvalue* for A if there is a non-zero vector v such that $Av = \lambda v$.

The vector v is called an *eigenvector* corresponding to λ. One sometimes finds eigenvalues by finding the roots of the polynomial $\det(A - \lambda I)$. We note that the roots of this equation can be complex even if A is real and we consider A to be an operator on $\mathbf{R}^n$.

In order to study the exponentiation of A, we first assume that A has n distinct eigenvalues. By linear algebra, shown below, there is an invertible matrix P and a diagonal matrix D such that $A = PDP^{-1}$. Since $(PDP^{-1})^k = PD^kP^{-1}$ for each k, it follows that

$$e^A = e^{PDP^{-1}} = Pe^D P^{-1}. \tag{20}$$

It is easy to find e^D; it is the diagonal matrix whose eigenvalues (the diagonal elements in this case) are the exponentials of the eigenvalues of D.

We recall how to find P. Given A with distinct eigenvalues, for each eigenvalue λ_j we find an eigenvector v_j. Thus v_j is a non-zero vector and $A(v_j) = \lambda_j v_j$. Then we may take P to be the matrix whose columns are these eigenvectors. We include the simple proof. First, the eigenvectors form a basis of $\mathbf{C}^n$ because the eigenvalues are distinct.

Let e_j be the j-th standard basis element of $\mathbf{R}^n$. Let D be the diagonal matrix with $D(e_j) = \lambda_j e_j$. By definition, $P(e_j) = v_j$. Therefore

$$PDP^{-1}(v_j) = PD(e_j) = P(\lambda_j e_j) = \lambda_j P(e_j) = \lambda_j v_j = A(v_j). \tag{21}$$

By (21), A and PDP^{-1} agree on a basis, and hence they define the same linear mapping. Thus $A = PDP^{-1}$.

We apply this reasoning to solve the general second order constant coefficient homogeneous differential equation $y'' = b_1 y' + b_0 y$. Let λ_1 and λ_2 be the roots of the polynomial $\lambda^2 - b_1\lambda - b_0 = 0$.

COROLLARY 1.4. *Assume $y : \mathbf{R} \to \mathbf{C}$ is twice differentiable, and*

$$y'' - (\lambda_1 + \lambda_2)y' + \lambda_1\lambda_2 y = 0 \tag{22}$$

for complex numbers λ_1 and λ_2. If $\lambda_1 \neq \lambda_2$, then there are complex constants c_1 and c_2 such that

$$y(t) = c_1 e^{\lambda_1 t} + c_2 e^{\lambda_2 t}. \tag{23}$$

In case $\lambda_1 = \lambda_2$, the answer is given by

$$y(t) = e^{\lambda t}y(0) + te^{\lambda t}(y'(0) - \lambda y(0)).$$

PROOF. Here the matrix A is given by

$$\begin{pmatrix} 0 & 1 \\ -\lambda_1\lambda_2 & \lambda_1 + \lambda_2 \end{pmatrix}. \tag{24}$$

Its eigenvalues are λ_1 and λ_2. When $\lambda_1 \neq \lambda_2$, we obtain e^{At} by the formula

$$e^{At} = Pe^{Dt}P^{-1},$$

where D is the diagonal matrix with eigenvalues λ_1 and λ_2, and P is the change of basis matrix. Here e^{Dt} is diagonal with eigenvalues $e^{\lambda_1 t}$ and $e^{\lambda_2 t}$:

$$e^{At} = \frac{1}{\lambda_2 - \lambda_1} \begin{pmatrix} 1 & 1 \\ \lambda_1 & \lambda_2 \end{pmatrix} \begin{pmatrix} e^{\lambda_1 t} & 0 \\ 0 & e^{\lambda_2 t} \end{pmatrix} \begin{pmatrix} \lambda_2 & -1 \\ -\lambda_1 & 1 \end{pmatrix}$$

The factor $\frac{1}{\lambda_2-\lambda_1}$ on the outside arises in finding P^{-1}. Performing the indicated matrix multiplications, introducing the values of y and y' at 0, and doing some tedious work gives

$$y(t) = \frac{(\lambda_2 e^{\lambda_1 t} - \lambda_1 e^{\lambda_2 t})y(0) + (e^{\lambda_2 t} - e^{\lambda_1 t})y'(0)}{\lambda_2 - \lambda_1}. \tag{25}$$

Formula (23) is a relabeling of (25). An ancillary advantage of writing the answer in the form (25) is that we can take the limit as λ_2 tends to λ_1 and obtain the solution in case these numbers are equal; write $\lambda = \lambda_1 = \lambda_2$ in this case. The result (See Exercise 1.24) is

$$y(t) = e^{\lambda t}y(0) + te^{\lambda t}(y'(0) - \lambda y(0)). \tag{26}$$

$$\square$$

A special case of this corollary arises often. For c real but not 0, the solutions to the differential equation $y'' = cy$ are (complex) exponentials. The behavior of the solutions depends in a significant way on the sign of c. When $c = k^2 > 0$, the solutions are linear combinations of $e^{\pm kt}$. Such exponentials either decay or grow at infinity. When $c = -k^2$, however, the solutions are linear combinations of $e^{\pm ikt}$, which we express instead in terms of sines and cosines. In this case the solutions oscillate.

EXERCISE 1.24. Show that (25) implies (26).

The assumption that A has distinct eigenvalues is used only to find e^{At} easily. Even when A has repeated eigenvalues and the eigenvectors do not span the space, the general solution to $Y' = AY$ remains $Y(t) = e^{At}Y(0)$. The Jordan normal form allows us to write $A = P(D + N)P^{-1}$, where D is diagonal and N is nilpotent of a particular form. If the eigenvectors do not span, then N is not 0. It is often easier in practice to exponentiate A by using the ideas of differential equations rather than by using linear algebra. The proof from Exercise 1.24 that (25) implies (26) nicely illustrates the general idea. See also Exercises 1.25 and 1.28.

EXERCISE 1.25. Find e^{At} if

$$A = \begin{pmatrix} \lambda & 1 \\ 0 & \lambda \end{pmatrix}.$$

If you know the Jordan normal form, then describe how to find e^{At} when the eigenvectors of A do not span the full space. Suppose first that A is a Jordan block of the form $\lambda I + N$, where N is nilpotent (as in the normal form) but not 0. What is e^{At}?

EXERCISE 1.26. Give an example of two-by-two matrices A and B such that $e^A e^B \neq e^{A+B}$.

4.1. Inhomogeneous linear differential equations. We also wish to solve inhomogeneous differential equations. To do so, we introduce natural notation. Let $p(z) = z^m - \sum_{j=0}^{m-1} c_j z^j$ be a monic polynomial of degree m. Let D represent the operation of differentiation with respect to x. We define the operator $p(D)$ by formally substituting D^j for z^j.

In Theorem 1.2, we solved the equation $p(D)y = 0$. In applications, however, one often needs to solve the equation $p(D)y = f$ for a given *forcing function* f. For example, one might turn on a switch at a given time x_0, in which case f could be the function that is 0 for $x < x_0$ and is 1 for $x \geq x_0$.

Since the operator $p(D)$ is linear, the general solution to $p(D)y = f$ can be written $y = y_0 + y_*$, where y_0 is a solution to the equation $p(D)y = 0$ and y_* is any particular solution to $p(D)y_* = f$. We already know how to find all solutions to $p(D)(y_0) = 0$. Thus we need only to find one particular solution. To do so, we proceed as follows. Using the fundamental theorem of algebra, we factor the polynomial p:

$$p(z) = \prod_{k=1}^{m}(z - \lambda_k),$$

where the λ_k can be repeated if necessary.

When $m = 1$ we can solve the equation $(D - \lambda_1)g_1 = f$ by the following method. We suppose that $g_1(x) = c(x)e^{\lambda_1 x}$ for some differentiable function c. Applying $D - \lambda_1$ to g_1 we get

$$(D - \lambda_1)g_1 = (c'(x) + c(x)\lambda_1)e^{\lambda_1 x} - \lambda_1 c(x)e^{\lambda_1 x} = c'(x)e^{\lambda_1 x} = f(x).$$

For an arbitrary real number a (often it is useful to take $a = \pm\infty$), we obtain

$$c(x) = \int_a^x f(t)e^{-\lambda_1 t}dt.$$

This formula yields the particular solution g_1 defined by

$$g_1(x) = e^{\lambda_1 x}\int_a^x e^{-\lambda_1 t}f(t)dt,$$

and amounts to finding the inverse of the operator $D - \lambda_1$.

The case $m > 1$ follows easily from the special case. We solve the equation

$$(D - \lambda_1)(D - \lambda_2)\ldots(D - \lambda_m)(y) = f$$

by solving $(D - \lambda_1)g_1 = f$, and then for $j > 1$ solving $(D - \lambda_j)g_j = g_{j-1}$. The function $y = g_m$ then solves the equation $p(D)y = f$.

REMARK 1.4. Why do we start with the term on the far left? The reason is that the inverse of the composition BA of operators is the composition $A^{-1}B^{-1}$ of the inverses in the *reverse* order. To take off our socks, we must first take off our shoes. The composition product $(D - \lambda_1)(D - \lambda_2)\ldots(D - \lambda_n)$ is independent of the ordering of the factors, and in this situation we could invert in any order.

EXAMPLE 1.2. We solve $(D - 5)(D - 3)y = e^x$. First we solve $(D - 5)g = e^x$, obtaining

$$g(x) = e^{5x}\int_\infty^x e^t e^{-5t}dt = \frac{-1}{4}e^x.$$

Then we solve $(D - 3)h = \frac{-1}{4}e^x$ to get

$$h(x) = e^{3x}\int_\infty^x \frac{-1}{4}e^t e^{-3t}dt = \frac{1}{8}e^x.$$

The general solution to the equation is $c_1 e^{5x} + c_2 e^{3x} + \frac{1}{8}e^x$, where c_1 and c_2 are constants. We put $a = \infty$ because $e^{-\lambda t}$ vanishes at ∞ if $\lambda > 0$.

EXERCISE 1.27. Find all solutions to $(D^2 + m^2)y = e^x$.

EXERCISE 1.28. Solve $(D - \lambda)y = e^{\lambda x}$. Use the result to solve $(D - \lambda)^2(y) = 0$. Compare the method with the result from Corollary 1.4, when $\lambda_1 = \lambda_2$.

EXERCISE 1.29. Find a particular solution to $(D - 5)y = 1 - 75x^2$.

EXERCISE 1.30. We wish to find a particular solution to $(D - \lambda)y = g$, when g is a polynomial of degree m. Identify the coefficients of g as a vector in $\mathbf{C}^{m+1}$. Assuming $\lambda \neq 0$, show that there is a unique particular solution y that is a polynomial of degree m. Write explicitly the matrix of the linear transformation that sends y to g and note that it is invertible. Explain precisely what happens when $\lambda = 0$.

EXERCISE 1.31. Consider the equation $(D - \lambda)^m y = 0$. Prove by induction that $x^j e^{\lambda x}$ for $0 \leq j \leq m - 1$ form a linearly independent set of solutions.

We conclude this section with some elementary remarks about solving systems of linear equations in finitely many variables; these remarks inform to a large degree the methods used throughout this book. The logical development enabling the passage from linear algebra to solving linear differential equations was one of the great achievements of 19-th century mathematics.

Consider a system of k linear equations in n real variables. We regard this system as a linear equation $Ly = w$, where $L : \mathbf{R}^n \to \mathbf{R}^k$. Things work out better (as we shall see) in terms of complex variables; thus we consider the linear equation $Lz = w$, where now $L : \mathbf{C}^n \to \mathbf{C}^k$. Let $\langle z, \zeta \rangle$ denote the usual Hermitian Euclidean inner product on both the domain and target spaces. Let L^* denote the adjoint of L. The matrix representation of L^* is the conjugate transpose of L. Then $Lz = w$ implies (for all ζ)

$$\langle w, \zeta \rangle = \langle Lz, \zeta \rangle = \langle z, L^* \zeta \rangle.$$

In order that the equation $Lz = w$ have a solution at all, the right-hand side w must be orthogonal to the nullspace of L^*.

Consider the case where the number of equations equals the number of variables. Using eigenvalues and orthonormal expansion (to be developed in Chapter 2 for Hilbert spaces), we can attempt to solve the equation $Lz = w$ as follows. Under the assumption that $L = L^*$, there is an orthonormal basis of $\mathbf{C}^n$ consisting of eigenvectors ϕ_j with corresponding real eigenvalues λ_j. We can then write both z and w in terms of this basis, obtaining

$$w = \sum \langle w, \phi_j \rangle \phi_j$$

$$z = \sum \langle z, \phi_j \rangle \phi_j.$$

Equating Lz to w, we get

$$\sum \langle z, \phi_j \rangle \lambda_j \phi_j = \sum \langle w, \phi_j \rangle \phi_j.$$

Now equating coefficients yields

$$\langle z, \phi_j \rangle \lambda_j = \langle w, \phi_j \rangle. \tag{27}$$

If $\lambda_j = 0$, then w must be orthogonal to ϕ_j. If w satisfies this condition for all appropriate j, then we can solve $Lz = w$ by division. On each eigenspace with $\lambda_j \neq 0$, we divide by λ_j to find $\langle z, \phi_j \rangle$ and hence we find a solution z. The solution is not unique in general; we can add to z any solution ζ to $L\zeta = 0$. These ideas recur throughout this book, both in the Fourier series setting and in differential equations.

5. The wave equation for a vibrating string

The wave equation discussed in this section governs the motion of a vibrating string. The solution of this equation naturally leads to Fourier series.

We are given a twice differentiable function u of two variables, x and t, with x representing position and t representing time. Using subscripts for partial derivatives, the wave equation is

$$u_{xx} = \frac{1}{c^2} u_{tt}. \tag{28}$$

Here c is a constant which equals the speed of propagation of the wave.

Recall that a function is *continuously differentiable* if it is differentiable and its derivative is continuous. It is *twice continuously differentiable* if it is twice differentiable and the second derivative is continuous. We have the following result about the partial differential equation (28). After the proof we discuss the appropriate initial conditions.

THEOREM 1.3. *Let* $u : \mathbf{R} \times \mathbf{R} \to \mathbf{R}$ *be twice continuously differentiable and satisfy (28). Then there are twice continuously differentiable functions F and G (of one variable) such that*

$$u(x,t) = F(x + ct) + G(x - ct).$$

PROOF. Motivated by writing $\alpha = x + ct$ and $\beta = x - ct$, we define a function ϕ by

$$\phi(\alpha, \beta) = u(\frac{\alpha + \beta}{2}, \frac{\alpha - \beta}{2c}).$$

We compute second derivatives by the chain rule, obtaining

$$\phi_{\alpha\beta} = \frac{d}{d\beta}\phi_\alpha = \frac{d}{d\beta}(\frac{u_x}{2} + \frac{u_t}{2c}) = \frac{u_{xx}}{4} - \frac{u_{xt}}{4c} + \frac{u_{tx}}{4c} - \frac{u_{tt}}{4c^2} = 0. \tag{29}$$

Note that we have used the equality of the mixed second partial derivatives u_{xt} and u_{tx}. It follows that ϕ_α is independent of β, hence a function h of α alone. Integrating again, we see that ϕ is the integral F of this function h plus an integration constant, say G, which will depend on β. We obtain

$$u(x,t) = \phi(\alpha, \beta) = F(\alpha) + G(\beta) = F(x + ct) + G(x - ct). \tag{30}$$

$\square$

This problem becomes more realistic if x lies in a fixed interval and u satisfies initial conditions. For convenience we let this interval be $[0, \pi]$. We can also choose units for time to make $c = 1$. The conditions $u(0,t) = u(\pi, t) = 0$ state that the string is fixed at the points 0 and π. The initial conditions for time are $u(x, 0) = f(x)$, and $u_t(x, 0) = g(x)$. The requirement $u(x, 0) = f(x)$ means that the initial displacement curve of the string is defined by the equation $y = f(x)$. The requirement on u_t means that the string is given the initial velocity $g(x)$.

Note that f and g are not the same functions as F and G from Theorem 1.3 above. We can, however, easily express F and G in terms of f and g.

First we derive d'Alembert's solution, Theorem 1.4. Then we attempt to solve the wave equation by way of *separation of variables*. That method leads to a Fourier series. In the next section we obtain the d'Alembert solution by regarding the wave equation as a constant coefficient ODE, and treating the second derivative operator D^2 as a number.

THEOREM 1.4. *Let u be twice continuously differentiable and satisfy $u_{xx} = u_{tt}$, together with the initial conditions $u(x,0) = f(x)$ and $u_t(x,0) = g(x)$. Then*

$$u(x,t) = \frac{f(x+t) + f(x-t)}{2} + \frac{1}{2} \int_{x-t}^{x+t} g(a)da. \tag{31}$$

PROOF. Using the F and G from Theorem 1.3, and assuming $c = 1$, we are given $F + G = f$ and $F' - G' = g$. Differentiating, we obtain the system

$$\begin{pmatrix} 1 & 1 \\ 1 & -1 \end{pmatrix} \begin{pmatrix} F' \\ G' \end{pmatrix} = \begin{pmatrix} f' \\ g \end{pmatrix}. \tag{32}$$

Solving (32) by linear algebra expresses F' and G' in terms of f and g: we obtain $F' = \frac{f'+g}{2}$ and $G' = \frac{f'-g}{2}$. Integrating and using (30) with $c = 1$ yields (31). □

We next attempt to solve the wave equation by separation of variables. The standard idea seeks a solution of the form $u(x,t) = A(x)B(t)$. Differentiating and using the equation $u_{xx} = u_{tt}$ leads to $A''(x)B(t) = A(x)B''(t)$, and therefore $\frac{A''(x)}{A(x)} = \frac{B''(t)}{B(t)}$. Since one side depends on x and the other on t, each must be constant. Thus we have $A''(x) = \xi A(x)$ and $B''(t) = \xi B(t)$ which we solve as in Corollary 1.4. For each ξ we obtain solutions. If we insist that the solution is a wave, then we must have $\xi < 0$ (as the roots are then purely imaginary). Thus

$$A(x) = a_1 \sin(\sqrt{|\xi|}x) + a_2 \cos(\sqrt{|\xi|}x)$$

for constants a_1, a_2. If the solution satisfies the condition $A(0) = 0$, then $a_2 = 0$. If the solution also satisfies the condition $A(\pi) = 0$, then $\sqrt{|\xi|}$ is an integer. Putting this information together, we obtain a collection of solutions u_m indexed by the integers:

$$u_m(x,t) = (d_{m_1} \cos(mt) + d_{m_2} \sin(mt)) \sin(mx),$$

for constants d_{m_1} and d_{m_2}. Adding these solutions (the superposition of these waves) we are led to a candidate for the solution:

$$u(x,t) = \sum_{m=0}^{\infty} (d_{m_1} \cos(mt) + d_{m_2} \sin(mt)) \sin(mx).$$

Given $u(x,0) = f(x)$, we now wish to solve the equation (where $d_m = d_{m_1}$)

$$f(x) = \sum_m d_m \sin(mx). \tag{33}$$

Again we encounter a series involving the terms $\sin(mx)$. At this stage, the basic question becomes, given a function f with $f(0) = f(\pi) = 0$, whether there are constants such that (33) holds. We are thus asking whether a given function can be represented as the superposition of (perhaps infinitely many) sine waves. This sort of question arises throughout applied mathematics.

EXERCISE 1.32. Give an example of a function on the real line that is differentiable (at all points) but not continuously differentiable.

EXERCISE 1.33. Suppose a given function f can be written in the form (33), where the sum is either finite or converges uniformly. How can we determine the constants d_m? (We will solve this problem in Section 9 of this chapter.)

We conclude this section with a few remarks about the inhomogeneous wave equation. Suppose that an external force acts on the vibrating string. The wave equation (28) then becomes

$$u_{xx} - \frac{1}{c^2} u_{tt} = h(x,t) \qquad\qquad (*)$$

for some function h determined by this force. Without loss of generality, we again assume that $c = 1$. We still have the initial conditions $u(0,t) = u(\pi,t) = 0$ as well as $u(x,0) = f(x)$ and $u_t(x,0) = g(x)$. We can approach this situation also by using sine series. We presume that both u and h are given by sine series:

$$u(x,t) = \sum d_m(t)\sin(mx)$$

$$h(x,t) = \sum k_m(t)\sin(mx).$$

Plugging into $(*)$, we then obtain a family of second order constant coefficient ODE relating d_m to k_m:

$$d_m''(t) + m^2 d_m(t) = -k_m(t).$$

We can then solve these ODEs by the method described just before Remark 1.4. This discussion indicates the usefulness of expanding functions in series such as (33) or, more generally, as series of the form $\sum_n d_n e^{inx}$.

6. Solving the wave equation via exponentiation

This section is not intended to be rigorous. Its purposes are to illuminate Theorem 1.4 and to glimpse some deeper ideas.

Consider the partial differential equation (PDE) on $\mathbf{R} \times \mathbf{R}$ given by $u_{xx} = u_{tt}$ with initial conditions $u(x,0) = f(x)$ and $u_t(x,0) = g(x)$. We regard it formally as a second order ordinary differential equation as follows:

$$\begin{pmatrix} u \\ u' \end{pmatrix}' = \begin{pmatrix} u' \\ u'' \end{pmatrix} = \begin{pmatrix} 0 & 1 \\ D^2 & 0 \end{pmatrix} \begin{pmatrix} u \\ u' \end{pmatrix}. \qquad\qquad (34)$$

Here D^2 is the operator of differentiating twice with respect to x, but we treat it formally as a number. Using the method of Corollary 1.4, we see formally that the answer is given by

$$\begin{pmatrix} u(x,t) \\ u_t(x,t) \end{pmatrix} = e^{\begin{pmatrix} 0 & 1 \\ D^2 & 0 \end{pmatrix} t} \begin{pmatrix} f(x) \\ g(x) \end{pmatrix} \qquad\qquad (35)$$

The eigenvalues of $\begin{pmatrix} 0 & 1 \\ D^2 & 0 \end{pmatrix}$ are $\pm D$ and the change of basis matrix is given by

$$P = \begin{pmatrix} 1 & -1 \\ D & D \end{pmatrix}.$$

Proceeding formally as if D were a nonzero number, we obtain by this method

$$u(x,t) = \frac{e^{Dt} + e^{-Dt}}{2} f(x) + \frac{e^{Dt} - e^{-Dt}}{2} (D^{-1}g)(x). \qquad\qquad (36)$$

We need to interpret the expressions $e^{\pm Dt}$ and D^{-1} in order for (36) to be useful. It is natural for D^{-1} to mean integration. We claim that $e^{Dt}f(x) = f(x+t)$. We do not attempt to prove the claim, as a rigorous discussion would take us far

from our aims, but we ask the reader to give a heuristic explanation in Exercise
1.34. The claim and (36) yield d'Alembert's solution, the same answer we obtained
in Theorem 1.4.

$$u(x,t) = \frac{f(x+t) + f(x-t)}{2} + \frac{1}{2} \int_{x-t}^{x+t} g(u)du. \tag{37}$$

EXERCISE 1.34. Give a suggestive argument why $e^{Dt}f(x) = f(x+t)$.

7. Integrals

We are now almost prepared to begin our study of Fourier series. In this section
we introduce some notation and say a few things about integration.

When we say that f is *integrable* on the circle, we mean that f is Riemann integrable on $[0, 2\pi]$ there and that $f(0) = f(2\pi)$. By definition, a Riemann integrable
function must be bounded. Each continuous function is Riemann integrable, but
there exist Riemann integrable functions that are discontinuous at an infinite (but
small, in the right sense) set of points.

Some readers may have seen the more general Lebesgue integral and measure
theory. We sometimes use notation and ideas usually introduced in that context.
For example, we can define *measure zero* without defining *measure*. A subset S of
R has *measure zero*, if for every $\epsilon > 0$, we can find a sequence $\{I_n\}$ of intervals
such that S is contained in the union of the I_n and the sum of the lengths of the
I_n is less than ϵ. A basic theorem we will neither prove nor use is that a bounded
function on a closed interval on **R** is Riemann integrable if and only if the set of
its discontinuities has measure zero.

In the theory of Lebesgue integration, we say that two functions are equivalent
if they agree except on a set of measure zero. We also say that f and g agree *almost
everywhere*. Then L^1 denotes the collection of (equivalence classes of measurable)
functions f on **R** such that $||f||_{L^1} = \int |f| < \infty$. For $1 \leq p < \infty$, L^p denotes the
collection of (equivalence classes of measurable) functions f such that $||f||_{L^p}^p =
\int |f|^p < \infty$. We are primarily interested in the case where $p = 2$. Finally, L^∞
denotes the collection of (equivalence classes of measurable) functions that agree
almost everywhere with a bounded function. For f continuous, $||f||_{L^\infty} = \sup |f|$.
We write $L^1(S)$, $L^2(S)$, and so on, when the domain of the function is some given
set S. We do not define measurability here, but we note that being measurable is
a weak condition on a function, satisfied by all functions encountered in this book.
See [F1] for detailed discussion of these topics.

Perhaps the fundamental result distinguishing Lebesgue integration from Riemann integration is that the L^p spaces are complete in the Lebesgue theory. In
other words, Cauchy sequences converge. We do not wish to make Measure Theory
a prerequisite for what follows. We therefore define $L^1(S)$ to be the *completion* of
the space of continuous functions on S in the topology defined by the L^1 norm.
We do the same for $L^2(S)$. In this approach, we do not ask what objects lie in the
completion. Doing so is analogous to defining **R** to be the (metric space) completion of **Q**, but never showing that a real number can be identified by an infinite
decimal expansion.

We mention a remarkable subtlety about integration theory. There exist sequences $\{f_n\}$ of functions on an interval such that each f_n is (Riemann or Lebesgue)
integrable, $\int f_n$ converges to 0, yet $\{f_n(x)\}$ diverges for every x in the interval.

EXAMPLE 1.3. For each positive integer n, we can find unique non-negative integers h and k such that $n = 2^h + k < 2^{h+1}$. Let f_n be the function on $[0, 1]$ that is 1 on the half-open interval I_n defined by

$$I_n = [\frac{k}{2^h}, \frac{k+1}{2^h})$$

and 0 off this interval. Then the integral of f_n is the length of I_n, hence $\frac{1}{2^h}$. As n tends to infinity, so does h, and thus $\int f_n$ also tends to 0. On the other hand, for each x, the terms of the sequence $\{f_n(x)\}$ equal both 0 and 1 infinitely often, and hence the sequence diverges.

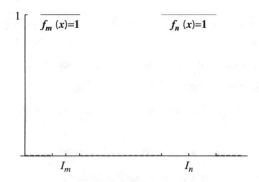

FIGURE 3. Example 1.3

In Example 1.3 there is a subsequence of the f_n converging almost everywhere to the 0 function, illustrating a basic result in integration theory.

We will use the following lemma about Riemann integrable functions. Since f is Riemann integrable, it is bounded, and hence we may use the notation $||f||_{L^\infty}$.

LEMMA 1.6. *Suppose f is Riemann integrable on the circle. Then there exists a sequence $\{f_n\}$ of continuous functions such that both hold:*

(1) *For all k, $\sup(|f_k(x)|) \le \sup|f(x)|$. That is, $||f_k||_{L^\infty} \le ||f||_{L^\infty}$.*
(2) *$\lim \int |f_k(x) - f(x)|dx = 0$. That is, $f_k \to f$ in L^1.*

We end this section by indicating why it is unreasonable to make the collection of Riemann integrable functions into a complete metric space. We first note that it is difficult to define a meaningful distance between two Riemann integrable functions. The natural distance might seem to be $d(f, g) = \int |f-g|$, but this definition violates one of the axioms for a distance. If f and g agree except on a (non-empty) set of measure zero, then $\int |f - g| = 0$, but f and g are not equal. Suppose instead we define f and g to be equivalent if they agree except on a set of Lebesgue measure zero. We then consider the space of equivalence classes of Riemann integrable functions. We define the distance between two equivalence classes F and G by choosing representatives f and g and putting $\delta(F, G) = \int |f - g|$. Then completeness fails. The next example shows, with this notion of distance, that the limit of a Cauchy sequence of Riemann integrable functions need not be itself Riemann integrable.

EXAMPLE 1.4. Define a sequence of functions $\{f_n\}$ on $[0,1]$ as follows: $f_n(x) = 0$ for $0 \leq x \leq \frac{1}{n}$ and $f_n(x) = -\log(x)$ otherwise. Each f_n is obviously Riemann integrable, and f_n converges pointwise to a limit f. Since f is unbounded, it is not Riemann integrable. This sequence is Cauchy in both the L^1 and the L^2 norms. To show that it is Cauchy in the L^1 norm, we must show that $||f_n - f_m||_{L^1}$ tends to 0 as m, n tend to infinity. But, for $n \geq m$,

$$\int_0^1 |f_n(x) - f_m(x)|dx = \int_{\frac{1}{n}}^{\frac{1}{m}} |\log(x)|dx.$$

The reader can easily show using calculus that the limit as n, m tend to infinity of this integral is 0. A similar but slightly harder calculus problem shows that $\{f_n\}$ is also Cauchy in the L^2 norm.

In this book we will use the language and notation from Lebesgue integration, but most of the time we work with Riemann integrable functions.

EXERCISE 1.35. Verify the statements in Example 1.4.

EXERCISE 1.36. Prove Lemma 1.6.

EXERCISE 1.37. Prove that each $n \in \mathbf{N}$ has a unique representation $n = 2^h + k$ where $0 \leq k < 2^h$.

8. Approximate identities

In his work on quantum mechanics, Paul Dirac introduced a mathematical object often called the Dirac delta function. This function $\delta : \mathbf{R} \to \mathbf{R}$ was supposed to have two properties: $\delta(x) = 0$ for $x \neq 0$, and $\int_{-\infty}^{\infty} \delta(x)f(x)dx = f(0)$ for all continuous functions f defined on the real line. No such function can exist, but it is possible to make things precise by defining δ to be a *linear functional*. That is, δ is the function $\delta : V \to \mathbf{C}$ defined by $\delta(f) = f(0)$, for V an appropriate space of functions. Note that

$$\delta(f + g) = (f + g)(0) = f(0) + g(0) = \delta(f) + \delta(g)$$
$$\delta(cf) = (cf)(0) = cf(0) = c\delta(f),$$

and hence δ is linear. We discuss linear functionals in Chapter 2. We provide a rigorous framework (known as distribution theory) in Chapter 3, for working with the Dirac delta function.

In this section we discuss *approximate identities*, often called *Dirac sequences*, which we use to approximate the behavior of the delta function.

DEFINITION 1.3. Let W denote either the natural numbers or an interval on the real line, and let S^1 denote the unit circle. An *approximate identity* on S^1 is a collection, for $t \in W$, of continuous functions $K_t : S^1 \to \mathbf{R}$ with the following properties:

(1) For all t, $\frac{1}{2\pi} \int_{-\pi}^{\pi} K_t(x)dx = 1$
(2) There is a constant C such that, for all t, $\frac{1}{2\pi} \int_{-\pi}^{\pi} |K_t(x)|dx \leq C$.
(3) For all $\epsilon > 0$, we have

$$\lim_{t \to T} \int_{\epsilon \leq |x| \leq \pi} |K_t(x)|dx = 0.$$

Here $T = \infty$ when W is the natural numbers, and $T = \sup(W)$ otherwise.

Often our approximate identity will be a sequence of functions K_n and we let n tend to infinity. In another fundamental example, called the Poisson kernel, our approximate identity will be a collection of functions P_r defined for $0 \le r < 1$. In this case we let r increase to 1. In the subsequent discussion we will write K_n for an approximate identity indexed by the natural numbers and P_r for the Poisson kernel, indexed by r with $0 \le r < 1$.

We note the following simple point. If $K_t \ge 0$, then the second property follows from the first property. We also note that the graphs of these functions K_t spike at 0. See Figures 4, 5, and 6. In some vague sense, Dirac's delta function is the limit of K_t. The crucial point, however, is not to consider the K_t on their own, but rather the operation of *convolution* with K_t.

We first state the definition of convolution and then prove a result clarifying why the sequence $\{K_n\}$ is called an approximate identity. In the next section we will observe another way in which convolution arises.

DEFINITION 1.4. Suppose f, g are integrable on the circle. Define $f * g$, the *convolution* of f and g, by

$$(f * g)(x) = \frac{1}{2\pi} \int_{-\pi}^{\pi} f(y)g(x - y)dy.$$

Note the normalizing factor of $\frac{1}{2\pi}$. One consequence, where 1 denotes the constant function equal to 1, is that $1 * 1 = 1$. The primary reason for the normalizing factor is the connection with probability. A non-negative function that integrates to 1 can be regarded as a *probability density*. The density of the sum of two random variables is the convolution of the given densities. See [HPS].

THEOREM 1.5. *Let $\{K_n\}$ be an approximate identity, and let f be Riemann integrable on the circle. If f is continuous at x, then*

$$\lim_{n \to \infty} (f * K_n)(x) = f(x).$$

If f is continuous on the circle, then $f * K_n$ converges uniformly to f. Also,

$$f(0) = \lim_{n \to \infty} \frac{1}{2\pi} \int_{-\pi}^{\pi} f(-y)K_n(y)dy = \lim_{n \to \infty} \frac{1}{2\pi} \int_{-\pi}^{\pi} f(y)K_n(-y)dy.$$

PROOF. The proof uses a simple idea, which is quite common in proofs in analysis. We estimate by treating $f(x)$ as a constant and integrating with respect to another variable y. Then $|(f * K_n)(x) - f(x)|$ is the absolute value of an integral, which is at most the integral of the absolute value of the integrand. We then break this integral into two pieces, where y is close to 0 and where y is far from 0. The first term can be made small because f is continuous at x. The second term is made small by choosing n large.

Here are the details. Since $\{K_n\}$ is an approximate identity, the integrals of $|K_n|$ are bounded by some number M. Assume that f is continuous at x. Given $\epsilon > 0$, we first find δ such that $|y| < \delta$ implies

$$|f(x - y) - f(x)| < \frac{\epsilon}{2M}.$$

If f is continuous on the circle, then f is uniformly continuous, and we can choose δ independent of x to make the same estimate hold. We next write

$$|(f * K_n)(x) - f(x)| = \left| \int K_n(y)(f(x - y) - f(x))dy \right|$$

$$\leq \int |K_n(y)(f(x-y)-f(x))|dy = I_1 + I_2 \tag{38}$$

Here I_1 denotes the integral over the set where y is close to 0. We have

$$I_1 = \int_{-\delta}^{\delta} |K_n(y)(f(x-y)-f(x))|dy \leq M\frac{\epsilon}{2M} = \frac{\epsilon}{2}. \tag{39}$$

Next, we estimate I_2, the integral over the set where $|y| \geq \delta$. Since f is integrable, it is bounded. For some constant C, I_2 is then bounded by

$$C \int_{|y|\geq\delta} |K_n(y)|dy. \tag{40}$$

The third defining property of an approximate identity enables us to choose N_0 sufficiently large such that, for $n \geq N_0$, we can bound (40) by $\frac{\epsilon}{2}$ as well. Both conclusions follow.   □

Each of the next three examples of approximate identities will arise in this book. The third example is defined on the real line rather than on the circle, but there is no essential difference.

EXAMPLE 1.5 (Fejer kernel). Let $D_N(x) = \sum_{-N}^{N} e^{ikx}$; this sequence is sometimes called the Dirichlet kernel. Although the integral of each D_N over the circle equals 1, the integral of the absolute value of D_N is not bounded independent of N. Hence the sequence $\{D_N\}$ does not form an approximate identity. Instead we average these functions; define F_N by

$$F_N(x) = \frac{1}{N} \sum_{n=0}^{N-1} D_n(x).$$

The sequence $\{F_N\}$ defines an approximate identity called the Fejer kernel. See Figure 4. See Theorem 1.8 both for the verification that $\{F_N\}$ defines an approximate identity and for a decisive consequence.

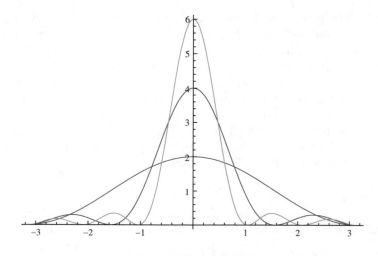

FIGURE 4. Fejer kernel

EXAMPLE 1.6 (Poisson kernel). For $0 \leq r < 1$, define $P_r(\theta)$ as follows. Put $z = re^{i\theta}$ and put $P_r(\theta) = \frac{1-|z|^2}{|1-z|^2}$. Then, as shown in Exercise 1.38, we have

$$P_r(\theta) = \sum_{n \in \mathbf{Z}} r^{|n|} e^{in\theta}. \tag{41}$$

It follows from (41) that the first property of an approximate identity holds. (The only term not integrating to 0 is the term when $n = 0$.) The second property is immediate, as $P_r(\theta) \geq 0$. The third property is also easy to check. Fix $\epsilon > 0$. If $|\theta| \geq \epsilon$ and $z = re^{i\theta}$, then $|1 - z|^2 \geq c_\epsilon > 0$. Hence $P_r(\theta) \leq \frac{1-r^2}{c_\epsilon}$. Thus the limit as r increases to 1 of $P_r(\theta)$ is 0. Hence the Poisson kernel defines an approximate identity on the circle. Figure 5 shows the Poisson kernel for the values $r = \frac{1}{3}$, $r = \frac{1}{2}$, and $r = \frac{2}{3}$.

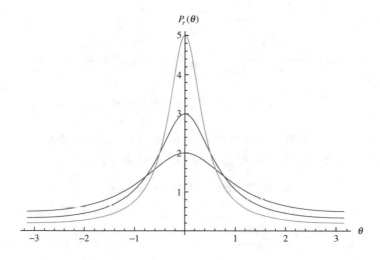

FIGURE 5. Poisson kernel

EXAMPLE 1.7 (Gaussian). For $0 < t < \infty$, put $\mathcal{G}_t(x) = \sqrt{\frac{t}{\pi}} e^{-tx^2}$. Then $\mathcal{G}_t$ defines an approximate identity on $\mathbf{R}$. Since $\mathcal{G}_t(x) > 0$, we need only to show that $\int_{-\infty}^{\infty} \mathcal{G}_t(x) dx = 1$ and, that for $\delta > 0$,

$$\lim_{t \to \infty} \int_{|x| \geq \delta} \mathcal{G}_t(x) dx = 0.$$

See Exercise 1.39. Figure 6 shows the Gaussian for three different values of t.

A *Gaussian* is any function of the form

$$G(x) = \frac{1}{\sigma\sqrt{2\pi}} e^{\frac{-(x-\mu)^2}{2\sigma^2}}.$$

Here $\sigma > 0$ and μ is an arbitrary real number. Gaussians are of crucial importance in probability and statistics. The function G represents the density function for a normal probability distribution with *mean* μ and *variance* σ^2. In Example 1.7, we are setting $\mu = 0$ and $\sigma^2 = \frac{1}{2t}$. When we let t tend to infinity, we are making the variance tend to zero and clustering the probability around the mean, thus giving an intuitive understanding of why $\mathcal{G}_t$ is an approximate identity. By contrast, when we let t tend to 0, the variance tends to infinity and the probability distribution spreads out. We will revisit this situation in Chapter 3.

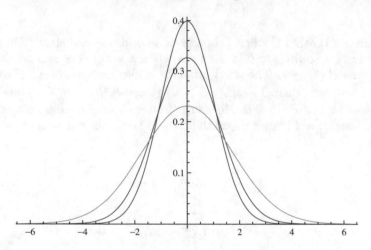

FIGURE 6. Gaussian kernel

EXERCISE 1.38. Verify (41). (Hint: Sum two geometric series.)

EXERCISE 1.39. Verify the statements in Example 1.7.

9. Definition of Fourier series

An infinite series of the form $\sum_{n=-\infty}^{n=\infty} c_n e^{in\theta}$ is called a *trigonometric series*. Such a series need not converge.

Let f be an integrable function on the circle. For $n \in \mathbf{Z}$ we define its Fourier coefficients by

$$\hat{f}(n) = \frac{1}{2\pi} \int_0^{2\pi} f(x) e^{-inx} dx.$$

Note by (9) that the coefficient c_j of a trig polynomial f is precisely the Fourier coefficient $\hat{f}(j)$. Sometimes we write $\mathcal{F}(f)(n) = \hat{f}(n)$. One reason is that this notation helps us think of $\mathcal{F}$ as an operator on functions. If f is integrable on the circle, then $\mathcal{F}(f)$ is a function on the integers, called the Fourier transform of f. Later we will consider Fourier transforms for functions defined on the real line. Another reason for the notation is that typographical considerations suggest it.

DEFINITION 1.5. The Fourier series of an integrable function f on the circle is the trigonometric series

$$\sum_{n \in \mathbf{Z}} \hat{f}(n) e^{inx}.$$

When considering convergence of a trigonometric series, we generally consider limits of the symmetric partial sums defined by

$$S_N(x) = \sum_{n=-N}^{N} a_n e^{inx}.$$

Considering the parts for n positive and n negative separately makes things more complicated. See [SS].

The Fourier series of an integrable function need not converge. Much of the subject of Fourier analysis arose from asking various questions about convergence. For example, under what conditions does the Fourier series of f converge pointwise to f, or when is the series summable to f using some more complicated summability method? We discuss some of these summability methods in the next section.

LEMMA 1.7 (properties of Fourier coefficients). *The following Fourier transform formulas hold:*

(1) $\mathcal{F}(f + g) = \mathcal{F}(f) + \mathcal{F}(g)$
(2) $\mathcal{F}(cf) = c\mathcal{F}(f)$
(3) *For all n,* $\mathcal{F}(\overline{f})(n) = \overline{\hat{f}}(n) = \hat{f}(-n) = \overline{\mathcal{F}(f)(-n)}$.
(4) *For all n,* $|\mathcal{F}(f)(n)| \le ||f||_{L^1}$. *Equivalently,* $||\mathcal{F}(f)||_{L^\infty} \le ||f||_{L^1}$.

PROOF. See Exercise 1.40. $\square$

The first two properties express the linearity of the integral. The third property generalizes the palindrome property of real trig polynomials. We will use the fourth property many times in the sequel.

We also note the relationship between anti-derivatives and Fourier coefficients.

LEMMA 1.8. *Let f be Riemann integrable (hence bounded). Assume that $\hat{f}(0) = \frac{1}{2\pi} \int f(u)du = 0$. Put $F(x) = \int_0^x f(u)du$. Then F is periodic and continuous; also, for $n \ne 0$,*

$$\hat{F}(n) = \frac{\hat{f}(n)}{in}. \tag{42}$$

PROOF. The following inequality implies the continuity of F:

$$|F(x) - F(y)| = \left| \int_y^x f(u)du \right| \le |x - y| \, ||f||_{L^\infty}.$$

Since $\hat{f}(0) = 0$, we see that $0 = F(2\pi) = F(0)$. Finally, formula (42) follows either by integration by parts (Exercise 1.40) or by interchanging the order of integration:

$$\hat{F}(n) = \frac{1}{2\pi} \int_0^{2\pi} \left(\int_0^x f(u)du \right) e^{-inx}dx = \frac{1}{2\pi} \int_0^{2\pi} \left(\int_u^{2\pi} e^{-inx}dx \right) f(u)du$$

$$= \frac{1}{2\pi} \int_0^{2\pi} \frac{1}{-in}(1 - e^{-inu})f(u)du = \frac{\hat{f}(n)}{in},$$

since $\int f(u)du = 0$. $\square$

See Exercise 1.43 for a generalization of this Lemma. The more times f is differentiable, the faster the Fourier coefficients must decay at infinity.

EXERCISE 1.40. Prove Lemma 1.7. Prove Lemma 1.8 using integration by parts. (Assume f is continuous almost everywhere.)

EXERCISE 1.41. Find the Fourier series for $\cos^{2N}(\theta)$. (Hint: Don't do any integrals!)

EXERCISE 1.42. Assume f is real-valued. Under what additional condition can we conclude that its Fourier coefficients are real? Under what condition are the coefficients purely imaginary?

EXERCISE 1.43. Assume that f is k times continuously differentiable. Show that there is a constant C such that

$$\left|\hat{f}(n)\right| \leq \frac{C}{n^k}.$$

EXERCISE 1.44. Assume that $f(x) = -1$ for $-\pi < x < 0$ and $f(x) = 1$ for $0 < x < \pi$. Compute the Fourier series for f.

EXERCISE 1.45. Put $f(x) = e^{ax}$ for $0 < x < 2\pi$. Compute the Fourier series for f.

EXERCISE 1.46. Put $f(x) = \sinh(x)$ for $-\pi < x < \pi$. Compute the Fourier series for f. Here $sinh$ is the hyperbolic sine defined by $\sinh(x) = \frac{e^x - e^{-x}}{2}$.

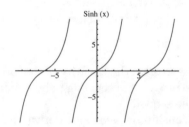

FIGURE 7. Periodic extension of hyperbolic sine

We next establish the fundamental relationship between Fourier series and convolution.

THEOREM 1.6. If f and g are integrable, then $f*g$ is continuous and $\mathcal{F}(f*g) = \mathcal{F}(f)\mathcal{F}(g)$. In other words, for all n we have

$$\mathcal{F}(f*g)(n) = (f*g)^{\hat{}}(n) = \hat{f}(n)\hat{g}(n) = \mathcal{F}(f)(n)\mathcal{F}(g)(n). \tag{43}$$

PROOF. The proof is computational when f and g are continuous. We compute the left-hand side of (43) as a double integral, and then interchange the order of integration. The general case then follows using the approximation Lemma 1.6.

Here are the details. Suppose first that f and g are continuous. Then

$$\mathcal{F}(f*g)(n) = \frac{1}{2\pi} \int_0^{2\pi} \left(\frac{1}{2\pi} \int_0^{2\pi} f(y)g(x-y)dy \right) e^{-inx}dx.$$

By continuity, we may interchange the order of integration, obtaining

$$\mathcal{F}(f*g)(n) = (\frac{1}{2\pi})^2 \int_0^{2\pi} f(y) \left(\int_0^{2\pi} g(x-y)e^{-inx}dx \right) dy.$$

Change variables by putting $x - y = t$. Then use $e^{-in(y+t)} = e^{-iny}e^{-int}$, and the result follows.

Next, assume f and g are Riemann integrable, hence bounded. By Lemma 1.6 we can find sequences of continuous functions f_k and g_k such that $||f_k||_{L^\infty} \leq ||f||_{L^\infty}$, also $||f - f_k||_{L^1} \to 0$, and similarly for g_k. By the usual adding and subtracting trick,

$$f*g - f_k*g_k = ((f - f_k)*g) + (f_k*(g - g_k)). \tag{44}$$

Since g and each f_k is bounded, both terms on the right-hand side of (44) tend to 0 uniformly. Therefore $f_k * g_k$ tends to $f * g$ uniformly. Since the uniform limit of continuous functions is continuous, $f * g$ is itself continuous. Since f_k tends to f in L^1 and (by property (4) from Lemma 1.7)

$$|\hat{f}_k(n) - \hat{f}(n)| \le \frac{1}{2\pi} \int_0^{2\pi} |f_k - f|, \tag{45}$$

it follows that $|\hat{f}_k(n) - \hat{f}(n)|$ converges to 0 for all n. Similarly $|\hat{g}_k(n) - \hat{g}(n)|$ converges to 0 for all n. Hence, for each n, $\hat{f}_k(n)\hat{g}_k(n)$ converges to $\hat{f}(n)\hat{g}(n)$. Since (43) holds for f_k and g_k, it holds for f and g. □

By the previous result, the function $f * g$ is continuous when f and g are assumed only to be integrable. Convolutions are often used to *regularize* a function. For example, if f is integrable and g is infinitely differentiable, then $f*g$ is infinitely differentiable. In Chapter 3 we will use this idea when g_n defines an approximate identity consisting of smooth (infinitely differentiable) functions.

10. Summability methods

We introduce two notions of summability, Cesàro summability and Abel summability, which arise in studying the convergence of Fourier series.

First we make an elementary remark. Let $\{A_n\}$ be a sequence of complex numbers. Let σ_N denote the average of the first N terms:

$$\sigma_N = \frac{A_1 + A_2 + ... + A_N}{N}.$$

If $A_N \to L$, then $\sigma_N \to L$ as well. We will prove this fact below. It applies in particular when A_N is the N-th partial sum of a sequence $\{a_n\}$. There exist examples where A_N does not converge but σ_N does converge. See Theorem 1.7. We therefore obtain a more general notion of summation for the infinite series $\sum a_n$.

Suppose next that $\sum a_n$ converges to L. For $0 \le r < 1$, put $f(r) = \sum a_n r^n$. We show in Corollary 1.5 that $\lim_{r \to 1} f(r) = L$. (Here we are taking the limit as r increases to 1.) There exist series $\sum a_n$ such that $\sum a_n$ diverges but this limit of $f(r)$ exists. A simple example is given by $a_n = (-1)^{n+1}$. A more interesting example is given by $a_n = n(-1)^{n+1}$.

DEFINITION 1.6. Let $\{a_n\}$ be a sequence of complex numbers. Let $A_N = \sum_{j=1}^N a_j$. Let $\sigma_N = \frac{1}{N} \sum_{j=1}^N A_j$. For $0 \le r < 1$ we put $F_N(r) = \sum_{j=1}^N a_j r^j$.

(1) $\sum_1^\infty a_j$ *converges* to L if $\lim_{N \to \infty} A_N = L$.
(2) $\sum_1^\infty a_j$ is *Cesàro summable* to L if $\lim_{N \to \infty} \sigma_N = L$.
(3) $\sum_1^\infty a_j$ is *Abel summable* to L if $\lim_{r \to 1} \lim_{N \to \infty} F_N(r) = L$.

THEOREM 1.7. *Let $\{a_n\}$ be a sequence of complex numbers.*

(1) *If $\sum a_n$ converges to L, then $\sum a_n$ is Cesàro summable to L. The converse fails.*
(2) *If $\sum a_n$ is Cesàro summable to L, then $\sum a_n$ is Abel summable to L. The converse fails.*

PROOF. We start by showing that the converse assertions are false. First put $a_n = (-1)^{n+1}$. The series $\sum a_n$ certainly diverges, because the terms do not tend to 0. On the other hand, the partial sum A_N equals 0 if N is even and equals 1 if

N is odd. Hence $\sigma_{2N} = \frac{1}{2}$ and $\sigma_{2N+1} = \frac{N+1}{2N+1} \to \frac{1}{2}$. Thus $\lim_{N\to\infty} \sigma_N = \frac{1}{2}$. Thus $\sum a_n$ is Cesàro summable but not convergent.

Next put $a_n = n(-1)^{n+1}$. Computation shows that $A_{2N} = -N$ and $A_{2N+1} = N + 1$. It follows that $\sigma_{2N} = 0$ and that $\sigma_{2N+1} = \frac{N+1}{2N+1}$. These expressions have different limits and hence $\lim_N \sigma_N$ does not exist. On the other hand, for $|r| < 1$,

$$\sum_{n=1}^{\infty} n(-1)^{n+1} r^n = r \sum_{1}^{\infty} n(-r)^{n-1} = \frac{r}{(1+r)^2}.$$

Letting r tend upwards to 1 gives the limiting value of $\frac{1}{4}$. Hence $\sum a_n$ is Abel summable to $\frac{1}{4}$ but not Cesàro summable.

Proof of (1): Suppose that $\sum a_n = L$. Replace a_1 with $a_1 - L$ and keep all the other terms the same. The new series sums to 0. Furthermore, each partial sum A_N is decreased by L. Hence the Cesàro means σ_N get decreased by L as well. It therefore suffices to consider the case where $\sum a_n = 0$. Fix $\epsilon > 0$. Since A_N tends to 0, we can find an N_0 such that $N \geq N_0$ implies $|A_N| < \frac{\epsilon}{2}$.

We have for $N \geq N_0$,

$$\sigma_N = \frac{1}{N} \sum_{j=1}^{N_0-1} A_j + \frac{1}{N} \sum_{j=N_0}^{N} A_j. \tag{46}$$

Since N_0 is fixed, the first term tends to 0 as N tends to infinity, and hence its absolute value is bounded by $\frac{\epsilon}{2}$ for large enough N. The absolute value of the second term is bounded by $\frac{\epsilon}{2} \frac{N-N_0+1}{N}$ and hence by $\frac{\epsilon}{2}$ because $N \geq N_0 \geq 1$. The conclusion follows.

Proof of (2): This proof is a bit elaborate and uses summation by parts. Suppose first that $\sigma_N \to 0$. For $0 \leq r < 1$ we claim that

$$(1 - r)^2 \sum_{n=1}^{\infty} n\sigma_n r^n = \sum_{n=1}^{\infty} a_n r^n. \tag{47}$$

We wish to show that the limit as r tends to 1 of the right-hand side of (47) exists and equals 0. Given the claim, consider $\epsilon > 0$. We can find N_0 such that $n \geq N_0$ implies $|\sigma_n| < \frac{\epsilon}{2}$. We break up the sum on the left-hand side of (47) into terms where $n \leq N_0 - 1$ and the rest. The absolute value of the first part is a finite sum times $(1-r)^2$ and hence can be made at most $\frac{\epsilon}{2}$ by choosing r close enough to 1. Note that $\sum_{n=1}^{\infty} nr^{n-1} = \frac{1}{(1-r)^2}$. The second term T can then be estimated by

$$|T| \leq (1-r)^2 \frac{\epsilon}{2} \sum_{N_0}^{\infty} nr^n \leq (1-r)^2 \frac{\epsilon}{2} \sum_{1}^{\infty} nr^n = r\frac{\epsilon}{2}.$$

Hence, given the claim, by choosing r close enough to 1 we can make the absolute value of (47) as small as we wish. Thus $\sum a_n$ is Abel summable to 0. As above, the case where σ_N tends to L reduces to the case where it tends to 0.

It remains to prove (47), which involves summation by parts twice.

$$\sum_{1}^{N} a_n r^n = A_N r^N - \sum_{1}^{N-1} A_n (r^{n+1} - r^n) = A_N r^N + (1-r) \sum_{1}^{N-1} A_n r^n.$$

Next we use summation by parts on $\sum_1^{N-1} A_n r^n$:

$$\sum_1^{N-1} A_n r^n = (N-1)\sigma_{N-1}r^{N-1} - \sum_1^{N-2} n\sigma_n(r^{n+1} - r^n)$$

$$= (N-1)\sigma_{N-1}r^{N-1} + (1-r)\sum_1^{N-2} n\sigma_n r^n.$$

Note that $A_N = N\sigma_N - (N-1)\sigma_{N-1}$. Hence we obtain

$$\sum_1^N a_n r^n =$$

$$(N\sigma_N - (N-1)\sigma_{N-1})r^N + (1-r)r^{N-1}(N-1)\sigma_{N-1} + (1-r)^2\sum_1^{N-2} n\sigma_n r^n. \quad (48)$$

Since $|r| < 1$, $\lim(Nr^N) = 0$. Since also σ_N is bounded, each of the terms in (48) other than the sum converges to 0. Thus $\sum_1^N a_n r^n$ converges to $(1-r)^2\sum_1^\infty n\sigma_n r^n$, as desired. $\qquad\square$

We will recall the notion of *radius of convergence* of a power series in Theorem 1.10 below. Here we note that Abel summability provides information about the behavior of a series on the circle of convergence.

COROLLARY 1.5 (Abel's theorem). *Suppose $\sum_{n=0}^\infty a_n x^n$ has radius of convergence 1, and assume $\sum_{n=0}^\infty a_n$ converges to L. Then the function f, defined on $(-1, 1)$ by this series and by $f(1) = L$, is continuous at $x = 1$. (The limit is taken as $x \to 1$ from the left.)*

PROOF. Combining the two conclusions of Theorem 1.7 shows that the series is Abel summable to L. But Abel summability is simply another way to state the conclusion of the corollary: $\lim_{x\to 1} f(x) = f(1)$. $\qquad\square$

COROLLARY 1.6. $\frac{\pi}{4} = \sum_{n=0}^\infty \frac{(-1)^n}{2n+1}$.

PROOF. By integrating the series for $\frac{1}{1+x^2}$ term by term on $(-1, 1)$, we obtain the power series for inverse tangent there:

$$\tan^{-1}(x) = \sum_{n=0}^\infty (-1)^n \frac{x^{2n+1}}{2n+1}.$$

The series converges at $x = 1$ by Corollary 1.2. By Corollary 1.5, it converges to $\tan^{-1}(1) = \frac{\pi}{4}$. $\qquad\square$

The series in Corollary 1.6 converges rather slowly, and hence it does not yield a good method for approximating π.

The reader should note the similarities in the proofs of Theorem 1.5 and Theorem 1.7. The same ideas appear also in one of the standard proofs of the Fundamental Theorem of Calculus.

Cesàro summability will be important in our analysis of the series $\sum_{n=1}^\infty \frac{\sin(nx)}{n}$. We will prove a general result about convergence to $f(x)$ of the Cesàro means of the Fourier series of the integrable function f at points x where f is continuous.

Then we will compute the Fourier series for the function x on the interval $[0, 2\pi]$. It then follows for $0 < x < 2\pi$ that

$$\sum_{n=1}^{\infty} \frac{\sin(nx)}{n} = \frac{\pi - x}{2}.$$

Note that equality fails at 0 and 2π. Figure 8 shows two partial sums of the series. See also Remark 1.5.

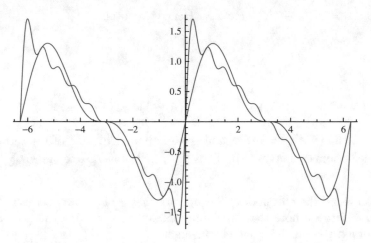

FIGURE 8. Approximations to the sawtooth function

Recall that $S_N(x) = \sum_{-N}^{N} \hat{f}(n) e^{inx}$ denotes the symmetric partial sums of the Fourier series of f.

THEOREM 1.8. *Suppose f is integrable on $[-\pi, \pi]$ and f is continuous at x. The Fourier series for f at x is Cesàro summable to $f(x)$.*

PROOF. Put $D_K(x) = \sum_{-K}^{K} e^{inx}$. Define F_N by

$$F_N(x) = \frac{D_0(x) + D_1(x) + \ldots + D_{N-1}(x)}{N}.$$

Note that $\sigma_N(f)(x) = (f * F_N)(x)$. We claim that $\{F_N\}$ defines an approximate identity.

Since each D_K integrates to 1, each F_N integrates to 1. The first property of an approximate identity therefore holds. A computation (Exercise 1.47) shows that

$$F_N(x) = \frac{1}{N} \frac{\sin^2(\frac{Nx}{2})}{\sin^2(\frac{x}{2})}. \tag{49}$$

Since $F_N \geq 0$, the second property of an approximate identity is automatic. The third is easy to prove. It suffices to show for each ϵ with $0 < \epsilon < \pi$ that

$$\lim_{N \to \infty} \int_{\epsilon}^{\pi} F_N(x) dx = 0.$$

But, for x in the interval $[\epsilon, \pi]$, the term $\frac{1}{\sin^2(\frac{x}{2})}$ is bounded above by a constant and the term $\sin^2(\frac{Nx}{2})$ is bounded above by 1. Hence $F_N \leq \frac{c}{N}$ and the claim follows.

The conclusion of the Theorem now follows by Theorem 1.5.   □

COROLLARY 1.7. *For $0 < x < 2\pi$, we have*

$$\sum_{n=1}^{\infty} \frac{\sin(nx)}{n} = \frac{\pi - x}{2}.$$

PROOF. Put $f(x) = x$ on $[0, 2\pi]$. Compute the Fourier coefficients of x. We have $\hat{f}(0) = \pi$ and $\hat{f}(n) = \frac{-i}{n}$. Hence the Fourier series for f is given by

$$\pi + \sum_{1}^{\infty} e^{-inx} \frac{-i}{n} + \sum_{1}^{\infty} e^{inx} \frac{i}{n}.$$

This series converges in the Cesàro sense to $f(x)$ at each x where f is continuous, namely everywhere except 0 and 2π. By Proposition 1.2 it converges for all x, and by Theorem 1.7, to $f(x)$ when f is continuous. For $0 < x < 2\pi$ we get

$$x = \pi + \sum_{1}^{\infty} e^{-inx} \frac{-i}{n} + \sum_{1}^{\infty} e^{inx} \frac{i}{n} = \pi - 2 \sum_{1}^{\infty} \frac{\sin(nx)}{n},$$

from which the conclusion follows. $\square$

REMARK 1.5. The sine series in Corollary 1.7 converges for all x. The limit function is continuous everywhere except at integer multiples of 2π. The value 0 at the jump discontinuities is the average of the limiting values obtained by approaching from the right and the left. This phenomenon holds rather generally. Figure 8 illustrates the situation in this case, and also displays the *Gibbs phenomenon*; the Fourier series overshoots (or undershoots) the value by a fixed percentage near a jump discontinuity. See [SS], [F2], and their references for more information.

COROLLARY 1.8. *Let f be continuous on the circle. Then there is a sequence of trig polynomials converging uniformly to f.*

PROOF. Each partial sum S_N is a trig polynomial, and hence each Cesàro mean also is. Therefore we need only to strengthen the theorem to show, when f is continuous, that $f * F_N$ converges uniformly to f. The strengthened version follows from the proof. $\square$

Thus trig polynomials are dense in the space of continuous functions on the circle with the norm $\|f\|_{L^\infty}$. Ordinary polynomials are dense in the space of continuous functions on a closed interval as well. This result, called the Weierstrass approximation theorem, has many proofs. In particular it can be derived from Corollary 1.8.

EXERCISE 1.47. Prove (49). Here is one possible approach. We have

$$F_N(x) = \frac{1}{N} \sum_{0}^{N-1} D_k(x) = \frac{1}{N} \sum_{0}^{N-1} \sum_{-k}^{k} e^{ijx} = \frac{1}{N} \sum_{0}^{N-1} \sum_{-k}^{k} w^j,$$

where $w = e^{ix}$. Hence $w^{-1} = \overline{w}$. After factoring out $\overline{w}^k$, rewrite the inner sum as a sum from 0 to $2k$. Sum the finite geometric series. Each of the two terms becomes itself a geometric series. Sum these, and simplify, canceling the common factor of $1 - w$, until you get

$$F_N(x) = \frac{1}{N|1 - w|^2} (2 - w^N - \overline{w}^N) = \frac{1}{N} \frac{2 - 2\cos(Nx)}{2 - 2\cos(x)}.$$

We finally obtain (49) after using the identity

$$\sin^2\left(\frac{\alpha}{2}\right) = \frac{1 - \cos(\alpha)}{2}.$$

EXERCISE 1.48. Verify the previous trig identity using complex exponentials.

EXERCISE 1.49. Put $f(x) = \frac{1}{\log(x)}$. Show that f is convex and decreasing. Conclude that $f(x + 2) + f(x) \geq 2f(x + 1)$.

EXERCISE 1.50. Find the following limit:

$$\lim_{x \to \infty} x \left(\frac{1}{\log(x + 2)} + \frac{1}{\log(x)} - \frac{2}{\log(x + 1)} \right).$$

EXERCISE 1.51. Use (49) and Exercise 1.49 to solve Exercise 1.6. (Again sum by parts twice.) Exercise 1.3 might also be useful.

REMARK 1.6. In solving Exercise 1.51, one must include the first term.

EXERCISE 1.52. Derive the Weierstrass approximation theorem from Corollary 1.8. Suggestion. First show that the Taylor polynomials of e^{ix} uniformly approximate it on any closed and bounded interval. Thus any trig polynomial can be uniformly approximated by ordinary polynomials.

EXERCISE 1.53. If $\{s_n\}$ is a monotone sequence of real numbers, show that the averages $\sigma_N = \frac{1}{N} \sum_{j=1}^{N} s_j$ also define a monotone sequence. Give an example where the converse assertion is false.

11. The Laplace equation

In this section we connect ideas from Abel summability to the Dirichlet problem for the Laplace equation.

We have defined Abel summability of a series $\sum_{n=0}^{\infty} z_n$ of complex numbers. For Fourier series, we make a small change and consider Abel summability of the series $\sum_{-\infty}^{\infty} a_n e^{in\theta}$. Thus we consider

$$\sum_{n \in \mathbf{Z}} a_n r^{|n|} e^{in\theta}, \tag{50}$$

and we let r increase to 1. The series in (50) has a simple expression; it is the convolution of the Poisson kernel with the function h, whose Fourier series is the given series. Since the Poisson kernel is an approximate identity, our next result will follow easily.

THEOREM 1.9. *Suppose h is integrable on the circle. Then the Fourier series of h is Abel summable to h at each point of continuity of h. If h is continuous on the circle, then the Abel means of its Fourier series converge uniformly to h.*

PROOF. Recall that $P_r(\theta) = \frac{1 - |z|^2}{|1 - z|^2}$ when $z = re^{i\theta}$. We have noted that P_r is an approximate identity and that

$$P_r(\theta) = \sum_{n \in \mathbf{Z}} r^{|n|} e^{in\theta}.$$

The Abel means of the Fourier series for h are then

$$\sum_{n \in \mathbf{Z}} \hat{h}(n) r^{|n|} e^{in\theta} = (P_r * h)(\theta).$$

By Theorem 1.5 the Abel means converge to $h(\theta)$ at each point where h is continuous. Also by Theorem 1.5, the convergence is uniform if h is continuous.   □

We recall that a twice differentiable function u of two real variables is called *harmonic* if $\Delta(u) = 0$; that is,

$$\Delta(u) = u_{xx} + u_{yy} = 0.$$

The Dirichlet problem is perhaps the most fundamental boundary-value problem in applied mathematics. We are given a continuous function h on the boundary of an open set Ω, and we wish to find a harmonic function u on Ω such that $u = h$ on the boundary.

For us, Ω will be the unit disk in $\mathbf{C}$; its boundary is the unit circle. Suppose h is given by a Fourier series,

$$h(\theta) = \sum_{n \in \mathbf{Z}} a_n e^{in\theta}.$$

Then the solution u to the Dirichlet problem is uniquely determined and satisfies

$$u(z) = u(re^{i\theta}) = \sum_{n \in \mathbf{Z}} r^{|n|} a_n e^{in\theta}. \tag{51}$$

Before proving this result (Theorem 1.11), we recall one of the basic ideas from complex variable theory, the notion of radius of convergence of a power series.

THEOREM 1.10. *Given a power series $\sum_{n=1}^{\infty} a_n z^n$, there is a non-negative real number R, or the value infinity, such that the series converges for $|z| < R$ and diverges for $|z| > R$. The number R can be computed by Hadamard's formula*

$$R = \sup\{r : |a_n| r^n \text{ is a bounded sequence}\}. \tag{52}$$

If $R = 0$, then the series converges only if $z = 0$. If $0 \le r < R$, the the series converges absolutely and uniformly for $|z| \le r$.

PROOF. Define R by (52). If $|z| > R$, then the terms of the series are unbounded, and hence the series diverges. Next assume that $0 \le \rho < r < R$. Assume that $|a_n| r^n \le M$. We claim that $\sum a_n z^n$ converges absolutely and uniformly in $\{|z| \le \rho\}$. The claim follows by the comparison test and the convergence of a geometric series with ratio $t = \frac{\rho}{r}$:

$$\sum |a_n z^n| \le \sum |a_n| \rho^n = \sum |a_n| r^n (\frac{\rho}{r})^n \le M \sum t^n.$$

Each assertion follows.   □

REMARK 1.7. We also can compute R by the root test. See Exercise 1.57.

The Laplacian Δ has a convenient expression using complex partial derivatives:

$$\Delta(u) = 4u_{z\bar{z}} = 4 \frac{\partial^2 u}{\partial z \partial \bar{z}}.$$

Here the complex partial derivatives are defined by

$$\frac{\partial}{\partial z} = \frac{1}{2} \left(\frac{\partial}{\partial x} - i \frac{\partial}{\partial y} \right)$$

$$\frac{\partial}{\partial \bar{z}} = \frac{1}{2}\left(\frac{\partial}{\partial x} + i\frac{\partial}{\partial y}\right).$$

EXERCISE 1.54. Verify the formula $\Delta(u) = 4u_{z\bar{z}}$.

EXERCISE 1.55. Show that the Laplacian in polar coordinates is given as follows:

$$\Delta(u) = u_{rr} + \frac{1}{r}u_r + \frac{1}{r^2}u_{\theta\theta}.$$

EXERCISE 1.56. Use the previous exercise to show that the real and imaginary parts of z^n are harmonic for n a positive integer.

EXERCISE 1.57. Given the series $\sum a_n z^n$, put $L = \lim \sup(|a_n|^{\frac{1}{n}})$. Show that the radius of convergence R satisfies $R = \frac{1}{L}$.

EXERCISE 1.58. Give three examples of power series with radius of convergence 1 with the following true. The first series converges at no points of the unit circle, the second series converges at some but not all points of the unit circle, and the third series converges at all points of the unit circle.

EXERCISE 1.59. Let p be a polynomial. Show that the series $\sum(-1)^n p(n)$ is Abel summable. More generally, for $|z| < 1$, show that $\sum_0^\infty p(n)z^n$ is a polynomial in $\frac{1}{1-z}$ with no constant term. Hence the limit, as we approach the unit circle from within, exists at every point except 1.

By analogy with the wave equation, the formula for the Laplacian in complex notation suggests that a function $u = u(x,y)$ is harmonic if and only if it can be written $u(x,y) = f(z) + g(\bar{z})$, for functions f and g of the indicated one complex variable. In particular, it suggests that a real-valued function is harmonic if and only if it is the real part of a complex analytic function.

We use these considerations to revisit the Dirichlet problem. Let h be continuous on the circle.

THEOREM 1.11. *Suppose h is continuous on the unit circle. Put*

$$u(re^{i\theta}) = (P_r * h)(\theta).$$

Then u is infinitely differentiable on the unit disk, u is harmonic, and $u = h$ on the circle.

PROOF. Since h is continuous, the Fourier coefficients $\hat{h}(n)$ are bounded. Hence for each $r < 1$ the series in (51) converges absolutely. Put $z = re^{i\theta}$ and write $u(re^{i\theta}) = (P_r * h)(\theta)$. We have

$$u(re^{i\theta}) = \sum_{n=0}^\infty \hat{h}(n)r^n e^{in\theta} + \sum_{n=1}^\infty \hat{h}(-n)r^n e^{-in\theta} = \sum_{n=0}^\infty \hat{h}(n)z^n + \sum_{n=1}^\infty \hat{h}(-n)\bar{z}^n$$

$$= f(z) + g(\bar{z}).$$

Each z^n or $\bar{z}^n$ is harmonic. The power series for f and g each converge absolutely and uniformly on compact subsets of the unit disk. Hence they represent infinitely differentiable functions. We can therefore differentiate term by term to conclude that u is harmonic. Since h is continuous, $h(\theta) = \lim_{r\to 1} u(re^{i\theta})$. $\square$

The Dirichlet problem for domains (open and connected sets) more general than the unit disk is of fundamental importance in applied mathematics. Amazingly enough, the solution for the unit disk extends to much more general situations. By the Riemann mapping theorem, each *simply connected* domain other than $\mathbf{C}$ is conformally equivalent to the unit disk. Hence one can transfer the problem to the disk and solve it there. See [A] or [D2] for additional discussion. Exercise 1.62 provides an important formula.

We make some additional remarks about the Cauchy-Riemann equations. Suppose f is complex analytic in an open set, and we write $f = u + iv$ there. Then u and v are harmonic and satisfy the system of PDE $u_x = v_y$ and $u_y = -v_x$. These two equations are equivalent to the simpler statement $f_{\bar{z}} = 0$ (which yields $\overline{f}_z = 0$ as well). Since $u = \frac{f+\bar{f}}{2}$ (and v has a similar formula), it follows from the formula $\Delta = 4\frac{\partial^2}{\partial z \partial \bar{z}}$ that u and v are harmonic. Furthermore, the Cauchy-Riemann equations guarantee that the level curves of u and v intersect orthogonally. This geometric fact partially explains why complex analytic functions are useful in applied subjects such as fluid flow.

The next exercise approaches the Laplace equation by way of polar coordinates and separation of variables. It presages spherical harmonics, discussed in Section 13 of Chapter 2, where the circle gets replaced with the sphere.

EXERCISE 1.60. Use Exercise 1.55 and separation of variables to find solutions of the Laplace equation $\Delta(f) = 0$. Your answers should be in the form $r^n e^{in\theta}$. Compare with Exercise 1.56.

EXERCISE 1.61. Graph some level sets of the real and imaginary parts of f, when $f(z) = z^2$, when $f(z) = e^z$, and when $f(z) = \log(z)$, for some branch of the logarithm.

EXERCISE 1.62. Assume that f is complex analytic and that h is twice differentiable in a neighborhood of the image of f. Compute the Laplacian of $h \circ f$. Suggestion: Use the formula from Exercise 1.54.

EXERCISE 1.63. Discuss the validity of the formula

$$\log(x + iy) = \frac{1}{2}\log(x^2 + y^2) + i\tan^{-1}(\frac{y}{x}).$$

EXERCISE 1.64. Assume $0 \leq r < 1$. Find formulas for

$$\sum_{n=1}^{\infty} \frac{r^n \cos(n\theta)}{n} \quad \text{and} \quad \sum_{n=1}^{\infty} \frac{r^n \sin(n\theta)}{n}.$$

Suggestion: Start with the geometric series

$$\sum_{n=0}^{\infty} z^n = \frac{1}{1-z},$$

valid for $|z| < 1$. Integrate to obtain a series for $-\log(1 - z)$. Replace z with $re^{i\theta}$, equate real and imaginary parts of the result, and use Exercise 1.63.

Fourier's original work considered the *heat equation* $u_t = \Delta(u)$. The Laplace equation can be regarded as the steady-state solution of the heat equation.

We pause to consider the heat equation by way of separation of variables. Put $u(x, y, t) = A(x, y)B(t)$. The heat equation becomes $\Delta A(x, y)B(t) = A(x, y)B'(t)$, and hence $\frac{\Delta A}{A} = \frac{B'}{B} = \lambda$ for some constant λ, and $B(t) = e^{\lambda t}$. To guarantee that B tends to zero at ∞, the constant λ must be negative. We also obtain the eigenvalue problem $\Delta A = \lambda A$. We then introduce polar coordinates and use the formula in Exercise 1.55. Doing so leads to the equation

$$r^2 A_{rr} + r A_r + A_{\theta\theta} = \lambda r^2 A.$$

We can attack this equation using separation of variables as well. Let us write $A(r,\theta) = g(r)h(\theta)$. We obtain two equations as usual. The equation for h has solutions $h(\theta) = e^{\pm ik\theta}$; we assume that k is an integer to ensure periodicity in θ. The equation for g becomes

$$r^2 g''(r) + r g'(r) - (\lambda r^2 - k^2) g(r) = 0.$$

The change of variables $x = \sqrt{|\lambda|} r$ yields the *Bessel* differential equation

$$x^2 f''(x) + x f'(x) + (x^2 - \nu^2) f(x) = 0. \tag{Bessel}$$

Here ν is the integer k, but the Bessel equation is meaningful for all real values of ν. We make only two related remarks about solutions to the Bessel equation. If we divide by x^2, and then think of $|x|$ as large, the equation tends to $f'' + f = 0$. Hence we might expect, for large $|x|$, that the solutions resemble cosine and sine. In fact, they more closely resemble (linear combinations of) $\frac{\cos(x)}{\sqrt{x}}$ and $\frac{\sin(x)}{\sqrt{x}}$. These statements can be made precise, and they are important in applications. The second remark is that a notion of Fourier-Bessel series exists, in which one expands functions in terms of scaled solutions to the Bessel equation. See [G], [GS], and [F2] for considerable information on Bessel functions and additional references. We note here only that Wilhelm Bessel (1784–1846) was an astronomer who encountered Bessel functions while studying planetary motion.

12. Uniqueness of Fourier coefficients for continuous functions

Suppose two functions have the same Fourier coefficients. Must the two functions be equal? We next show that the answer is yes when the given functions are continuous. This conclusion follows from Theorem 1.5, but we give a somewhat different proof here in order to illustrate the power of approximate identities. The answer is certainly no when the given functions fail to be continuous; if a function is zero except at a finite set of points, for example, then all its Fourier coefficients vanish, but it is not the zero function. Thus continuity is a natural hypothesis in the following theorem and its corollaries. See Remarks 1.8 and 1.9 below for additional information.

THEOREM 1.12. *Suppose f is integrable on the circle and $\hat{f}(n) = 0$ for all n. If f is continuous at p, then $f(p) = 0$. In particular, if f is continuous on the circle, and $\hat{f}(n) = 0$ for all n, then f is the zero function.*

PROOF. Assume first that f is real-valued. Represent the circle as $[-\pi, \pi]$ and suppose without loss of generality that $p = 0$. Assuming that f is continuous at 0 and that $f(p) > 0$, we will show that some Fourier coefficients must be nonzero, thereby proving the contrapositive statement.

In the proof we consider the integrals

$$\int_{-\pi}^{\pi} (c + \cos(\theta))^k f(\theta) d\theta.$$

Here c is a suitable positive constant and k is chosen sufficiently large that this integral is positive. Let $\chi_k(x) = (c + \cos(x))^k$. Since χ_k is a trig polynomial, the positivity of this integral guarantees that $\hat{f}(n) \neq 0$ for some n. Note that, as k tends to infinity, the functions χ_k concentrate at 0, and hence the idea of the proof is one we have seen several times.

We divide the interval $[-\pi, \pi]$ into several parts. See Figure 9. These parts will be given by $|\theta| \leq \eta$, by $\eta \leq |\theta| \leq \delta$, and by $\delta \leq |\theta| \leq \pi$. Since we are assuming $f(0) > 0$, there is a δ with $0 < \delta < \frac{\pi}{2}$ such that $f(\theta) \geq \frac{f(0)}{2}$ for $|\theta| \leq \delta$. Once δ is chosen, we find a small positive c such that $\cos(\theta) \leq 1 - \frac{3c}{2}$ when $|\theta| \geq \delta$. Doing so is possible because $\cos(\theta)$ is bounded away from 1 there. The inequality $|c + \cos(\theta)| \leq 1 - \frac{c}{2}$ for $|\theta| \geq \delta$ follows.

We want $\chi_k(\theta)$ big near 0. We next find η with $0 < \eta < \delta$ such that $c + \cos(\theta) \geq \frac{c}{2} + 1$ for $|\theta| \leq \eta$. Doing so is possible because $1 - \cos(\theta)$ is small near 0.

On the part where θ is close to 0, $\int \chi_k f \geq C(1 + (\frac{c}{2}))^k$. On the part where $\eta \leq |\theta| \leq \delta$, $\int \chi_k f \geq 0$. On the part where $|\theta| \geq \delta$,

$$\left| \int \chi_k \, f \right| \leq C\left(1 - \frac{c}{2}\right)^k. \tag{53}$$

We gather this information. The integral

$$\int_{|\theta| \leq \eta} \chi_k(x) f(x) dx$$

actually tends to infinity with k. The integral over the set where $\eta \leq |\theta| \leq \delta$ yields a positive number. By (53), the integral over the remaining part is bounded as k tends to infinity. Thus the sum of the three pieces tends to infinity and hence

$$\int_{-\pi}^{\pi} \chi_k(x) f(x) dx > 0$$

for large k. Hence some Fourier coefficient of f must be nonzero.

The case when f is complex-valued follows by applying the above reasoning to its real and imaginary parts. Note that the real part of f is $\frac{f + \bar{f}}{2}$ and the imaginary part of f is $\frac{f - \bar{f}}{2i}$. By Lemma 1.7 we know that

$$\hat{\bar{f}}(n) = \overline{\hat{f}(-n)}.$$

Hence all the Fourier coefficients of f vanish if and only if all the Fourier coefficients of its real and imaginary parts vanish. □

COROLLARY 1.9. *If both f and g are continuous on the circle, and $\hat{f}(n) = \hat{g}(n)$ for all n, then $f = g$.*

PROOF. The function $f - g$ is continuous and all its Fourier coefficients vanish. Hence $f - g = 0$ and thus $f = g$. □

COROLLARY 1.10. *Suppose $\sum |\hat{f}(n)|$ converges and f is continuous. Then S_N converges uniformly to f.*

PROOF. Recall that $S_N(f)(x) = \sum_{-N}^{N} e^{inx} \hat{f}(n)$. Each $S_N(f)$ is continuous, and the hypothesis guarantees that $S_N(f)$ converges uniformly. Hence it has a continuous limit g. But g and f have the same Fourier coefficients. By Corollary 1.9, $f = g$. □

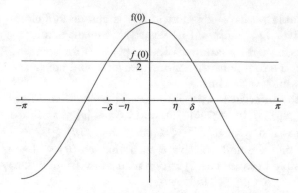

FIGURE 9. Proof of Theorem 1.12

Assuming the continuity of f is not adequate to ensure that the Fourier series of f converges absolutely. If f is twice differentiable, however, then a simple computation shows that $|\hat{f}(n)| \leq \frac{C}{n^2}$ and hence the Fourier series for f does converge absolutely, as $\sum \frac{1}{n^2}$ converges. Since $\sum \frac{1}{n^p}$ converges for $p > 1$, we see that an inequality of the form

$$|\hat{f}(n)| \leq \frac{C}{n^p} \tag{54}$$

for $p > 1$ also guarantees absolute convergence. In Chapter 2 we prove a related but more difficult result involving Hölder continuous functions.

REMARK 1.8. The following fact follows immediately from Theorem 1.12. If all the Fourier coefficients of a Riemann integrable function f vanish, then $f(x) = 0$ at all points x at which f is continuous. In the theory of integration, one establishes also that the set of points at which a Riemann integrable function fails to be continuous has measure zero. Thus we can conclude, when $\hat{f}(n) = 0$ for all n, that f is zero almost everywhere.

REMARK 1.9. There exist continuous functions whose Fourier series do not converge at all points. Constructing such a function is a bit difficult. See [K] or pages 83–87 in [SS].

13. Inequalities

The primary purpose of this section is to use Fourier series to establish some inequalities. Before doing so, we briefly mention versions of the Cauchy-Schwarz inequality. This famous inequality is discussed in detail in Chapter 2. The more abstract formulation there is easier to understand than the following specific special cases. See Chapter 2 for the proof.

PROPOSITION 1.3 (Cauchy-Schwarz inequalities). *Several versions.*

- *Let $a_1, ..., a_n$ and $b_1, ..., b_n$ be complex numbers. Then*

$$\left| \sum_{j=1}^{n} a_j \bar{b}_j \right|^2 \leq \sum_{j=1}^{n} |a_j|^2 \sum_{j=1}^{n} |b_j|^2.$$

- Let $\{a_j\}$ and $\{b_j\}$ be sequences of complex numbers such that $\sum_{j=1}^{\infty} |a_j|^2$ and $\sum_{j=1}^{\infty} |b_j|^2$ converge. Then $\sum_{j=1}^{\infty} a_j \bar{b}_j$ converges and

$$\left| \sum_{j=1}^{\infty} a_j \bar{b}_j \right|^2 \leq \sum_{j=1}^{\infty} |a_j|^2 \sum_{j=1}^{\infty} |b_j|^2.$$

- Suppose that $f, g : [a, b] \to \mathbf{C}$ are such that $\int_a^b |f|^2$ and $\int_a^b |g|^2$ are finite. Then

$$\left| \int_a^b f\bar{g} \right|^2 \leq \int_a^b |f|^2 \int_a^b |g|^2.$$

We now wish to use Fourier series to establish inequalities. The first example is a famous inequality of Hilbert. Rather than proving it directly, we derive it from a general result. There is a vast literature on generalizations of Hilbert's inequality. See [HLP], [B], [S] and their references.

THEOREM 1.13 (Hilbert's inequality). *Let $\{z_n\}$ (for $n \geq 0$) be a sequence of complex numbers with $\sum |z_n|^2$ finite. Then*

$$\left| \sum_{j,k=0}^{\infty} \frac{z_j \bar{z}_k}{1+j+k} \right| \leq \pi \sum_{k=0}^{\infty} |z_k|^2. \tag{55}$$

Furthermore the constant π is the smallest possible.

Hilbert's inequality follows by choosing $g(t) = i(\pi - t)e^{-it}$ in Theorem 1.14 below. With this choice of g we have $\hat{g}(n) = \frac{1}{n+1}$ and hence the j, k entry of the infinite matrix C is $\frac{1}{1+j+k}$. Furthermore $|g|$ is bounded by π.

The inequality can be stated in equivalent ways. For example, by choosing $g(t) = i(\pi - t)$ we have $\hat{g}(n) = \frac{1}{n}$ and we obtain the following:

$$\left| \sum_{j,k=1}^{\infty} \frac{z_j \bar{z}_k}{j+k} \right| \leq \pi \sum_{k=1}^{\infty} |z_k|^2.$$

Polarization yields, in case both sequences $\{z_k\}$ and $\{w_k\}$ are square-summable,

$$\left| \sum_{j,k=1}^{\infty} \frac{z_j \bar{w}_k}{j+k} \right| \leq \pi \left(\sum_{k=1}^{\infty} |z_k|^2 \right)^{\frac{1}{2}} \left(\sum_{k=1}^{\infty} |w_k|^2 \right)^{\frac{1}{2}}. \tag{56}$$

We omit the proof that π is the smallest possible constant. [HLP] has several proofs. We do remark however that equality holds in (55) only if z is the zero sequence. In other words, unless z is the zero sequence, one can replace $\leq$ with $<$ in (55).

We can also write (55) or (56) in terms of integrals rather than sums:

$$\int_0^{\infty} \int_0^{\infty} \frac{f(x)g(y)}{x+y} \, dx dy \leq \pi \|f\|_{L^2} \|g\|_{L^2}, \tag{57.1}$$

where the L^2 norm is taken over $[0, \infty)$. See Exercise 1.67. This formulation suggests a generalization due to Hardy. Let $\frac{1}{p} + \frac{1}{q} = 1$. Then p, q are called conjugate exponents. For $1 < p < \infty$, Hardy's result gives the following inequality:

$$\int_0^\infty \int_0^\infty \frac{f(x)g(y)}{x+y} \, dx \, dy \leq \frac{\pi}{\sin(\frac{\pi}{p})} ||f||_{L^p} ||g||_{L^q}. \tag{57.2}$$

Again the constant is the smallest possible, and again strict inequality holds unless f or g is zero almost everywhere. We will verify (57.2) in Chapter 3 after we prove Hölder's inequality.

THEOREM 1.14. *Let g be integrable on $[0, 2\pi]$ with $\sup(|g|) = M$ and Fourier coefficients $\hat{g}(k)$. Let C denote an infinite matrix whose entries c_{jk} for $0 \leq j, k$ satisfy*

$$c_{jk} = \hat{g}(j+k).$$

Let $\{z_k\}$ and $\{w_k\}$ be square-summable sequences. The following inequalities hold:

$$\left| \sum_{j,k=0}^\infty c_{jk} z_j \overline{z}_k \right| \leq M \sum_{j=0}^\infty |z_j|^2. \tag{58.1}$$

$$\left| \sum_{j,k=0}^\infty c_{jk} z_j \overline{w}_k \right| \leq M \left(\sum_{j=0}^\infty |z_j|^2 \right)^{\frac{1}{2}} \left(\sum_{j=0}^\infty |w_j|^2 \right)^{\frac{1}{2}}. \tag{58.2}$$

PROOF. Since (58.1) is the special case of (58.2) when the sequences are equal, it suffices to prove (58.2). Put $u_N = \sum_{j=0}^N z_j e^{-ijt}$ and $v_N = \sum_{k=0}^N \overline{w}_k e^{-ikt}$. For each N we have

$$\sum_{j,k=0}^N c_{jk} z_j \overline{w}_k = \frac{1}{2\pi} \int_0^{2\pi} \sum_{j,k=0}^N z_j \overline{w}_k e^{-i(j+k)t} g(t) dt$$

$$= \frac{1}{2\pi} \int_0^{2\pi} \sum_{j=0}^N z_j e^{-ijt} \sum_{k=0}^N \overline{w}_k e^{-ikt} g(t) dt = \frac{1}{2\pi} \int_0^{2\pi} u_N(t) v_N(t) \, g(t) dt. \tag{59}$$

Since $|g| \leq M$ we obtain

$$\left| \sum_{j,k=0}^N c_{jk} z_j \overline{w}_k \right| \leq \frac{M}{2\pi} \int_0^{2\pi} |u_N(t)| \, |v_N(t)| dt \tag{60}$$

The Cauchy-Schwarz inequality for integrals (see Proposition above or Chapter 2) implies that

$$\int |u_N| \, |v_N| \leq \left(\int |u_N|^2 \right)^{\frac{1}{2}} \left(\int |v_N|^2 \right)^{\frac{1}{2}}.$$

By the orthogonality of the functions $t \to e^{int}$, we also see that

$$\frac{1}{2\pi} \int |u_N|^2 = \sum_{j=0}^N |z_j|^2 \leq \sum_{j=0}^\infty |z_j|^2 = ||z||_2^2, \tag{61.1}$$

$$\frac{1}{2\pi} \int |v_N|^2 = \sum_{k=0}^N |w_k|^2 \leq \sum_{k=0}^\infty |w_k|^2 = ||w||_2^2. \tag{61.2}$$

We can therefore continue estimating (60) to get

$$| \sum_{j,k=0}^{N} c_{jk} z_j \overline{w}_k | \leq M \left(\sum_{j=0}^{N} |z_j|^2 \right)^{\frac{1}{2}} \left(\sum_{j=0}^{N} |w_j|^2 \right)^{\frac{1}{2}}.$$

The desired inequality (58.2) follows by letting N tend to infinity.   □

The computation in the proof of this theorem differs when the coefficients of the matrix C are instead given by $\hat{g}(j - k)$. Suppose the sequences z and w are equal. Then we obtain

$$\sum_{j,k=0}^{\infty} c_{jk} z_j \overline{z}_k = \frac{1}{2\pi} \int_0^{2\pi} \sum_{j,k=0}^{\infty} z_j \overline{z}_k e^{-i(j-k)t} g(t) dt = \frac{1}{2\pi} \int_0^{2\pi} | \sum_{j=0}^{\infty} z_j e^{-ijt} |^2 g(t) dt.$$

To this point no inequality is used. We obtain information from both upper and lower bounds for g. When g is non-negative, we conclude that the infinite matrix $\hat{g}(j - k)$ is non-negative definite. This result is the easy direction of Herglotz's theorem: The matrix whose entries satisfy $C_{jk} = c_{j-k}$ is non-negative definite if and only if there is a positive measure μ such that $c_j = \hat{\mu}(j)$. In our case the measure is simply $\frac{g dt}{2\pi}$. See [K] for a proof of Herglotz's theorem.

We sketch another proof of Hilbert's inequality. We change notation slightly; the coefficients a_n play the role of the sequence $\{z_n\}$ in (55).

PROOF. Consider a power series $f(z) = \sum_{n=0}^{\infty} a_n z^n$ that converges in a region containing the closed unit disk. Squaring f yields the series

$$(f(z))^2 = \sum_{m=0}^{\infty} \sum_{n=0}^{\infty} a_n a_m z^{m+n}.$$

To obtain the expression in Hilbert's inequality, we integrate from 0 to 1:

$$\int_0^1 (f(z))^2 dz = \sum_{m,n=0}^{\infty} \frac{a_n a_m}{1 + m + n}. \tag{62}$$

By the Cauchy integral theorem, Theorem 6.8 of the appendix, the integral of a complex analytic function around a closed loop vanishes. Hence the integral $\int_{-1}^{1} (f(z))^2 dz$ along the real axis equals the integral $- \int_\gamma (f(z))^2 dz$, where γ denotes the semi-circle (of radius 1) from 1 to -1. We perform this "deformation of contour" in order to involve polar coordinates. When the coefficients are non-negative and at least one is positive, the integral from -1 to 1 exceeds the integral from 0 to 1.

Assume that all coefficients are non-negative and that f is not identically 0. Using (62) and the orthogonality of the $e^{in\theta}$ we obtain

$$\sum_{m,n=0}^{\infty} \frac{a_n a_m}{1 + m + n} = \int_0^1 (f(z))^2 dz < \int_{-1}^{1} (f(z))^2 dz$$

$$= - \int_\gamma (f(z))^2 dz \leq \frac{1}{2} \int_{-\pi}^{\pi} |f(e^{i\theta})|^2 d\theta = \pi \sum |a_n|^2. \tag{63}$$

Assuming the series converges in a region containing the closed unit disk is a bit too strong. It we follow the same proof for f a polynomial of degree N, and then let N tend to infinity, we obtain Hilbert's inequality, but with the strict $<$ replaced by $\leq$ in (63). The computation in the last step in (63) recurs throughout this book.   □

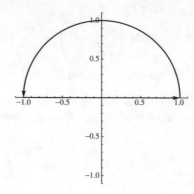

FIGURE 10. Contour used in second proof of Hilbert's inequality

EXERCISE 1.65. Give an example of a sequence $\{a_n\}$ such that the radius of convergence of $\sum a_n z^n$ equals 1 but $\sum |a_n|^2$ diverges. Give another example where $\sum |a_n|^2$ converges.

EXERCISE 1.66. Verify for $n \neq -1$ that $\hat{g}(n) = \frac{1}{n+1}$ when $g(t) = i(\pi - t)e^{-it}$.

EXERCISE 1.67. Show that (57.1) is equivalent to (55). Suggestion: Given sequences $(x_0, x_1, \dots)$ and $(y_0, y_1, \dots)$ in l^2, define $f(x)$ by $f(x) = x_0$ on $[0, 1)$, by x_1 on $[1, 2)$ and so on, and give a similar definition of g.

EXERCISE 1.68. The trig polynomial $p(\theta) = 1 + a\cos(\theta)$ is non-negative on the circle if and only if $|a| \leq 1$. By Herglotz's theorem (see the discussion preceding the second proof of Hilbert's inequality) a certain infinite matrix is therefore non-negative definite if and only if $|a| \leq 1$. Find this matrix, and verify directly that it is non-negative definite if and only if $|a| \leq 1$. Suggestion: Find an explicit formula for the determinant of the N-th principal minor, and then let N tend to infinity.

EXERCISE 1.69. (Difficult) Generalize Exercise 1.68 to the polynomial considered in Exercise 1.8.

EXERCISE 1.70. Use the Cauchy-Schwarz inequality to show the following:

$$\left| \sum_{j=1}^{n} a_j \right|^2 \leq n \sum_{j=1}^{n} |a_j|^2$$

$$\left| \sum_{j=1}^{n} j a_j \right|^2 \leq \left(\frac{n^3}{3} + \frac{n^2}{2} + \frac{n}{6} \right) \sum_{j=1}^{n} |a_j|^2$$

CHAPTER 2

Hilbert spaces

1. Introduction

Fourier series played a significant role in the development of Hilbert spaces and other aspects of abstract analysis. The theory of Hilbert spaces returns the favor by illuminating much of the information about Fourier series. We first develop enough information about Hilbert spaces to allow us to regard Fourier series as orthonormal expansions. We prove that (the symmetric partial sums of) the Fourier series of a square integrable function converges in L^2. From this basic result we obtain corollaries such as Parseval's formula and the Riemann-Lebesgue lemma. We prove Bernstein's theorem: the Fourier series of a Hölder continuous function (with exponent greater than $\frac{1}{2}$) converges absolutely. We prove the spectral theorem for compact Hermitian operators. We include Sturm-Liouville theory to illustrate orthonormal expansion. We close by discussing spherical harmonics, indicating one way to pass from the circle to the sphere. These results leave one in awe at the strength of 19-th century mathematicians.

The ideas of real and complex geometry combine to make Hilbert spaces a beautiful and intuitive topic. A Hilbert space is a complex vector space with a Hermitian inner product and corresponding norm making it into a complete metric space. Completeness enables a deep connection between analytic and geometric ideas. Polarization also plays a significant role. Several of the results in Section 6 apply only to complex inner product spaces.

2. Norms and inner products

Let V be a vector space over the complex numbers. In order to discuss convergence in V, it is natural to use *norms* to compute the lengths of vectors in V. In Chapter 3 we will see the more general concept of a *semi-norm*.

DEFINITION 2.1 (norm). A *norm* on a (real or) complex vector space V is a function $v \mapsto ||v||$ satisfying the following three properties:

(1) $||v|| > 0$ for all nonzero v.
(2) $||cv|| = |c| \, ||v||$ for all $c \in \mathbf{C}$ and all $v \in V$.
(3) (The triangle inequality) $||v + w|| \leq ||v|| + ||w||$ for all $v, w \in V$.

Given a norm $|| \ ||$, we define its corresponding *distance function* by

$$d(u, v) = ||u - v||. \tag{1}$$

The function d is symmetric in its arguments u and v, its values are non-negative, and its values are positive when $u \neq v$. The triangle inequality

$$||u - \zeta|| \leq ||u - v|| + ||v - \zeta||$$

© Springer Nature Switzerland AG 2019
J. P. D'Angelo, *Hermitian Analysis*, Cornerstones,
https://doi.org/10.1007/978-3-030-16514-7_2

follows immediately from the triangle inequality for the norm. Therefore d defines a distance function in the metric space sense (defined in the appendix) and (V, d) is a metric space.

DEFINITION 2.2. A sequence $\{z_n\}$ in a normed vector space V converges to z if $||z_n - z||$ converges to 0. A series $\sum z_k$ converges to w if the sequence $\{\sum_{k=1}^{n} z_k\}$ of partial sums converges to w.

Many of the proofs from elementary real analysis extend to the setting of metric spaces and even more of them extend to normed vector spaces. The norm in the Hilbert space setting arises from an inner product. The norm is a much more general concept. Before we give the definition of Hermitian inner product, we recall the basic example of complex Euclidean space.

EXAMPLE 2.1. Let $\mathbf{C}^n$ denote complex Euclidean space of dimension n. As a set, $\mathbf{C}^n$ consists of all n-tuples of complex numbers; we write $z = (z_1, \ldots, z_n)$ for a point in $\mathbf{C}^n$. This set has the structure of a complex vector space with the usual operations of vector addition and scalar multiplication. The notation $\mathbf{C}^n$ includes the vector space structure, the Hermitian inner product defined by (2.1), and the squared norm defined by (2.2). The *Euclidean inner product* is given by

$$\langle z, w \rangle = \sum_{j=1}^{n} z_j \overline{w}_j \tag{2.1}$$

and the *Euclidean squared norm* is given by

$$||z||^2 = \langle z, z \rangle. \tag{2.2}$$

Properties (1) and (2) of a norm are evident. We establish property (3) below.

The Euclidean norm on $\mathbf{C}^n$ determines by (1) the usual Euclidean distance function. A sequence of vectors in $\mathbf{C}^n$ converges if and only if each component sequence converges; hence $\mathbf{C}^n$ is a complete metric space. See Exercise 2.6.

DEFINITION 2.3 (Hermitian inner product). Let V be a complex vector space. A *Hermitian inner product* on V is a function $\langle \, , \, \rangle$ from $V \times V$ to $\mathbf{C}$ satisfying the following four properties. For all $u, v, w \in V$, and for all $c \in \mathbf{C}$:

(1) $\langle u + v, w \rangle = \langle u, w \rangle + \langle v, w \rangle$.
(2) $\langle cu, v \rangle = c \langle u, v \rangle$.
(3) $\langle u, v \rangle = \overline{\langle v, u \rangle}$. (Hermitian symmetry)
(4) $\langle u, u \rangle > 0$ for $u \neq 0$. (Positive definiteness)

Three additional properties are consequences:

- $\langle u, v + w \rangle = \langle u, v \rangle + \langle u, w \rangle$.
- $\langle u, cv \rangle = \overline{c} \langle u, v \rangle$.
- $\langle 0, w \rangle = 0$ for all $w \in V$. In particular $\langle 0, 0 \rangle = 0$.

Positive definiteness provides a technique for verifying that a given z equals 0. We see from the above that $z = 0$ if and only if $\langle z, w \rangle = 0$ for all w in V.

DEFINITION 2.4. The *norm* $|| \, ||$ corresponding to the Hermitian inner product $\langle \, , \, \rangle$ is defined by

$$||v|| = \sqrt{\langle v, v \rangle}.$$

A Hermitian inner product determines a norm, but most norms do not come from inner products. See Exercise 2.5.

EXERCISE 2.1. Verify the three additional properties of the inner product.

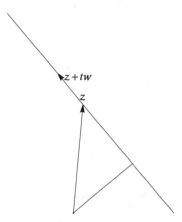

FIGURE 1. Proof of the Cauchy-Schwarz inequality

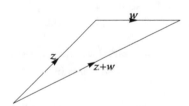

FIGURE 2. Triangle inequality

THEOREM 2.1 (The Cauchy-Schwarz and triangle inequalities). *Let V be a complex vector space, let $\langle\ ,\ \rangle$ be a Hermitian inner product on V, and let $||v|| = \sqrt{\langle v, v\rangle}$. The function $||\ ||$ defines a norm on V and the following inequalities hold for all $z, w \in V$:*

$$|\langle z, w\rangle| \leq ||z||\ ||w|| \tag{3}$$

$$||z + w|| \leq ||z|| + ||w||. \tag{4}$$

PROOF. The first two properties of a norm are evident. The first follows from the positive definiteness of the inner product. To prove the second, it suffices to show that $|c|^2||v||^2 = ||cv||^2$. This conclusion follows from

$$||cv||^2 = \langle cv, cv\rangle = c\langle v, cv\rangle = |c|^2\langle v, v\rangle = |c|^2\ ||v||^2.$$

Note that we have used the linearity in the first slot and the conjugate linearity in the second slot. The third property of a norm is the triangle inequality (4).

We first prove the Cauchy-Schwarz inequality (3). For all $t \in \mathbf{C}$, and for all z and w in V,

$$0 \leq ||z + tw||^2 = ||z||^2 + 2\mathrm{Re}\langle z, tw \rangle + |t|^2 ||w||^2. \tag{5}$$

Think of z and w as fixed, and let ϕ be the quadratic Hermitian polynomial in t and $\bar{t}$ defined by the right-hand side of (5). The values of ϕ are non-negative; we seek its minimum value by setting its differential equal to 0. (Compare with Exercise 1.13) We use subscripts to denote the derivatives with respect to t and $\bar{t}$. Since ϕ is real-valued, we have $\phi_t = 0$ if and only if $\phi_{\bar{t}} = 0$. From (5) we find

$$\phi_{\bar{t}} = \langle z, w \rangle + t||w||^2.$$

When $w = 0$ we get no useful information, but inequality (3) is true when $w = 0$. To prove (3) when $w \neq 0$, we may set

$$t = \frac{-\langle z, w \rangle}{||w||^2}$$

in (5) and conclude that

$$0 \leq ||z||^2 - 2\frac{|\langle z, w \rangle|^2}{||w||^2} + \frac{|\langle z, w \rangle|^2}{||w||^2} = ||z||^2 - \frac{|\langle z, w \rangle|^2}{||w||^2}. \tag{6}$$

Inequality (6) yields

$$|\langle z, w \rangle|^2 \leq ||z||^2 ||w||^2,$$

from which (3) follows by taking square roots.

To establish the triangle inequality (4), we begin by squaring its left-hand side:

$$||z + w||^2 = ||z||^2 + 2\mathrm{Re}\langle z, w \rangle + ||w||^2. \tag{7}$$

Since $\mathrm{Re}\langle z, w \rangle \leq |\langle z, w \rangle|$, the Cauchy-Schwarz inequality yields

$$||z + w||^2 = ||z||^2 + 2\mathrm{Re}\langle z, w \rangle + ||w||^2 \leq ||z||^2 + 2||z||\,||w|| + ||w||^2 = (||z|| + ||w||)^2.$$

Taking the square root of each side gives the triangle inequality and completes the proof that $\sqrt{\langle v, v \rangle}$ defines a norm on V. $\qquad\square$

In the proof we noted the identity (7). This (essentially trivial) identity has two significant corollaries.

THEOREM 2.2. *Let V be a complex inner product space. The following hold:*
Pythagorean theorem: $\langle z, w \rangle = 0$ *implies* $||z + w||^2 = ||z||^2 + ||w||^2$.
Parallelogram law: $||z + w||^2 + ||z - w||^2 = 2(||z||^2 + ||w||^2)$.

PROOF. The Pythagorean theorem is immediate from (7), because $\langle z, w \rangle = 0$ implies that $\mathrm{Re}(\langle z, w \rangle) = 0$. The parallelogram law follows from (7) by adding the result in (7) to the result of replacing w by $-w$ in (7). $\qquad\square$

The two inequalities from Theorem 2.1 have many consequences. We use them here to show that the inner product and norm on V are (sequentially) continuous functions.

PROPOSITION 2.1. (Continuity of the inner product and the norm) *Let V be a complex vector space with Hermitian inner product and corresponding norm. Let $\{z_n\}$ be a sequence that converges to z in V. Then, for all $w \in V$, the sequence of inner products $\langle z_n, w \rangle$ converges to $\langle z, w \rangle$. Furthermore $||z_n||$ converges to $||z||$.*

PROOF. By the linearity of the inner product and the Cauchy-Schwarz inequality, we have

$$|\langle z_n, w \rangle - \langle z, w \rangle| = |\langle z_n - z, w \rangle| \leq ||z_n - z||\, ||w||. \tag{8}$$

Thus, when z_n converges to z, the right-hand side of (8) converges to 0, and therefore so does the left-hand side. Thus the inner product (with w) is continuous.

The proof of the second statement uses the triangle inequality. From it we obtain the inequality $||z|| \leq ||z - z_n|| + ||z_n||$ and hence

$$||z|| - ||z_n|| \leq ||z - z_n||.$$

Interchanging the roles of z_n and z gives the same inequality with a negative sign on the left-hand side. Combining these inequalities yields

$$|\,||z|| - ||z_n||\,| \leq ||z - z_n||,$$

from which the second statement follows.   □

Suppose that $\sum v_n$ converges in V. For all $w \in V$, we have

$$\left\langle \sum_n v_n, w \right\rangle = \sum_n \langle v_n, w \rangle.$$

This conclusion follows by applying Proposition 2.1 to the partial sums of the series. We will often apply this result when working with orthonormal expansions.

Finite-dimensional complex Euclidean spaces are complete in the sense that Cauchy sequences have limits. Infinite-dimensional complex vector spaces with Hermitian inner products need not be complete. By definition, Hilbert spaces are complete.

DEFINITION 2.5. A *Hilbert space* $\mathcal{H}$ is a complex vector space, together with a Hermitian inner product whose corresponding distance function makes $\mathcal{H}$ into a complete metric space.

EXERCISE 2.2. Prove the Cauchy-Schwarz inequality in $\mathbf{R}^n$ by writing $||x||^2||y||^2 - |\langle x, y \rangle|^2$ as a sum of squares. Give the analogous proof in $\mathbf{C}^n$.

EXERCISE 2.3. Prove the Cauchy-Schwarz inequality in $\mathbf{R}^n$ using Lagrange multipliers.

EXERCISE 2.4. Let $\mathcal{H}$ be an inner product space. We showed, for all z and w in $\mathcal{H}$, that (9) holds:

$$||z + w||^2 + ||z - w||^2 = 2||z||^2 + 2||w||^2. \tag{9}$$

Why is this identity called the *parallelogram law*?

EXERCISE 2.5. (Difficult) Let V be a real or complex vector space with a norm. Show that this norm comes from an inner product if and only if the norm satisfies the parallelogram law (9). Comment: Given the norm, one has to define the inner product somehow, and then prove that the inner product satisfies all the necessary properties. Use a polarization identity such as (19) to get started.

We give several examples of Hilbert spaces. We cannot verify completeness in the last example without developing the Lebesgue integral. We do, however, make the following remark. Suppose we are given a metric space that is not complete. We may form its completion by considering equivalence classes of Cauchy sequences in a manner similar to defining the real numbers $\mathbf{R}$ as the completion of the rational numbers $\mathbf{Q}$. Given an inner product space, we may complete it into a Hilbert space. The problem is that we wish to have a concrete realization of the limiting objects.

EXAMPLE 2.2. (Hilbert Spaces)

(1) Complex Euclidean space $\mathbf{C}^n$ is a complete metric space with the distance function given by $d(z,w) = ||z - w||$, and hence it is a Hilbert space.
(2) l^2. Let $a = \{a_\nu\}$ denote a sequence of complex numbers. We say that a is *square-summable*, and we write $a \in l^2$, if $||a||_2^2 = \sum_\nu |a_\nu|^2$ is finite. When $a, b \in l^2$ we write

$$\langle a, b \rangle_2 = \sum_\nu a_\nu \bar{b}_\nu$$

for their Hermitian inner product. Exercise 2.6 requests a proof that l^2 is a complete metric space; here $d(a,b) = ||a - b||_2$.
(3) $\mathcal{A}^2(B_1)$. This space consists of all complex analytic functions f on the unit disk B_1 in $\mathbf{C}$ such that $\int_{B_1} |f|^2 dxdy$ is finite. The inner product is given by

$$\langle f, g \rangle = \int_{B_1} f\bar{g} dxdy.$$

(4) $L^2(\Omega)$. Let Ω be an open subset of $\mathbf{R}^n$. Let dV denote Lebesgue measure in $\mathbf{R}^n$. We write $L^2(\Omega)$ for the complex vector space of (equivalence classes of) measurable functions $f : \Omega \to \mathbf{C}$ for which $\int_\Omega |f(x)|^2 dV(x)$ is finite. When f and g are elements of $L^2(\Omega)$, we define their inner product by

$$\langle f, g \rangle = \int_\Omega f(x)\overline{g(x)} dV(x).$$

The corresponding norm and distance function make $L^2(\Omega)$ into a complete metric space, so $L^2(\Omega)$ is a Hilbert space. See [F1] for a proof of completeness.

EXERCISE 2.6. Verify that $\mathbf{C}^n$ and l^2 are complete.

EXERCISE 2.7. Let V be a normed vector space. Show that V is complete if and only if, whenever $\sum_n ||v_n||$ converges, then $\sum_n v_n$ converges. Compare with Exercise 1.5.

3. Subspaces and linear maps

A *subspace* of a vector space is a subset that is itself a vector space under the same operations of addition and scalar multiplication. A finite-dimensional subspace of a Hilbert space is necessarily closed (in the metric space sense) whereas infinite-dimensional subspaces need not be closed. A closed linear subspace of a Hilbert space is complete and therefore also a Hilbert space. Let B be a bounded domain in $\mathbf{C}^n$. Then $\mathcal{A}^2(B)$ is a closed subspace of $L^2(B)$, and thus a Hilbert space. See Theorem 4.12 and Remark 4.20 from Chapter 4.

Next we define *bounded linear transformations* or *operators*. These mappings are the continuous functions between Hilbert spaces that preserve the vector space structure.

DEFINITION 2.6. Let $\mathcal{H}$ and $\mathcal{H}'$ be Hilbert spaces. A function $L : \mathcal{H} \to \mathcal{H}'$ is called *linear* if it satisfies properties (1) and (2). Also, L is called a *bounded linear transformation* from $\mathcal{H}$ to $\mathcal{H}'$ if L satisfies all three of the following properties:
   (1) $L(z_1 + z_2) = L(z_1) + L(z_2)$ for all z_1 and z_2 in $\mathcal{H}$.
   (2) $L(cz) = cL(z)$ for all $z \in \mathcal{H}$ and all $c \in \mathbf{C}$.
   (3) There is a constant C such that $||L(z)|| \leq C||z||$ for all $z \in \mathcal{H}$.

We write $\mathcal{L}(\mathcal{H}, \mathcal{H}')$ for the collection of bounded linear transformations from $\mathcal{H}$ to $\mathcal{H}'$, and $\mathcal{L}(\mathcal{H})$ in the important special case when $\mathcal{H} = \mathcal{H}'$. In this case I denotes the identity linear transformation, given by $I(z) = z$. Elements of $\mathcal{L}(\mathcal{H})$ are often called *bounded operators* on $\mathcal{H}$. The collection of bounded operators is an algebra, where composition plays the role of multiplication.

Properties (1) and (2) define the *linearity* of L. Property (3) guarantees the continuity of L; see Lemma 2.1 below. The infimum of the set of constants C that work in (3) provides a measurement of the size of the transformation L; it is called the norm of L, and is written $||L||$. Exercise 2.9 justifies the terminology. An equivalent way to define $||L||$ is the formula

$$||L|| = \sup_{\{z \neq 0\}} \frac{||L(z)||}{||z||}.$$

The set $\mathcal{L}(\mathcal{H}, \mathcal{H}')$ becomes a complete normed vector space. See Exercise 2.9.

We next discuss the relationship between boundedness and continuity for linear transformations.

LEMMA 2.1. *Assume $L : \mathcal{H} \to \mathcal{H}'$ is linear. The following three statements are equivalent:*

   (1) *There is a constant $C > 0$ such that, for all z,*

   $$||Lz|| \leq C||z||.$$

   (2) *L is continuous at the origin.*
   (3) *L is continuous at every point.*

PROOF. It follows from the ϵ-δ definition of continuity at a point and the linearity of L that statements (1) and (2) are equivalent. Statement (3) implies statement (2). Statement (1) and the linearity of L imply statement (3) because

$$||Lz - Lw|| = ||L(z - w)|| \leq C||z - w||.$$

$\square$

We associate two natural subspaces with a linear mapping.

DEFINITION 2.7. For $L \in \mathcal{L}(\mathcal{H}, \mathcal{H}')$, the *nullspace* $\mathcal{N}(L)$ is the set of $v \in \mathcal{H}$ for which $L(v) = 0$. The *range* $\mathcal{R}(L)$ is the set of $w \in \mathcal{H}'$ for which there is a $v \in \mathcal{H}$ with $L(v) = w$.

DEFINITION 2.8. An operator $P \in \mathcal{L}(\mathcal{H})$ is a *projection* if $P^2 = P$.

Observe (see Exercise 2.11) that $P^2 = P$ if and only if $(I - P)^2 = I - P$. Thus $I - P$ is also a projection if P is. Furthermore, in this case $\mathcal{R}(P) = \mathcal{N}(I - P)$ and $\mathcal{H} = \mathcal{R}(P) + \mathcal{N}(P)$.

Bounded linear functionals, that is, elements of $\mathcal{L}(\mathcal{H}, \mathbf{C})$, are especially important. The vector space of bounded linear functionals on $\mathcal{H}$ is called the *dual space* of $\mathcal{H}$. We characterize this space in Theorem 2.4 below.

DEFINITION 2.9. A *bounded linear functional* on a Hilbert space $\mathcal{H}$ is a bounded linear transformation from $\mathcal{H}$ to $\mathbf{C}$.

One of the major results in pure and applied analysis is the Riesz lemma, Theorem 2.4 below. A bounded linear functional on a Hilbert space is always given by an inner product. In order to prove this basic result we develop material on orthogonality that also particularly illuminates our work on Fourier series.

EXERCISE 2.8. For $L \in \mathcal{L}(\mathcal{H}, \mathcal{H}')$, verify that $\mathcal{N}(L)$ is a subspace of $\mathcal{H}$ and $\mathcal{R}(L)$ is a subspace of $\mathcal{H}'$.

EXERCISE 2.9. With $||L||$ defined as above, show that $\mathcal{L}(\mathcal{H})$ is a complete normed vector space.

EXERCISE 2.10. Show by using a basis that a linear functional on $\mathbf{C}^n$ is given by an inner product.

EXERCISE 2.11. Let P be a projection. Verify that $I - P$ is a projection, that $\mathcal{R}(P) = \mathcal{N}(I - P)$, and that $\mathcal{H} = \mathcal{R}(P) + \mathcal{N}(P)$.

4. Orthogonality

Let $\mathcal{H}$ be a Hilbert space, and suppose $z, w \in \mathcal{H}$. We say that z and w are *orthogonal* if $\langle z, w \rangle = 0$. The Pythagorean theorem indicates that orthogonality generalizes perpendicularity and provides geometric insight in the general Hilbert space setting. The term "orthogonal" applies also for subspaces. Subspaces V and W of $\mathcal{H}$ are orthogonal if $\langle v, w \rangle = 0$ for all $v \in V$ and $w \in W$. We say that z is orthogonal to V if $\langle z, v \rangle = 0$ for all v in V, or equivalently, if the one-dimensional subspace generated by z is orthogonal to V.

Let V and W be orthogonal closed subspaces of a Hilbert space; $V \oplus W$ denotes their orthogonal sum. It is the subspace of $\mathcal{H}$ consisting of those z that can be written $z = v + w$, where $v \in V$ and $w \in W$. We sometimes write $z = v \oplus w$ in order to emphasize orthogonality. By the Pythagorean theorem, $||v \oplus w||^2 = ||v||^2 + ||w||^2$. Thus $v \oplus w = 0$ if and only if both $v = 0$ and $w = 0$.

We now study the geometric notion of orthogonal projection onto a closed subspace. The next theorem guarantees that we can *project* a vector w in a Hilbert space onto a closed subspace. This existence and uniqueness theorem has diverse corollaries.

THEOREM 2.3. *Let V be a closed subspace of a Hilbert space $\mathcal{H}$. For each w in $\mathcal{H}$ there is a unique $z \in V$ that minimizes $||z - w||$. This z is the orthogonal projection of w onto V.*

PROOF. Fix w. If $w \in V$ then the conclusion holds with $z = w$. In general let $d = \inf_{z \in V} ||z - w||$. Choose a sequence $\{z_n\}$ such that $z_n \in V$ for all n and $||z_n - w||$ tends to d. We will show that $\{z_n\}$ is a Cauchy sequence, and hence it converges to some z. Since V is closed, z is in V. By continuity of the norm (Proposition 2.1), $||z - w|| = d$.

By the parallelogram law, we express $||z_n - z_m||^2$ as follows:

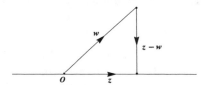

FIGURE 3. Orthogonal projection

$$||z_n - z_m||^2 = ||(z_n - w) + (w - z_m)||^2 = 2||z_n - w||^2 + 2||w - z_m||^2 - ||(z_n - w) - (w - z_m)||^2.$$

The last term on the right-hand side is

$$4||\frac{z_n + z_m}{2} - w||^2.$$

Since V is a subspace, the midpoint $\frac{z_n + z_m}{2}$ lies in V as well. Therefore this term is at least $4d^2$, and we obtain

$$0 \leq ||z_n - z_m||^2 \leq 2||z_n - w||^2 + 2||w - z_m||^2 - 4d^2. \tag{10}$$

As m and n tend to infinity the right-hand side of (10) tends to $2d^2 + 2d^2 - 4d^2 = 0$. Thus $\{z_n\}$ is a Cauchy sequence in $\mathcal{H}$ and hence converges to some z in V.

It remains only to show uniqueness. Given a pair of minimizers z and ζ, let d_m^2 denote the squared distance from their midpoint to w. By the parallelogram law, we may write

$$2d^2 = ||z - w||^2 + ||\zeta - w||^2 = 2||\frac{z + \zeta}{2} - w||^2 + 2||\frac{z - \zeta}{2}||^2 = 2d_m^2 + 2||\frac{z - \zeta}{2}||^2.$$

Thus $d^2 \geq d_m^2$. But d is minimal. Hence $d_m = d$ and thus $\zeta = z$. $\square$

COROLLARY 2.1. *Let V be a closed subspace of a Hilbert space $\mathcal{H}$. For each $w \in \mathcal{H}$, there is a unique way to write $w = v + \zeta = v \oplus \zeta$, where $v \in V$ and ζ is orthogonal to V.*

PROOF. Let v be the projection of w onto V guaranteed by Theorem 2.3. Since $w = v + (w - v)$, the existence result follows if we can show that $w - v$ is orthogonal to V. To see the orthogonality choose $u \in V$. Then consider the function f of one complex variable defined by

$$f(\lambda) = ||v + \lambda u - w||^2.$$

By Theorem 2.3, f achieves its minimum at $\lambda = 0$. Therefore for all λ

$$0 \leq f(\lambda) - f(0) = 2\mathrm{Re}\langle v - w, \lambda u \rangle + |\lambda|^2 ||u||^2. \tag{11}$$

We claim that (11) forces $\langle v - w, u \rangle = 0$. Granted the claim, we note that u is an arbitrary element of V. Therefore $v - w$ is orthogonal to V, as required.

To prove the claim, thereby completing the proof of existence, we note that $\langle v - w, u \rangle$ is the (partial) derivative of f with respect to $\overline{\lambda}$ at 0, and hence vanishes at a minimum of f.

The uniqueness assertion is easy; we use the notation for orthogonal sum. Suppose $w = v \oplus \zeta = v' \oplus \zeta'$, as in the statement of the Corollary. Then

$$0 = w - w = (v - v') \oplus (\zeta - \zeta')$$

from which we obtain $v = v'$ and $\zeta = \zeta'$. □

COROLLARY 2.2. *Let V be a closed subspace of a Hilbert space $\mathcal{H}$. For each $w \in \mathcal{H}$, let Pw denote the unique $z \in V$ guaranteed by Theorem 2.3; Pw is also the v guaranteed by Corollary 2.1. Then the mapping $w \to P(w)$ is a bounded linear transformation satisfying $P^2 = P$. Thus P is a projection.*

PROOF. Both the existence and uniqueness assertions in Corollary 2.1 matter in this proof. Given w_1 and w_2 in $\mathcal{H}$, by existence we may write $w_1 = Pw_1 \oplus \zeta_1$ and $w_2 = Pw_2 \oplus \zeta_2$. Adding gives

$$w_1 + w_2 = (Pw_1 \oplus \zeta_1) + (Pw_2 \oplus \zeta_2) = (Pw_1 + Pw_2) \oplus (\zeta_1 + \zeta_2). \tag{12}$$

The uniqueness assertion and (12) show that $Pw_1 + Pw_2$ is the unique element of V corresponding to $w_1 + w_2$ guaranteed by Corollary 2.1; by definition this element is $P(w_1 + w_2)$. By uniqueness $Pw_1 + Pw_2 = P(w_1 + w_2)$, and P is additive. In a similar fashion we write $w = Pw \oplus \zeta$ and hence

$$cw = c(Pw) \oplus c\zeta.$$

Again by uniqueness, $c(Pw)$ must be the unique element corresponding to cw guaranteed by Corollary 2.1; by definition this element is $P(cw)$. Hence $cP(w) = P(cw)$. We have now shown that P is linear.

To show that P is bounded, we note from the Pythagorean theorem that $||w||^2 = ||Pw||^2 + ||\zeta||^2$, and hence $||Pw|| \leq ||w||$.

Finally we show that $P^2 = P$. For $z = v \oplus \zeta$, we have $P(z) = v = v \oplus 0$. Hence

$$P^2(z) = P(P(z)) = P(v \oplus 0) = v = P(z).$$

□

Theorem 2.3 and its consequences are among the most powerful results in the book. The theorem guarantees that we can solve a minimization problem in diverse infinite-dimensional settings, and it implies the Riesz representation lemma.

Fix $w \in \mathcal{H}$, and consider the function from $\mathcal{H}$ to $\mathbf{C}$ defined by $Lz = \langle z, w \rangle$. Then L is a bounded linear functional. The linearity is evident. The boundedness follows from the Cauchy-Schwarz inequality; setting $C = ||w||$ yields $|L(z)| \leq C||z||$ for all $z \in \mathcal{H}$.

The following fundamental result of F. Riesz characterizes bounded linear functionals on a Hilbert space; a bounded linear functional must be given by an inner product. The proof relies on projection onto a closed subspace.

THEOREM 2.4 (Riesz Lemma). *Let $\mathcal{H}$ be a Hilbert space. Assume $L \in \mathcal{L}(\mathcal{H}, \mathbf{C})$. Then there is a unique $w \in \mathcal{H}$ such that*

$$L(z) = \langle z, w \rangle$$

for all $z \in \mathcal{H}$. The norm $||L||$ of the linear transformation L equals $||w||$.

PROOF. Since L is bounded, its nullspace $\mathcal{N}(L)$ is closed. If $\mathcal{N}(L) = \mathcal{H}$, we take $w = 0$ and the result is true.

Suppose that $\mathcal{N}(L)$ is not $\mathcal{H}$. Theorem 2.3 implies that there is a nonzero element w_0 orthogonal to $\mathcal{N}(L)$. To find such a w_0, choose any nonzero element not in $\mathcal{N}(L)$ and subtract its orthogonal projection onto $\mathcal{N}(L)$.

Let z be an arbitrary element of $\mathcal{H}$. For a complex number α we can write

$$z = (z - \alpha w_0) + \alpha w_0.$$

Note that $L(z - \alpha w_0) = 0$ if and only if $\alpha = \frac{L(z)}{L(w_0)}$. For each z we therefore let $\alpha_z = \frac{L(z)}{L(w_0)}$.

Since w_0 is orthogonal to $\mathcal{N}(L)$, computing the inner product with w_0 yields

$$\langle z, w_0 \rangle = \alpha_z ||w_0||^2 = \frac{L(z)}{L(w_0)} ||w_0||^2. \tag{13}$$

From (13) we see that

$$L(z) = \langle z, \frac{w_0}{||w_0||^2} \overline{L(w_0)} \rangle$$

and the existence result is proved. An explicit formula for w holds:

$$w = \frac{w_0}{||w_0||^2} \overline{L(w_0)}.$$

The uniqueness for w is immediate from the test we mentioned earlier. If $\langle \zeta, w - w' \rangle$ vanishes for all ζ, then $w - w' = 0$.

It remains to show that $||L|| = ||w||$. The Cauchy-Schwarz inequality yields

$$||L|| = \sup_{||z||=1} |\langle z, w \rangle| \leq ||w||.$$

Choosing $\frac{w}{||w||}$ for z yields

$$||L|| \geq |L(\frac{w}{||w||})| = \frac{\langle w, w \rangle}{||w||} = ||w||.$$

Combining the two inequalities shows that $||L|| = ||w||$. □

EXERCISE 2.12. Fix w with $w \neq 0$. Define $P(v)$ by

$$P(v) = \frac{\langle v, w \rangle}{||w||^2} w.$$

Verify that $P^2 = P$.

EXERCISE 2.13. Let $\mathcal{H} = L^2[-1, 1]$. Recall that f is *even* if $f(-x) = f(x)$ and f is *odd* if $f(-x) = -f(x)$. Let V_e be the subspace of even functions, and V_o the subspace of odd functions. Show that V_e is orthogonal to V_o.

EXERCISE 2.14. A hyperplane in $\mathcal{H}$ is a level set of a non-trivial linear functional. Assume that $w \neq 0$. Find the distance between the parallel hyperplanes given by $\langle z, w \rangle = c_1$ and $\langle z, w \rangle = c_2$.

EXERCISE 2.15. Let $b = \{b_j\}$ be a sequence of complex numbers, and suppose there is a positive number C such that

$$\left| \sum_{j=1}^{\infty} a_j \bar{b}_j \right| \leq C \left(\sum_{j=1}^{\infty} |a_j|^2 \right)^{\frac{1}{2}}$$

for all $a \in l^2$. Show that $b \in l^2$ and that $\sum |b_j|^2 \leq C^2$. Suggestion: Consider the map that sends a to $\sum a_j \bar{b}_j$.

5. Orthonormal expansion

We continue our general discussion of Hilbert spaces by studying orthonormal expansions. The simplest example comes from basic physics. Let $v = (a, b, c)$ be a point or vector in $\mathbf{R}^3$. Physicists write $v = a\mathbf{i} + b\mathbf{j} + c\mathbf{k}$, where $\mathbf{i}, \mathbf{j}, \mathbf{k}$ are mutually perpendicular vectors of unit length. Mathematicians write the same equation as $v = ae_1 + be_2 + ce_3$; here $e_1 = (1, 0, 0) = \mathbf{i}$, $e_2 = (0, 1, 0) = \mathbf{j}$, and $e_3 = (0, 0, 1) = \mathbf{k}$. This equation expresses v in terms of an *orthonormal expansion*:

$$ae_1 + be_2 + ce_3 = (a, b, c) = v = \langle v, e_1 \rangle e_1 + \langle v, e_2 \rangle e_2 + \langle v, e_3 \rangle e_3.$$

Orthonormal expansion in a Hilbert space abstracts this idea. Fourier series provide the basic example, where the functions $x \to e^{inx}$ are analogous to mutually perpendicular unit vectors.

We assume here that a Hilbert space is *separable*. This term means that the Hilbert space has a countable dense set; separability implies that the orthonormal systems we are about to define are either finite or countably infinite sets. All the specific Hilbert spaces mentioned or used in this book are separable. Some of the proofs given tacitly use separability even when the result holds more generally.

DEFINITION 2.10. Let $S = \{z_n\}$ be a finite or countably infinite collection of elements in a Hilbert space $\mathcal{H}$. We say that S is an *orthonormal system* in $\mathcal{H}$ if, for each n we have $||z_n||^2 = 1$, and for each n, m with $n \neq m$, we have $\langle z_n, z_m \rangle = 0$. We say that S is a *complete orthonormal system* if, in addition, $\langle z, z_n \rangle = 0$ for all n implies $z = 0$.

PROPOSITION 2.2 (Bessel's inequality). *Let $S = \{z_n\}$ be a countably infinite orthonormal system in $\mathcal{H}$. For each $z \in \mathcal{H}$ we have*

$$\sum_{n=1}^{\infty} |\langle z, z_n \rangle|^2 \leq ||z||^2. \tag{14}$$

PROOF. Choose $z \in \mathcal{H}$. By orthonormality, for each positive integer N we have

$$0 \leq ||z - \sum_{n=1}^{N} \langle z, z_n \rangle z_n||^2 = ||z||^2 - \sum_{n=1}^{N} |\langle z, z_n \rangle|^2. \tag{15}$$

Define a sequence of real numbers $r_N = r_N(z)$ by

$$r_N = \sum_{n=1}^{N} |\langle z, z_n \rangle|^2.$$

By (15), r_N is bounded above by $||z||^2$ and nondecreasing. Therefore it has a limit $r = r(z)$. Bessel's inequality follows.   □

PROPOSITION 2.3 (Best approximation lemma). *Let $S = \{z_n\}$ be an orthonormal system (finite or countable) in $\mathcal{H}$. Let V be the span of S. Then, for each $z \in \mathcal{H}$ and each $w \in V$,*

$$||z - \sum \langle z, z_n \rangle z_n|| \leq ||z - w||.$$

PROOF. The expression $\sum \langle z, z_n \rangle z_n$ equals the orthogonal projection of z onto V. Hence the result follows from Theorem 2.3.   □

The limit $r(z)$ of the sequence in the proof of Bessel's inequality equals $||z||^2$ for each z if and only if the orthonormal system S is *complete*. This statement is the content of the following fundamental theorem. In general $r(z)$ is the squared norm of the projection of z onto the span of the z_j.

THEOREM 2.5 (Orthonormal Expansion). *An orthonormal system $S = \{z_n\}$ is complete if and only if, for each $z \in \mathcal{H}$, we have*

$$z = \sum_n \langle z, z_n \rangle z_n. \tag{16}$$

PROOF. The cases where S is a finite set or where $\mathcal{H}$ is finite-dimensional are evident. Assume then that $\mathcal{H}$ is infinite-dimensional and S is a countably infinite set. We first verify that the series in (16) converges. Fix $z \in \mathcal{H}$, and put

$$T_N = \sum_{n=1}^{N} \langle z, z_n \rangle z_n.$$

Define r_N as in the proof of Bessel's inequality. For $N > M$, observe that

$$||T_N - T_M||^2 = ||\sum_{n=M+1}^{N} \langle z, z_n \rangle z_n||^2 = \sum_{n=M+1}^{N} |\langle z, z_n \rangle|^2 = r_N - r_M. \tag{17}$$

Since $\{r_N\}$ converges, it is a Cauchy sequence of real numbers. By (17), $\{T_N\}$ is a Cauchy sequence in $\mathcal{H}$. Since $\mathcal{H}$ is complete, T_N converges to some element w of $\mathcal{H}$, and $w = \sum \langle z, z_n \rangle z_n$, the right-hand side of (16). Note that $\langle w, z_n \rangle = \langle z, z_n \rangle$ for each n, so $z - w$ is orthogonal to each z_n.

We can now establish both implications. Suppose first that S is a complete system. Since $z - w$ is orthogonal to each z_n, we have $z - w = 0$. Thus (16) holds. Conversely, suppose that (16) holds. To show that S is a complete system, we assume that $\langle z, z_n \rangle = 0$ for all n, and hope to show that $z = 0$. This conclusion follows immediately from (16).   □

EXERCISE 2.16. Verify (15).

EXERCISE 2.17. Let $\mathcal{H} = L^2([0,1])$ with the usual inner product. Let V be the span of 1 and x. Find the orthogonal projection of x^2 onto V. Do the same problem if $\mathcal{H} = L^2([-1,1])$.

EXERCISE 2.18. Let $\mathcal{H} = L^2([-1,1])$ with the usual inner product. Apply the Gram-Schmidt process (see [G]) to orthonormalize the polynomials $1, x, x^2, x^3$.

EXERCISE 2.19. A sequence $\{f_n\}$ in a Hilbert space $\mathcal{H}$ *converges weakly* to f if, for each $g \in \mathcal{H}$, the sequence $\{\langle f_n, g \rangle\}$ converges to $\langle f, g \rangle$. Put $\mathcal{H} = L^2([0, 2\pi])$. Put $f_n(x) = \sin(nx)$. Show that $\{f_n\}$ converges weakly to 0, but does not converge to 0.

EXERCISE 2.20. Assume $\mathcal{H}$ is infinite-dimensional. Show that a sequence of orthonormal vectors does not converge, but does converge weakly to 0.

6. Polarization

In a Hilbert space we can recover the Hermitian inner product from the squared norm. In addition, for each linear operator L we can recover $\langle Lz, w \rangle$ for all z, w from knowing $\langle Lz, z \rangle$ for all z. See Theorem 2.6. The corresponding result for real vector spaces with inner products fails. See Exercise 2.21.

To introduce these ideas, let m be an integer with $m \geq 2$. Recall, for a complex number $a \neq 1$, the sum of the finite geometric series:

$$1 + a + a^2 + \cdots + a^{m-1} = \frac{1 - a^m}{1 - a}.$$

When a is an m-th root of unity, the sum is zero. A *primitive m-th root of unity* is a complex number ω such that $\omega^m = 1$, but no smaller positive power equals 1. The set of powers ω^j for $j = 0, 1, \ldots, m-1$ forms a cyclic group Γ of order m.

Let z, ζ be elements of a Hilbert space $\mathcal{H}$. Let ω be a primitive m-th root of unity and consider averaging the m complex numbers $\gamma \|z + \gamma \zeta\|^2$ as γ varies over Γ. Since each group element is a power of ω, this average equals

$$\frac{1}{m} \sum_{j=0}^{m-1} \omega^j \|z + \omega^j \zeta\|^2.$$

The next proposition gives a simple expression for the average.

PROPOSITION 2.4 (Polarization identities). *Let ω be a primitive m-th root of unity. For $m \geq 3$ we have*

$$\langle z, \zeta \rangle = \frac{1}{m} \sum_{j=0}^{m-1} \omega^j \|z + \omega^j \zeta\|^2. \tag{18}$$

For $m = 2$ the right-hand side of (18) equals $2\mathrm{Re}\langle z, \zeta \rangle$.

PROOF. We prove (18) below when $m = 4$, leaving the general case to the reader. □

For $m \geq 3$, each identity in (18) expresses the inner product in terms of squared norms. It is both beautiful and useful to recover the inner product from the squared norm. The special case of (18) where $m = 4$, and thus $\omega = i$, arises often. We state it explicitly and prove it.

$$4\langle z, \zeta \rangle = ||z + \zeta||^2 + i||z + i\zeta||^2 - ||z - \zeta||^2 - i||z - i\zeta||^2. \tag{19}$$

To verify (19), observe that expanding the squared norms gives both equations:

$$4\mathrm{Re}\langle z, \zeta \rangle = ||z + \zeta||^2 - ||z - \zeta||^2$$

$$4\mathrm{Re}\langle z, i\zeta \rangle = ||z + i\zeta||^2 - ||z - i\zeta||^2.$$

Observe for $a \in \mathbf{C}$ that $\mathrm{Re}(-ia) = \mathrm{Im}(a)$. Thus multiplying the second equation by i, using $i(-i) = 1$, and then adding the two equations, gives (19).

In addition to polarizing the inner product, we often polarize expressions involving linear transformations.

THEOREM 2.6 (Polarization identities for operators). *Let* $L \in \mathcal{L}(\mathcal{H})$. *Let* ω *be a primitive* m-*th root of unity.*

(1) For $m \geq 3$ *we have*

$$\langle Lz, \zeta \rangle = \frac{1}{m} \sum_{j=0}^{m-1} \omega^j \langle L(z + \omega^j \zeta), z + \omega^j \zeta \rangle. \tag{20}$$

(2) For $m = 2$ *we have*

$$\langle Lz, \zeta \rangle + \langle L\zeta, z \rangle = \frac{1}{2}(\langle L(z + \zeta), z + \zeta \rangle - \langle L(z - \zeta), z - \zeta \rangle). \tag{21}$$

(3) Suppose in addition that $\langle Lv, v \rangle$ *is real for all* $v \in \mathcal{H}$. *Then, for all* z *and* ζ,

$$\langle Lz, \zeta \rangle = \overline{\langle L\zeta, z \rangle}.$$

(4) Suppose $\langle Lz, z \rangle = 0$ *for all* z. *Then* $L = 0$.

PROOF. To prove (20) and (21), expand each $\langle L(z + \omega^j \zeta), z + \omega^j \zeta \rangle$ using the linearity of L and the defining properties of the inner product. Collect similar terms, and use the above comment about roots of unity. For $m \geq 3$, all terms inside the sum cancel except for m copies of $\langle Lz, \zeta \rangle$. The result gives (20). For $m = 2$, the coefficient of $\langle L\zeta, z \rangle$ does not vanish, and we obtain (21). Thus statements (1) and (2) hold.

To prove the third statement, we apply the first for some m with $m \geq 3$ and $\omega^m = 1$; the result is

$$\langle Lz, \zeta \rangle = \frac{1}{m} \sum_{j=0}^{m-1} \omega^j \langle L(z + \omega^j \zeta), z + \omega^j \zeta \rangle = \frac{1}{m} \sum_{j=0}^{m-1} \omega^j \langle L(\omega^{m-j} z + \zeta), \omega^{m-j} z + \zeta \rangle. \tag{22}$$

Change the index of summation by setting $l = m - j$. Also observe that $\omega^{-1} = \overline{\omega}$. Combining gives the first equality in (23) below. Finally, because $\langle Lv, v \rangle$ is real, and $\omega^0 = \omega^m$ we obtain the second equality in (23):

$$\langle Lz, \zeta \rangle = \frac{1}{m} \sum_{l=1}^{m} \overline{\omega}^l \langle L(\zeta + \omega^l z), \zeta + \omega^l z \rangle = \overline{\langle L\zeta, z \rangle}. \tag{23}$$

We have now proved the third statement.

The fourth statement follows from (20); each term in the sum on the right-hand side of (20) vanishes if $\langle Lw, w \rangle = 0$ for all w. Thus $\langle Lz, \zeta \rangle = 0$ for all ζ. Hence $Lz = 0$ for all z, and thus $L = 0$. $\qquad \square$

The reader should compare these results about polarization with our earlier results about Hermitian symmetric polynomials.

EXERCISE 2.21. Give an example of a linear map of $\mathbf{R}^2$ such that $\langle Lu, u \rangle = 0$ for all u but L is not 0.

7. Adjoints and unitary operators

Let I denote the identity linear transformation on a Hilbert space $\mathcal{H}$. Let $L \in \mathcal{L}(\mathcal{H})$. Then L is called *invertible* if there is a bounded linear mapping T such that $LT = TL = I$. If such a T exists, then T is unique, and written L^{-1}. We warn the reader (see the exercises) that, in infinite dimensions, $LT = I$ does not imply that L is invertible. When L is bounded, injective, and surjective, the usual set-theoretic inverse is also linear and bounded.

Given a bounded linear mapping L, the adjoint of L is written L^*. It is defined as follows. Fix $v \in \mathcal{H}$. Consider the map $u \to \langle Lu, v \rangle = \phi_v(u)$. It is obviously a linear functional. It is also continuous because

$$|\phi_v(u)| = |\langle Lu, v \rangle| \le ||Lu|| \, ||v|| \le ||u|| \, ||L|| \, ||v|| = c||u||, \qquad (24)$$

where the constant c is independent of u. By Theorem 2.4, there is a unique $w_v \in \mathcal{H}$ for which $\phi_v(u) = \langle u, w_v \rangle$. We denote w_v by L^*v. It is easy to prove that L^* is itself a bounded linear mapping on $\mathcal{H}$, called the adjoint of L.

The following properties of adjoints are left as exercises.

PROPOSITION 2.5. *Let* $L, T \in \mathcal{L}(\mathcal{H})$. *The following hold:*

(1) $L^* : \mathcal{H} \to \mathcal{H}$ *is linear.*
(2) L^* *is bounded. (In fact* $||L^*|| = ||L||$.)
(3) $(L^*)^* = L$.
(4) $\langle Lu, v \rangle = \langle u, L^*v \rangle$ *for all* u, v.
(5) $(LT)^* = T^*L^*$.

PROOF. See Exercise 2.22. $\qquad \square$

EXERCISE 2.22. Prove Proposition 2.5.

DEFINITION 2.11. A bounded linear transformation L on a Hilbert space $\mathcal{H}$ is called *Hermitian* or *self adjoint* if $L = L^*$. It is called *unitary* if it is invertible and $L^* = L^{-1}$.

The following simple but beautiful result characterizes unitary transformations.

PROPOSITION 2.6. *The following are equivalent for* $L \in \mathcal{L}(\mathcal{H})$.

(1) L *is surjective and preserves norms:* $||Lu||^2 = ||u||^2$ *for all* u.
(2) L *is surjective and preserves inner products:* $\langle Lu, Lv \rangle = \langle u, v \rangle$ *for all* u, v.
(3) L *is unitary:* $L^* = L^{-1}$.

PROOF. If $L \in \mathcal{L}(\mathcal{H})$, then $\langle Lu, Lv \rangle = \langle u, v \rangle$ for all u, v if and only if $\langle u, L^*Lv \rangle = \langle u, v \rangle$ for all u, v and thus if and only if $\langle u, (L^*L - I)v \rangle = 0$ for all u, v. This last statement holds if and only if $(L^*L - I)v = 0$ for all v. Thus $L^*L = I$. If L is also surjective, then $L^* = L^{-1}$, and therefore the second and third statements are equivalent.

The second statement obviously implies the first. It remains to prove the subtle point that the first statement implies the second or third statement. We are given $\langle L^*Lz, z \rangle = \langle z, z \rangle$ for all z. Hence $\langle (L^*L - I)z, z \rangle = 0$. By part 4 of Theorem 2.6, $L^*L - I = 0$, and the second statement holds. If L is also surjective, then L is invertible and hence unitary. □

The equivalence of the first two statements does not require L to be surjective. See the exercises for examples where L preserves inner products but L is not surjective and hence not unitary.

PROPOSITION 2.7. *Let* $L \in \mathcal{L}(\mathcal{H})$. *Then*

$$\mathcal{N}(L) = \mathcal{R}(L^*)^{\perp}$$

$$\mathcal{N}(L^*) = \mathcal{R}(L)^{\perp}.$$

PROOF. Note that $L^*(z) = 0$ if and only if $\langle L^*z, w \rangle = 0$ for all w, if and only if $\langle z, Lw \rangle = 0$ for all w, if and only if $z \perp \mathcal{R}(L)$. Thus the second statement holds. When $L \in \mathcal{L}(\mathcal{H})$, it is easy to check that $(L^*)^* = L$. See Exercise 2.22. The first statement then follows from the second statement by replacing L with L^*. □

EXERCISE 2.23. If $L : \mathbf{C}^n \to \mathbf{C}^n$ and $L = L^*$, what can we conclude about the matrix of L with respect to the usual basis $(1, 0, \ldots, 0)$, ..., $(0, 0, \ldots, 1)$?

EXERCISE 2.24. Suppose U is unitary and $Uz = \lambda z$ for $z \neq 0$. Prove that $|\lambda| = 1$. Suppose L is Hermitian and $Lz = \lambda z$ for $z \neq 0$. Prove that λ is real.

EXERCISE 2.25. Let $L : l^2 \to l^2$ be defined by

$$L(z_1, z_2, \ldots) = (0, z_1, z_2, \ldots).$$

Show that $||Lz||_2 = ||z||_2$ for all z but that L is not unitary.

EXERCISE 2.26. Give an example of a bounded linear $L : \mathcal{H} \to \mathcal{H}$ that is injective but not surjective, and an example that is surjective but not injective.

EXERCISE 2.27. Let V be the vector space of all polynomials in one variable. Let D denote differentiation, and J denote integration (with integration constant 0). Show that $DJ = I$ but that $JD \neq I$. Explain.

EXERCISE 2.28. Give an example of an operator L for which $||L^2|| \neq ||L||^2$. Suppose $L = L^*$; show that $||L^2|| = ||L||^2$.

We close this section with an interesting difference between real and complex vector spaces, related to inverses, polarization, and Exercise 2.21. The formula $(*)$ below interests the author partly because, although no real numbers satisfy the equation, teachers often see it on exams.

DEFINITION 2.12. A real vector space V *admits a complex structure* if there is a linear map $J : V \to V$ such that $J^2 = -I$.

It is easy to show (Exercise 2.30) that a finite-dimensional real vector space admits a complex structure if and only if its dimension is even. The linear transformation $J : \mathbf{R}^2 \to \mathbf{R}^2$ corresponding to the complex structure is given by the matrix

$$J = \begin{pmatrix} 0 & -1 \\ 1 & 0 \end{pmatrix}.$$

PROPOSITION 2.8. *Let V be a vector space over $\mathbf{R}$. Then there are invertible linear transformations A, B on V satisfying*

$$(A + B)^{-1} = A^{-1} + B^{-1} \tag{$*$}$$

if and only if V admits a complex structure.

PROOF. Invertible A, B satisfying $(*)$ exist if and only if

$$I = (A + B)(A^{-1} + B^{-1}) = I + BA^{-1} + I + AB^{-1}.$$

Put $C = BA^{-1}$. The condition $(*)$ is therefore equivalent to finding C such that $0 = I + C + C^{-1}$, which is equivalent to $0 = I + C + C^2$. Suppose such C exists. Put $J = \frac{1}{\sqrt{3}}(I + 2C)$. Then we have

$$J^2 = \frac{1}{3}(I + 2C)^2 = \frac{1}{3}(I + 4C + 4C^2) = \frac{1}{3}(-3I + 4(I + C + C^2)) = -I.$$

Hence V admits a complex structure. Conversely, if V admits a complex structure, then J exists with $J^2 = -I$. Put $C = \frac{-I + \sqrt{3}J}{2}$; then $I + C + C^2 = 0$. $\qquad \square$

COROLLARY 2.3. *There exist n by n matrices satisfying $(*)$ if and only if n is even.*

EXERCISE 2.29. Explain the proof of Proposition 2.8 in terms of cube roots of unity.

EXERCISE 2.30. Prove that a finite-dimensional real vector space with a complex structure must have even dimension. Hint: Consider the determinant of J.

8. A return to Fourier series

The specific topic of Fourier series motivated many of the abstract results about Hilbert spaces and it provides one of the best examples of the general theory. In return, the general theory clarifies the subject of Fourier series.

Let h be (Riemann) integrable on the circle and consider its Fourier series $\sum \hat{h}(n)e^{inx}$. Recall that its symmetric partial sums S_N are given by

$$S_N(h)(x) = \sum_{n=-N}^{N} \hat{h}(n)e^{inx}.$$

When h is sufficiently smooth, $S_N(h)$ converges to h. See for example Theorem 2.8. We show next that $S_N(h)$ converges to h in L^2. Rather than attempting to prove convergence at each point, this result considers an integrated form of convergence.

THEOREM 2.7. *Suppose f is integrable on the circle. Then $||S_N(f) - f||_{L^2} \to 0$.*

PROOF. Given $\epsilon > 0$ and an integrable f, we first approximate f to within $\frac{\epsilon}{2}$ in the L^2 norm by a continuous function g. Then we approximate g by a trig polynomial p to within $\frac{\epsilon}{2}$. See below for details. These approximations yield

$$||f - p||_{L^2} \leq ||f - g||_{L^2} + ||g - p||_{L^2} < \frac{\epsilon}{2} + \frac{\epsilon}{2} = \epsilon. \tag{25}$$

Once we have found this p, we use orthogonality as in Theorem 2.3. Let N be at least as large as the degree of p. Let V_N denote the $(2N+1)$-dimensional (hence closed) subspace spanned by the functions e^{inx} for $|n| \leq N$. By Theorem 2.3, there is a unique element w of V_N minimizing $||f - w||_{L^2}$. That w is the partial sum $S_N(f)$, namely the orthogonal projection of f onto V_N.

By Proposition 2.3, we have

$$||f - S_N(f)||_{L^2} \leq ||f - p||_{L^2} \tag{26}$$

for all elements p of V_N. Take p to be the polynomial in (25) and take N at least the degree of p. Combining (26) and (25) then gives

$$||f - S_N(f)||_{L^2} \leq ||f - p||_{L^2} \leq ||f - g||_{L^2} + ||g - p||_{L^2} < \epsilon. \tag{27}$$

It therefore suffices to verify that the two above approximations are valid. Given f integrable, by Lemma 1.6 we can find a continuous g such that $\sup(|g|) \leq \sup(|f|) = M$ and such that $||f - g||_{L^1}$ is as small as we wish. Since

$$||f - g||_{L^2}^2 = \frac{1}{2\pi} \int_0^{2\pi} |f - g|^2 dx \leq \frac{\sup(|f - g|)}{2\pi} \int_0^{2\pi} |f - g| dx \leq 2M||f - g||_{L^1}, \tag{28}$$

we may choose g to bound the expression in (28) by $\frac{\epsilon}{2}$.

Now g is given and continuous on the circle. By Corollary 1.8, there is a trig polynomial p such that $||g - p||_{L^\infty} < \frac{\epsilon}{2}$. Therefore

$$||g - p||_{L^2}^2 = \frac{1}{2\pi} \int_0^{2\pi} |g(x) - p(x)|^2 dx \leq ||g - p||_{L^\infty}^2.$$

Hence $||g - p||_{L^2} < \frac{\epsilon}{2}$ as well. We have established both approximations used in (25), and hence the conclusion of the theorem. $\square$

COROLLARY 2.4 (Parseval's formula). *If f is integrable on the circle, then*

$$\sum_{-\infty}^{\infty} |\hat{f}(n)|^2 = ||f||_{L^2}^2. \tag{29}$$

PROOF. By the orthonormality properties of the functions $x \to e^{inx}$, $f - S_N(f)$ is orthogonal to V_N. By the Pythagorean theorem, we have

$$||f||_{L^2}^2 = ||f - S_N(f)||_{L^2}^2 + ||S_N(f)||_{L^2}^2 = ||f - S_N(f)||_{L^2}^2 + \sum_{-N}^{N} |\hat{f}(n)|^2. \tag{30}$$

Letting N tend to infinity in (30) and using Theorem 2.7 gives (29). $\square$

COROLLARY 2.5 (Riemann-Lebesgue lemma). *If f is integrable on the circle, then* $\lim_{|n| \to \infty} \hat{f}(n) = 0$.

PROOF. The series in (29) converges; hence its terms tend to 0. $\square$

Polarization has several applications to Fourier series. By (29), if f and g are integrable on the circle S^1, then $\sum |\hat{f}|^2 = ||f||_{L^2}^2$ and similarly for g. It follows by polarization that

$$\langle \hat{f}, \hat{g} \rangle_2 = \sum_{-\infty}^{\infty} \hat{f}(n)\overline{\hat{g}(n)} = \frac{1}{2\pi} \int_0^{2\pi} f(x)\overline{g(x)}dx = \langle f, g \rangle_{L^2}. \tag{31}$$

COROLLARY 2.6. *If f and g are integrable on the circle, then (31) holds.*

COROLLARY 2.7. *The map $f \to \mathcal{F}(f)$ from $L^2(S^1)$ to l^2 satisfies the relation*

$$\langle \mathcal{F}f, \mathcal{F}g \rangle_2 = \langle f, g \rangle_{L^2}.$$

The analogue of this corollary holds for Fourier transforms on $\mathbf{R}$, $\mathbf{R}^n$, or in even more abstract settings. Such results, called Plancherel theorems, play a crucial role in extending the definition of Fourier transform to objects (called distributions) more general than functions. See Chapter 3.

THEOREM 2.8. *Suppose f is continuously differentiable on the circle. Then its Fourier series converges absolutely to f.*

PROOF. By Lemma 1.8, we have $\hat{f}(n) = \frac{\hat{f}'(n)}{in}$ for $n \neq 0$. We first apply the Parseval identity to the Fourier series for f', getting

$$\frac{1}{2\pi} \int |f'(x)|^2 dx = \sum |\hat{f}'(n)|^2 = \sum n^2 |\hat{f}(n)|^2. \tag{32}$$

Then we use the Cauchy-Schwarz inequality on $\sum |\hat{f}(n)|$ to get

$$\sum |\hat{f}(n)| = |\hat{f}(0)| + \sum \frac{1}{n} n|\hat{f}(n)| \leq |\hat{f}(0)| + (\sum \frac{1}{n^2})^{\frac{1}{2}} (\sum n^2 |\hat{f}(n)|^2)^{\frac{1}{2}}. \tag{33}$$

By (32), the second sum on the right-hand side of (33) converges. The sum $\sum_{n \neq 0} \frac{1}{n^2}$ also converges and can be determined exactly using Fourier series. See Exercise 2.31.

Since each partial sum is continuous and the partial sums converge uniformly, the limit is continuous. By Corollary 1.10, the Fourier series converges absolutely to f. $\qquad\square$

EXERCISE 2.31. Compute the Fourier series for the function f defined by $f(x) = (\pi - x)^2$ on $(0, 2\pi)$. Use this series to show that $\sum_{n=1}^{\infty} \frac{1}{n^2} = \frac{\pi^2}{6}$.

EXERCISE 2.32. Find $\sum_{n=1}^{\infty} \frac{(-1)^n}{n^2}$. Suggestion. Find the Fourier series for x^2 on $(-\pi, \pi)$.

9. Bernstein's theorem

We continue by proving a fairly difficult result. We include it to illustrate circumstances more general than Theorem 2.8 in which Fourier series converge absolutely and uniformly.

DEFINITION 2.13. Let $f : S^1 \to \mathbf{C}$ be a function and suppose $\alpha > 0$. We say that f satisfies a Hölder condition of order α if there is a constant C such that

$$|f(x) - f(y)| \leq C|x - y|^\alpha \tag{34}$$

for all x, y. Sometimes we say f is Hölder continuous of order α.

By the mean-value theorem from calculus, a differentiable function satisfies the inequality

$$|f(x) - f(y)| \leq \sup |f'(t)| \, |x - y|.$$

Hence, if f' is bounded, f satisfies a Hölder condition with $\alpha = 1$. Note also that a function satisfying (34) must be uniformly continuous.

THEOREM 2.9. *Suppose f is Hölder continuous on the circle of order α, and $\alpha > \frac{1}{2}$. Then the Fourier series for f converges absolutely and uniformly.*

PROOF. The Hölder condition means that there is a constant C such that inequality (34) holds. We must somehow use this condition to study

$$\sum_{n \in \mathbf{Z}} |\hat{f}(n)|.$$

The remarkable idea here is to break up this sum into dyadic parts, and estimate differently in different parts. For p a natural number let R_p denote the set of $n \in \mathbf{Z}$ for which $2^{p-1} \leq |n| < 2^p$. Note that there are 2^p integers in R_p. We have

$$\sum_{n \in \mathbf{Z}} |\hat{f}(n)| = |\hat{f}(0)| + \sum_p \sum_{n \in R_p} |\hat{f}(n)|. \qquad (35)$$

In each R_p we can use the Cauchy-Schwarz inequality to write

$$\sum_{n \in R_p} |\hat{f}(n)| \leq \left(\sum_{n \in R_p} |\hat{f}(n)|^2 \right)^{\frac{1}{2}} (2^p)^{\frac{1}{2}}. \qquad (36)$$

At first glance the factor $2^{\frac{p}{2}}$ looks troublesome, but we will nonetheless verify convergence of the Fourier series.

Let g_h be defined by $g_h(x) = f(x + h) - f(x - h)$. The Hölder condition gives

$$|g_h(x)|^2 \leq C^2 |2h|^{2\alpha} = C' |h|^{2\alpha},$$

and integrating we obtain

$$\|g_h\|_{L^2}^2 \leq C' |h|^{2\alpha}.$$

By the Parseval-Plancherel theorem (Corollary 2.7), for any h we have

$$\sum_{n \in \mathbf{Z}} |\hat{g}_h(n)|^2 = \|g_h\|_{L^2}^2 \leq C' |h|^{2\alpha}. \qquad (37)$$

Now we compute the Fourier coefficients of g_h, relating them to f. Using the definition directly, we get

$$\hat{g}_h(n) = \frac{1}{2\pi} \int_0^{2\pi} (f(x + h) - f(x - h)) e^{-inx} dx.$$

Changing variables in each term and recollecting gives

$$\hat{g}_h(n) = \frac{1}{2\pi} \int_0^{2\pi} f(y) e^{-iny} e^{inh} dy - \frac{1}{2\pi} \int_0^{2\pi} f(y) e^{-iny} e^{-inh} dy = 2i \sin(nh) \hat{f}(n).$$

Hence we have

$$|\hat{g}_h(n)|^2 = 4 \sin^2(nh) |\hat{f}(n)|^2.$$

Putting things together we obtain, with a new constant c,

$$\sin^2(nh)|\hat{f}(n)|^2 = \frac{1}{4}|\hat{g}_h(n)|^2 \le \frac{1}{4}\sum_n |\hat{g}_h(n)|^2 \le c|h|^{2\alpha}. \tag{38}$$

Also we have

$$\sum_{n\in R_p} |\hat{f}(n)|^2 = \sum_{n\in R_p} |\hat{f}(n)|^2 \sin^2(nh)\frac{1}{\sin^2(nh)} = \sum_{n\in R_p} |\hat{g}_h(n)|^2 \frac{1}{4\sin^2(nh)}. \tag{39}$$

Put $h = \frac{\pi}{2^p+1}$. Then $\frac{\pi}{4} \le |n|h \le \frac{\pi}{2}$ and hence $\frac{1}{2} \le \sin^2(nh) \le 1$. Using $\sin^2(nh) \ge \frac{1}{2}$ in (39), we get

$$\sum_{n\in R_p} |\hat{f}(n)|^2 \le \frac{1}{2}\sum_{n\in R_p} |\hat{g}_h(n)|^2. \tag{40}$$

For $h = \frac{\pi}{2^p+1}$, we have

$$|\hat{g}_h(n)|^2 \le C_1 \left|\frac{\pi}{2^p+1}\right|^{2\alpha} \le C_2 2^{-2\alpha p}. \tag{41}$$

Combining (40), (41), and (36) (note the exponent $\frac{1}{2}$ there) gives

$$\sum_{n\in\mathbf{Z}} |\hat{f}(n)| = |\hat{f}(0)| + \sum_p \sum_{n\in R_p} |\hat{f}(n)| \le |\hat{f}(0)| + C_2 \sum_p 2^{-\alpha p} 2^{\frac{p}{2}}. \tag{42}$$

The series on the right-hand side of (42) is of the form $\sum x^p$ where $x = 2^{\frac{1-2\alpha}{2}}$. If $\alpha > \frac{1}{2}$, then $|x| < 1$ and this series converges. $\square$

The conclusion of the theorem fails if f satisfies a Hölder condition of order $\frac{1}{2}$. See [K].

10. Compact Hermitian operators

Fourier series give but one of many examples of orthonormal expansions. In this section we establish the spectral theorem for compact Hermitian operators. Such operators determine complete orthonormal systems consisting of eigenvectors. In the next section we apply this result to Sturm-Liouville equations. These second order ordinary differential equations with homogeneous boundary conditions played a major role in the historical development of operator theory and remain significant in current applied mathematics, engineering, and physics.

An operator on a Hilbert space is *compact* if it can be approximated (in norm) arbitrarily well by operators with finite-dimensional range. We mention this characterization for the intuition it provides. The precise definition, which also applies in the context of complete normed vector spaces, involves subsequences. In older literature, compact operators are called *completely continuous*.

DEFINITION 2.14. Suppose $L \in \mathcal{L}(\mathcal{H})$. Then L is *compact* if, whenever $\{z_n\}$ is a bounded sequence in $\mathcal{H}$, then $\{L(z_n)\}$ has a convergent subsequence.

By the Bolzano-Weierstrass theorem (see Theorem 6.2), each bounded sequence in $\mathbf{C}^d$ has a convergent subsequence. Hence an operator with finite-dimensional range must be compact. A constant multiple of a compact operator is compact. The sum of two compact operators is compact. We check in Proposition 2.10 that the composition (on either side) of a compact operator with a bounded operator is compact. On the other hand, the identity operator is compact only when the

Hilbert space is finite-dimensional. Proposition 2.13 gives one of many possible proofs of this last statement.

We will use the following simple characterization of compact operators. See [D1] for many uses of the method. The two statements in the proof are equivalent, with different values of ϵ. In the statement we write f for an element of $\mathcal{H}$, to remind us that we are typically working on function spaces.

PROPOSITION 2.9. *Suppose* $L \in \mathcal{L}(\mathcal{H})$. *Then* L *is compact if and only if, for each* $\epsilon > 0$, *there are compact operators* K_ϵ *and* T_ϵ *such that either of the following (equivalent) statements holds:*

$$||Lf|| \leq \epsilon||f|| + ||K_\epsilon f|| \tag{$*1$}$$

$$||Lf||^2 \leq \epsilon||f||^2 + ||T_\epsilon f||^2. \tag{$*2$}$$

PROOF. When L is compact both ($*1$) and ($*2$) are obvious; we take K_ϵ and T_ϵ equal to L. The issue is to use either inequality to establish compactness.

Assuming the inequality ($*1$), we prove that L is compact. The proof assuming the inequality ($*2$) is similar. Let $\{f_n\}$ be a bounded sequence; we may assume that $||f_n|| \leq 1$. We wish to extract a Cauchy subsequence of $L(f_n)$. For each positive integer m, we set $\epsilon = \frac{1}{2m}$ in ($*1$). We obtain a sequence $\{L_m\}$ of compact operators. Thus, for fixed m, each sequence $\{L_m(f_n)\}$ has a convergent (hence Cauchy) subsequence. Hence we can find a subsequence, still labeled $\{f_n\}$, such that (for n, k large)

$$||L(f_n) - L(f_k)|| = ||L(f_n - f_k)|| \leq \frac{1}{2m} + ||L_m(f_n - f_k)|| \leq \frac{1}{2m} + \frac{1}{2m} = \frac{1}{m}. \tag{$*$}$$

Let $\{f_n^0\}$ denote the original sequence. Using ($*$), we choose a subsequence $\{f_n^1\}$ of $\{f_n^0\}$, and inductively for each j a subsequence $\{f_n^{j+1}\}$ of $\{f_n^j\}$, such that the following holds. For any pair of elements f, g in the sequence $\{f_n^j\}$ we have

$$||L(f) - L(g)|| \leq \frac{1}{j}.$$

Using the Cantor diagonalization trick, we extract the diagonal subsequence $\{f_j^j\}$. This sequence is a subsequence of the original sequence $\{f_n^0\}$ and its image under L is Cauchy. Since $\mathcal{H}$ is complete, $\{L(f_j^j)\}$ converges, and thus L is compact. $\square$

As noted in the proof, when L is compact, we may choose K_ϵ or T_ϵ equal to L. The point of Proposition 2.9 is the converse. It enables us to establish compactness by proving an inequality, instead of dealing with subsequences. We illustrate with several examples, which can of course also be proved using subsequences.

PROPOSITION 2.10. *Suppose* $L \in \mathcal{L}(\mathcal{H})$ *and* L *is compact. If* $M, T \in \mathcal{L}(\mathcal{H})$, *then* ML *and* LT *are compact.*

PROOF. That LT is compact follows directly from the definition of compactness. If $\{z_n\}$ is a bounded sequence, then $\{Tz_n\}$ also is, and hence $\{L(Tz_n)\}$ has a convergent subsequence. Similarly, ML is compact.

That ML is compact can also be proved using Proposition 2.9 as follows. Given $\epsilon > 0$, put $\epsilon' = \frac{\epsilon}{1+||M||}$. Put $K = ||M|| \, L$; then K is compact. We have

$$||MLz|| \leq ||M|| \, ||Lz|| \leq ||M||(\epsilon'||z|| + ||Lz||) \leq \epsilon||z|| + ||Kz||.$$

By Proposition 2.9, ML is also compact. $\square$

PROPOSITION 2.11. *Let $\{L_n\}$ be a sequence of operators with $\lim_n ||L_n - L|| = 0$. If each L_n is compact, then L is also compact.*

PROOF. Given $\epsilon > 0$, we can find an n such that $||L - L_n|| < \epsilon$. Then we write

$$||Lf|| \leq ||(L - L_n)f|| + ||L_n(f)|| \leq \epsilon ||f|| + ||L_n(f)||.$$

The result therefore follows from Proposition 2.9.   □

A converse of Proposition 2.11 also holds; each compact operator is the limit in norm of a sequence of operators with finite-dimensional ranges. We can also use Proposition 2.9 to prove the following result.

THEOREM 2.10. *Assume $L \in \mathcal{L}(\mathcal{H})$. If L is compact, then L^* is compact. Furthermore, L is compact if and only if $L^* L$ is compact.*

PROOF. See Exercise 2.35.   □

EXERCISE 2.33 (Small constant large constant trick). Given $\epsilon > 0$, prove that there is a $C_\epsilon > 0$ such that

$$|\langle x, y \rangle| \leq \epsilon ||x||^2 + C_\epsilon ||y||^2.$$

EXERCISE 2.34. Prove that the second inequality in Proposition 2.9 implies compactness.

EXERCISE 2.35. Prove Theorem 2.10. Use Proposition 2.9 and Exercise 2.33 to verify the *if* part of the implication.

Before turning to the spectral theorem for compact Hermitian operators, we give one of the classical types of examples. The function K in this example is called the *integral kernel* of the operator T. Such integral operators arise in the solutions of differential equations such as the Sturm-Liouville equation.

PROPOSITION 2.12. *Let $\mathcal{H} = L^2([a, b])$. Assume that $(x, t) \to K(x, t)$ is continuous on $[a, b] \times [a, b]$. Define an operator T on $\mathcal{H}$ by*

$$Tf(x) = \int_a^b K(x, t) f(t) dt.$$

Then T is compact. (The conclusion holds under weaker assumptions on K.)

PROOF. Let $\{f_n\}$ be a bounded sequence in $L^2([a, b])$. The following estimate follows from the Cauchy-Schwarz inequality:

$$|T(f_n)(x) - T(f_n)(y)|^2 \leq \sup |K(x, t) - K(y, t)|^2 ||f_n||_{L^2}^2.$$

Since K is continuous on the compact set $[a, b] \times [a, b]$, it is uniformly continuous. It follows that the sequence $\{T(f_n)\}$ is equi-continuous and uniformly bounded. By the Arzela-Ascoli theorem, there is a subsequence of $\{T(f_n)\}$ that converges uniformly. In particular, this subsequence converges in L^2. Hence, $\{T(f_n)\}$ has a convergent subsequence, and thus T is compact.   □

EXERCISE 2.36. Suppose that the integral kernel in Proposition 2.12 satisfies $\int_a^b |K(x, t)| dt \leq C$ and $\int_a^b |K(x, t)| dx \leq C$. Show that $T \in \mathcal{L}(\mathcal{H})$ and that $||T|| \leq C$.

A compact operator need not have any eigenvalues or eigenvectors.

EXAMPLE 2.3. Let $L : l^2 \to l^2$ be defined by

$$L(z_1, z_2, \dots) = (0, z_1, \frac{z_2}{2}, \frac{z_3}{3}, \dots).$$

Think of L as given by an infinite matrix with sub-diagonal entries $1, \frac{1}{2}, \frac{1}{3}, \dots$. Then L is compact but has no eigenvalues.

For a second example, consider the integral operator T on continuous functions in $L^2([0,1])$ defined by

$$Tf(x) = \int_0^x f(t)dt.$$

Then T is compact but it has no eigenvalues. It is obvious that 0 is not an eigenvalue. For $\lambda \neq 0$, the relation $Tf = \lambda f$ forces f to be a constant times an exponential but also $f(0)$ must be 0. Hence f must be 0. We will see this operator again in Theorem 4.2 of Chapter 4.

EXERCISE 2.37. Verify the conclusions of Example 2.3.

Compact Hermitian operators, however, have many eigenvectors. In fact, by the spectral theorem, there is a complete orthonormal system of eigenvectors. Before proving the spectral theorem, we note two easy results about eigenvectors and eigenvalues.

PROPOSITION 2.13. *An eigenspace of a compact operator corresponding to a non-zero eigenvalue must be finite-dimensional.*

PROOF. Assume that L is compact and $L(z_j) = \lambda z_j$ for a sequence of orthogonal unit vectors z_j. Since L is compact, $L(z_j) = \lambda z_j$ has a convergent subsequence. If $\lambda \neq 0$, then z_j has a convergent subsequence. But no sequence of orthogonal unit vectors can converge. Thus $\lambda = 0$. ◻

PROPOSITION 2.14. *The eigenvalues of a Hermitian operator are real and the eigenvectors corresponding to distinct eigenvalues are orthogonal.*

PROOF. Assume $Lf = \lambda f$ and $f \neq 0$. We then have

$$\lambda \|f\|^2 = \langle Lf, f \rangle = \langle f, L^*f \rangle = \langle f, Lf \rangle = \langle f, \lambda f \rangle = \overline{\lambda} \|f\|^2.$$

Since $\|f\|^2 \neq 0$, we conclude that $\lambda = \overline{\lambda}$.

The proof of the second statement amounts to polarizing the first. Thus we suppose $Lf = \lambda f$ and $Lg = \mu g$ where $\lambda \neq \mu$. We have, as μ is real,

$$\lambda \langle f, g \rangle = \langle Lf, g \rangle = \langle f, Lg \rangle = \mu \langle f, g \rangle.$$

Hence $0 = (\lambda - \mu) \langle f, g \rangle$ and the second conclusion follows. ◻

PROPOSITION 2.15. *Suppose $L \in \mathcal{L}(\mathcal{H})$ is Hermitian. Then*

$$\|L\| = \sup_{\|z\|=1} |\langle Lz, z \rangle|. \tag{43}$$

PROOF. Let α equal the right-hand side of (43). We prove both inequalities: $\alpha \leq \|L\|$ and $\|L\| \leq \alpha$. Since $|\langle Lz, z \rangle| \leq \|Lz\| \, \|z\|$, we see that

$$\alpha = \sup_{\|z\|=1} |\langle Lz, z \rangle| \leq \sup_{\|z\|=1} \|Lz\| = \|L\|.$$

The opposite inequality is harder. It uses the polarization identity (21) and the parallelogram law (9). We first note, by Theorem 2.6, that $\alpha = 0$ implies $L = 0$. Hence we may assume $\alpha \neq 0$. Since L is Hermitian, it follows that

$$\langle Lz, w \rangle = \langle z, Lw \rangle = \overline{\langle Lw, z \rangle}.$$

Applying this equality in (21), we obtain, for all z, w,

$$4\text{Re}\langle Lz, w \rangle = \langle L(z+w), z+w \rangle - \langle L(z-w), z-w \rangle.$$

Using $\langle L\zeta, \zeta \rangle \leq \alpha ||\zeta||^2$ and the parallelogram law, we obtain

$$4\text{Re}\langle Lz, w \rangle \leq \alpha(||z+w||^2 + ||z-w||^2) = 2\alpha(||z||^2 + ||w||^2). \tag{44}$$

Set $w = \frac{Lz}{\alpha}$ in (44) to get

$$\frac{4||Lz||^2}{\alpha} \leq 2\alpha(||z||^2 + \frac{||Lz||^2}{\alpha^2}).$$

Simplifying shows that this inequality is equivalent to $2\frac{||Lz||^2}{\alpha} \leq 2\alpha ||z||^2$, which implies $||Lz||^2 \leq \alpha^2 ||z||^2$. Hence $||L|| \leq \alpha$. $\qquad \square$

THEOREM 2.11 (Spectral Theorem). *Suppose $L \in \mathcal{L}(\mathcal{H})$ is compact and Hermitian. Then there is a complete orthonormal system consisting of eigenvectors of L. Each eigenspace corresponding to a non-zero eigenvalue is finite-dimensional.*

PROOF. The conclusion holds if L is the zero operator; we therefore ignore this case and assume $||L|| > 0$.

The first fact needed is that there is an eigenvalue λ with $|\lambda| = ||L||$. Note also, since L is Hermitian, that in this case λ is real and thus $\lambda = \pm ||L||$. In the proof we write α for $\pm ||L||$; in general only one of the two values works.

Because L is Hermitian, the subtle formula (43) for the norm of L holds. We let $\{z_\nu\}$ be a sequence on the unit sphere such that $|\langle Lz_\nu, z_\nu \rangle|$ converges to $||L||$. Since L is compact, we can find a subsequence (still labeled $\{z_\nu\}$) such that $L(z_\nu)$ converges to some w.

We will show that $||w|| = ||L||$ and also that αz_ν converges to w. It follows that z_ν converges to $z = \frac{w}{\alpha}$. Then we have a unit vector z for which $Lz = w = \alpha z$, and hence the first required fact will hold.

To see that $||w|| = ||L||$, we prove both inequalities. Since the norm is continuous and $||z_\nu|| = 1$, we obtain

$$||w|| = \lim_\nu ||Lz_\nu|| \leq ||L||.$$

To see the other inequality, note that $|\langle Lz_\nu, z_\nu \rangle|$ converges to $||L||$ and $L(z_\nu)$ converges to w. Hence $|\langle w, z_\nu \rangle|$ converges to $||L||$ as well. We then have

$$||L|| = \lim_\nu |\langle w, z_\nu \rangle| \leq ||w||.$$

Thus $||w|| = ||L||$.

Next we show that αz_ν converges to w. Consider the squared norm

$$||L(z_\nu) - \alpha z_\nu||^2 = ||L(z_\nu)||^2 - \alpha 2\text{Re}\langle Lz_\nu, z_\nu \rangle + ||L||^2.$$

The right-hand side converges to $||w||^2 - 2||L||^2 + ||L||^2 = 0$. Therefore the left-hand side converges to 0 as well, and hence $w = \lim(\alpha z_\nu)$. Thus z_ν itself converges to $z = \frac{w}{\alpha}$. Finally

$$L(z) = \lim(L(z_\nu)) = w = \alpha z.$$

We have found an eigenvector z with eigenvalue $\alpha = \pm||L||$. By Proposition 2.13, the eigenspace E_α corresponding to α is finite-dimensional and thus a closed subspace of $\mathcal{H}$.

Once we have found one eigenvalue λ_1, we consider the orthogonal complement $E_{\lambda_1}^\perp$ of the eigenspace E_{λ_1}. Then $E_{\lambda_1}^\perp$ is invariant under L, and the restriction of L to this subspace remains compact and Hermitian. We repeat the procedure, obtaining an eigenvalue λ_2. The eigenspaces E_{λ_1} and E_{λ_2} are orthogonal. Continuing in this fashion, we obtain a non-increasing sequence of (absolute values of) eigenvalues and corresponding eigenvectors. Each eigenspace is finite-dimensional and the eigenspaces are orthogonal. We normalize the eigenvectors to have norm 1; hence there is a bounded sequence $\{z_j\}$ of eigenvectors. By compactness $\{L(z_j)\}$ has a convergent subsequence. Since $L(z_j) = \lambda_j z_j$, also $\{\lambda_j z_j\}$ has a convergent subsequence. A sequence of orthonormal vectors cannot converge; the subsequence cannot be eventually constant because each eigenspace is of finite dimension. The only possibilities are that there are only finitely many nonzero eigenvalues, or that the eigenvalues λ_j tend to 0.

Finally we establish completeness. Let M denote a maximal collection of orthonormal eigenvectors, including those with eigenvalue 0. Since we are assuming $\mathcal{H}$ is separable, we may assume the eigenvectors are indexed by the positive integers. Let P_n denote the projection onto the span of the first n eigenvectors. We obtain

$$P_n(\zeta) = \sum_{j=1}^{n} \langle \zeta, z_j \rangle z_j.$$

Therefore

$$||L(P_n(\zeta)) - L(\zeta)|| \leq \max_{(j \geq n+1)} |\lambda_j|\, ||\zeta||. \tag{45}$$

Since the eigenvalues tend to zero, (45) shows that $L(P_n(\zeta))$ converges to $L(\zeta)$. Hence we obtain the orthonormal expansion for w in the range $\mathcal{R}(L)$ of L:

$$w = L(\zeta) = \sum_{j=1}^{\infty} \langle \zeta, z_j \rangle \lambda_j z_j. \tag{46}$$

The nullspace $\mathcal{N}(L)$ is the eigenspace corresponding to eigenvalue 0, and hence any element of $\mathcal{N}(L)$ has an expansion in terms of vectors in M. Finally, for any bounded linear map L, Proposition 2.7 guarantees that $\mathcal{N}(L) \oplus \mathcal{R}(L^*) = \mathcal{H}$. If also $L = L^*$, then $\mathcal{N}(L) \oplus \mathcal{R}(L) = \mathcal{H}$. Therefore 0 is the only vector orthogonal to M, and M is complete. $\qquad\square$

EXERCISE 2.38. Try to give a different proof of (43). (In finite dimensions one can use Lagrange multipliers.)

EXERCISE 2.39. Show that L^*L is compact and Hermitian if L is compact.

REMARK 2.1. The next several exercises concern *commutators* of operators.

DEFINITION 2.15. Let A, B be bounded operators. Their commutator $[A, B]$ is defined by $AB - BA$.

EXERCISE 2.40. Let A, B, C be bounded operators, and assume that $[C, A]$ and $[C, B]$ are compact. Prove that $[C, AB]$ is also compact. Suggestion: Do some easy algebra and then use Proposition 2.10.

EXERCISE 2.41. For a positive integer n, express $[A, B^n]$ as a sum of n terms involving $[A, B]$. What is the result when $[A, B] = I$?

EXERCISE 2.42. Use the previous exercise to show that there are no bounded operators satisfying $[A, B] = I$. Suggestion: Compute the norm of $[A, B^n]$ in two ways and let n tend to infinity.

EXERCISE 2.43. Suppose that $\langle Lz, z \rangle \geq 0$ for all z and that $\|L\| \leq 1$. Show that $\|I - L\| \leq 1$.

EXERCISE 2.44. Assume $L \in \mathcal{L}(\mathcal{H})$. Show that L is a linear combination of two Hermitian operators.

EXERCISE 2.45. Fill in the following outline to show that a Hermitian operator A is a linear combination of two unitary operators. Without loss of generality, we may assume $\|A\| \leq 1$. If $-1 \leq a \leq 1$, put $b = \sqrt{1 - a^2}$. Then $a = \frac{1}{2}((a + ib) + (a - ib))$ is the average of two points on the unit circle. We can analogously write the operator A as the average of unitary operators $A + iB$ and $A - iB$, if we can find a square root of $I - A^2$. Put $L = I - A^2$. We can find a square root of L as follows. We consider the power series expansion for $\sqrt{1 - z}$, and replace z by A^2. In other words, $\sqrt{I - C}$ makes sense if $\|C\| \leq 1$. You will need to know the sign of the coefficients in the expansion to verify convergence. Hence $\sqrt{L} = \sqrt{I - (I - L)}$ makes sense.

We close this section with a few words about *unbounded operators*. This term refers to linear mappings, defined on dense subsets of a Hilbert space, but not continuous.

Suppose $\mathcal{D}$ is a dense subset of a Hilbert space $\mathcal{H}$ and L is defined and linear on $\mathcal{D}$. If L were continuous, then L would extend to a linear mapping on $\mathcal{H}$. Many important operators are not continuous. Differentiation $\frac{d}{dx}$ is defined and linear on a dense set in $L^2([0, 2\pi])$, but it is certainly not continuous. For example, $\{\frac{e^{inx}}{in}\}$ converges to 0 in L^2, but $\frac{d}{dx}(\frac{e^{inx}}{in}) = e^{inx}$, whose L^2 norm equals 1 for each n. To apply Hilbert space methods to differential operators, we must be careful.

Let $L : \mathcal{D}(L) \subseteq \mathcal{H} \to \mathcal{H}$ be an unbounded operator. The domain $\mathcal{D}(L^*)$ of the adjoint of L is the set of $v \in \mathcal{H}$ such that the mapping $u \to \langle Lu, v \rangle$ is a continuous linear functional. By the Riesz Lemma, there is then a unique w such that $\langle Lu, v \rangle = \langle u, w \rangle$. We then put $L^*(v) = w$. It can happen that the domain of L^* is not dense in $\mathcal{H}$.

We say that an unbounded (but densely defined) operator L is *Hermitian* if

$$\langle Lz, w \rangle = \langle z, Lw \rangle$$

for all z and w in the domain of L. We say that L is *self-adjoint* if $\mathcal{D}(L) = \mathcal{D}(L^*)$ and the two maps agree there. Thus L is Hermitian if $Lz = L^*z$ when both are defined, and self-adjoint if also $\mathcal{D}(L) = \mathcal{D}(L^*)$. It often happens, with a given definition of $\mathcal{D}(L)$, that L^* agrees with L on $\mathcal{D}(L)$, but L is not self-adjoint. One must increase the domain of L, thereby decreasing the domain of L^*, until these domains are equal, before one can use without qualification the term self-adjoint.

EXERCISE 2.46. (Subtle) Put $L = i\frac{d}{dx}$ on the subspace of differentiable functions f in $L^2([0,1])$ for which $f(0) = f(1) = 0$. Show that $\langle Lf, g \rangle = \langle f, Lg \rangle$, but that L is not self-adjoint. Can you state precisely a domain for L making it self-adjoint? Comment: Look up the term *absolutely continuous* and weaken the boundary condition.

11. Sturm-Liouville theory

Fourier series provide the most famous example of orthonormal expansion, but many other orthonormal systems arise in applied mathematics and engineering. We illustrate by considering certain differential equations known as Sturm-Liouville equations. Mathematicians from the 19-th century were well-aware that many properties of the functions sine and cosine have analogues when these functions are replaced by linearly independent solutions of a second-order linear ordinary differential equation. In addition to orthonormal expansions, certain oscillation issues generalize as well. We prove the Sturm separation theorem, an easy result, to illustrate this sort of generalization, before we turn to the more difficult matter of orthonormal expansion.

Consider a second order linear ordinary differential equation $y'' + qy' + ry = 0$. Here q and r are continuous functions of x. What can we say about the zeroes of solutions? Figure 4 illustrates the situation for cosine and sine. Theorem 2.12 provides a general result.

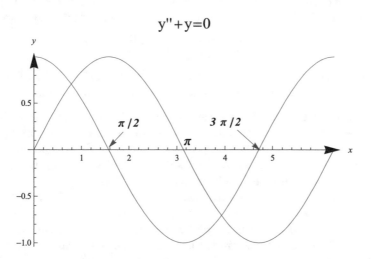

FIGURE 4. Sturm Separation

THEOREM 2.12 (Sturm separation theorem). *Let y_1 and y_2 be linearly independent (twice differentiable) solutions of $y'' + qy' + ry = 0$. Suppose that $\alpha < \beta$ and α, β are consecutive zeroes of y_1. Then there is a unique x in the interval (α, β) with $y_2(x) = 0$. Hence the zeroes of y_1 and y_2 alternate.*

PROOF. Consider the expression $W(x) = y_1(x)y_2'(x) - y_2(x)y_1'(x)$, called the Wronskian. We claim that it does not vanish. Assuming the claim, W has only one sign. We evaluate W at α and β, obtaining $-y_2(\alpha)y_1'(\alpha)$ and $-y_2(\beta)y_1'(\beta)$; these expressions must have the same sign. In particular, y_1' does not vanish at these points. Also, the values $y_1'(\alpha)$ and $y_1'(\beta)$ must have opposite signs because α and β are consecutive zeroes of y_1. Hence the values of $y_2(\alpha)$ and $y_2(\beta)$ have opposite signs. By the intermediate value theorem, there is an x in between α and β with $y_2(x) = 0$. This x must be unique, because otherwise the same reasoning would find a zero of y_1 in between the two zeroes of y_2. Since α and β are consecutive zeroes of y_1, we would get a contradiction.

It remains to show that W is of one sign. We show more in Lemma 2.2.   □

LEMMA 2.2. *Suppose y_1 and y_2 both solve $L(y) = y'' + qy' + ry = 0$. Then y_1 and y_2 are linearly dependent if and only if W vanishes identically. Also y_1 and y_2 are linearly independent if and only if W vanishes nowhere.*

PROOF. Suppose first that $W(x_0) = 0$. Since $W(x_0)$ is the determinant of the matrix of coefficients, the system of equations

$$\begin{pmatrix} y_1(x_0) & y_2(x_0) \\ y_1'(x_0) & y_2'(x_0) \end{pmatrix} \begin{pmatrix} c_1 \\ c_2 \end{pmatrix} = \begin{pmatrix} 0 \\ 0 \end{pmatrix}$$

has a non-trivial solution (c_1, c_2). Since L is linear, the function $y = c_1 y_1 + c_2 y_2$ also satisfies $L(y) = 0$. Since $y(x_0) = y'(x_0) = 0$, this solution y is identically 0. (See the paragraph after the proof.) Therefore the matrix equation holds at all x, the functions y_1 and y_2 are linearly dependent, and W is identically 0.

Suppose next that W is never zero. Consider a linear combination $c_1 y_1 + c_2 y_2$ that vanishes identically. Then also $c_1 y_1' + c_2 y_2'$ vanishes identically, and hence

$$\begin{pmatrix} y_1 & y_2 \\ y_1' & y_2' \end{pmatrix} \begin{pmatrix} c_1 \\ c_2 \end{pmatrix} = \begin{pmatrix} 0 \\ 0 \end{pmatrix}.$$

Since W is the determinant of the matrix here and $W(x) \neq 0$ for all x, the only solution is $c_1 = c_2 = 0$. Therefore y_1 and y_2 are linearly independent.   □

In the proof of Lemma 2.2, we used the following standard fact. The second order linear equation $Ly = 0$, together with initial conditions $y(x_0)$ and $y'(x_0)$, has a unique solution. This result can be proved by reducing the second order equation to a first order system. Uniqueness for the first order system can be proved using the contraction mapping principle in metric spaces. See [Ro].

We now turn to the more sophisticated Sturm-Liouville theory. Consider the following second-order differential equation on a real interval $[a, b]$. Here y is the unknown function; p, q, w are fixed real-valued functions, and the α_j and β_j are real constants. These constants are subject only to the constraint that both (SL.1) and (SL.2) are non-trivial. In other words, neither $\alpha_1^2 + \alpha_2^2$ nor $\beta_1^2 + \beta_2^2$ is 0. This condition makes the equation into a *boundary-value problem*. Both endpoints of the interval $[a, b]$ matter. The functions p', q, w are assumed to be continuous and the functions p and w are assumed positive.

$$(py')' + qy + \lambda wy = 0 \tag{SL}$$

$$\alpha_1 y(a) + \alpha_2 y'(a) = 0 \tag{SL.1}$$

$$\beta_1 y(b) + \beta_2 y'(b) = 0. \tag{SL.2}$$

REMARK 2.2. It is natural to ask how general the Sturm-Liouville equation is among second order linear equations. Consider any second order ODE of the form $Py'' + Qy' + Ry = 0$, where $P \neq 0$. We can always put it into the Sturm-Liouville form by the following typical trick from ODE, called an *integrating factor*. We multiply the equation by an unknown function u, and figure out what u must be to put the equation in Sturm-Liouville form:

$$0 = uPy'' + uQy' + uRy = (py')' + ry.$$

To make this equation hold, we need $uP = p$ and $uQ = p'$. Hence we require $\frac{p'}{p} = \frac{Q}{P}$, which yields $p = e^{\int \frac{Q}{P}}$. Hence, if we choose $u = \frac{1}{P} e^{\int \frac{Q}{P}}$, we succeed in putting the equation in the form (SL).

The following lemma involving the Wronskian gets used in an important integration by parts below, and it also implies that each eigenspace is one-dimensional. Note that the conclusion also holds if we replace g by $\bar{g}$, because all the parameters in (SL), (SL.1), and (SL.2) are real.

LEMMA 2.3. *If f and g both satisfy (SL.1) and (SL.2), then*

$$f(a)g'(a) - f'(a)g(a) = f(b)g'(b) - f'(b)g(b) = 0. \tag{47}$$

PROOF. Assume both f and g satisfy the conditions in (SL). We then can write

$$\begin{pmatrix} f(b) & f'(b) \\ g(b) & g'(b) \end{pmatrix} \begin{pmatrix} \beta_1 \\ \beta_2 \end{pmatrix} = \begin{pmatrix} 0 \\ 0 \end{pmatrix}, \tag{48}$$

and similarly for the values at a and the α_j. Equations (SL.1) and (SL.2) are non-trivial; hence (48) and its analogue for a have non-trivial solutions, and each of the matrices

$$\begin{pmatrix} f(a) & f'(a) \\ g(a) & g'(a) \end{pmatrix}$$

$$\begin{pmatrix} f(b) & f'(b) \\ g(b) & g'(b) \end{pmatrix}$$

has a non-trivial nullspace. Hence each determinant vanishes. □

COROLLARY 2.8. *Suppose f and g both solve the same (SL) equation. Then f and g are linearly dependent.*

PROOF. By Lemma 2.3, the two expressions in (47) vanish. But these expressions are Wronskian determinants. By Lemma 2.2, the two solutions are linearly independent if and only if their Wronskian determinant is (everywhere) non-zero. □

Later we use one more fact about the Wronskian.

LEMMA 2.4. *Assume u, v both solve the Sturm-Liouville equation $(py')' + qy = 0$. Let $W = uv' - u'v$. Then pW is constant. If u, v are linearly independent, then this constant is non-zero.*

PROOF. We want to show that $(p(uv' - u'v))' = 0$. Computing the expression, without any assumptions on u, v, gives

$$p(uv'' - u''v) + p'(uv' - u'v).$$

Since u and v satisfy the equation we also have

$$pu'' + p'u' + qu = 0$$

$$pv'' + p'v' + qv = 0.$$

Multiply the first equation by v, the second by u and then subtract. We get

$$p(u''v - uv'') + p'(u'v - uv') = 0,$$

which is what we need. The last statement follows immediately from Lemma 2.2.

$\square$

Each λ for which (SL) admits a non-zero solution is called an eigenvalue of the problem, and each non-zero solution is called an eigenfunction corresponding to this eigenvalue. The terminology is consistent with the standard notions of eigenvalue and eigenvector, as noted in Lemma 2.5 below. In general, when the elements of a vector space are functions, one often says *eigenfunction* instead of *eigenvector*. Corollary 2.8 thus says that the eigenspace corresponding to each eigenvalue is one-dimensional.

To connect the Sturm-Liouville setting with Fourier series, take $p = 1$, $q = 0$, and $w = 1$. We get the familiar equation

$$y'' + \lambda y = 0,$$

whose solutions are sines and cosines. For example, if the interval is $[0, \pi]$, and we assume that (SL.1) and (SL.2) give $y(0) = y(\pi) = 0$, then the eigenvalues are m^2 for positive integers m. The solutions are $y_m(x) = \sin(mx)$.

Sturm-Liouville theory uses the Hilbert space $\mathcal{H} = (L^2([a, b]), w)$, consisting of (equivalence classes of) square-integrable measurable functions with respect to the weight function w. The inner product is defined by

$$\langle f, g \rangle_w = \int_a^b f(x)\overline{g(x)}w(x)dx.$$

Although the Sturm-Liouville situation is much more general than the equation $y'' + \lambda y = 0$, the conclusions in the following theorem are remarkably similar to the results we have proved about Fourier series.

THEOREM 2.13. *Consider the Sturm-Liouville equation (SL) with boundary conditions (SL.1) and (SL.2). There is a countable collection of real eigenvalues λ_j tending to ∞ with $\lambda_1 < \lambda_2 < \ldots$. For each eigenvalue the corresponding eigenspace is one-dimensional. The corresponding eigenfunctions ϕ_j are orthogonal. After dividing each ϕ_j by a constant, we assume that these eigenfunctions are orthonormal. These eigenfunctions form a complete orthonormal system for $\mathcal{H}$. If f is continuously differentiable on $[a, b]$, then the series*

$$\sum_{j=1}^{\infty} \langle f, \phi_j \rangle_w \phi_j(x) \tag{49}$$

converges to $f(x)$ at each point of (a, b).

Proving this theorem is not easy, but we will give a fairly complete proof. We begin by rephrasing everything in terms of an unbounded operator L on $\mathcal{H}$. On an appropriate domain, L is defined by

$$L = \frac{-1}{w}\left(\frac{d}{dx}\left(p\frac{d}{dx}\right) + q\right). \tag{50}$$

The domain $\mathcal{D}(L)$ contains all twice continuously differentiable functions satisfying the (SL) boundary conditions. Eigenvalues of the Sturm-Liouville problem correspond to eigenvalues of this operator L.

LEMMA 2.5. *Equation (SL) is equivalent to $Ly = \lambda y$.*

PROOF. Left to the reader. □

PROPOSITION 2.16. *The operator L is Hermitian. In other words, if f and g are twice continuously differentiable functions on $[a, b]$ and satisfy (SL.1) and (SL.2), then*

$$\langle Lf, g \rangle_w = \langle f, Lg \rangle_w. \tag{51}$$

PROOF. The proof amounts to integrating by parts twice and using the boundary conditions. One integration by parts gives

$$\langle Lf, g \rangle_w = \int_a^b \frac{-1}{w(x)} \left(\frac{d}{dx} (p(x)f'(x)) + q(x)f(x) \right) \overline{g(x)} w(x) dx$$

$$= - \int_a^b \left(\frac{d}{dx}(p(x)f'(x)) + q(x)f(x) \right) \overline{g(x)} dx$$

$$= -p(x)f'(x)\overline{g(x)}\Big|_a^b + \int_a^b (p(x)f'(x))\overline{g'(x)} dx - \int_a^b q(x)f(x)\overline{g(x)} dx. \tag{52}$$

We integrate the middle term by parts, and stop writing the variable x, to obtain

$$\langle Lf, g \rangle_w = -p\, f'\overline{g}\Big|_a^b + p\, f\overline{g'}\Big|_a^b - \int_a^b f \frac{d}{dx}(p\overline{g'})\, dx - \int_a^b q f\overline{g}\, dx. \tag{53}$$

After multiplying and dividing by w, the integrals in (53) become

$$\int_a^b \left(\frac{-f}{w} \left(\frac{d}{dx}(p\overline{g'}) + q\overline{g} \right) \right) w\, dx = \langle f, Lg \rangle_w. \tag{54}$$

The boundary terms in (53) become

$$p(x) \left(f(x)\overline{g'(x)} - f'(x)\overline{g(x)} \right) \Big|_a^b. \tag{55}$$

Since both f and g satisfy the homogeneous boundary conditions, the term in (55) vanishes by Lemma 2.3 (using $\overline{g}$ instead of g). Hence $\langle Lf, g \rangle_w = \langle f, Lg \rangle_w$. □

In order to proceed with Sturm-Liouville theory, we must introduce some standard ideas in operator theory. These ideas are needed because differential operators such as L are defined on only a dense subspace of the Hilbert space, and they cannot be extended continuously to the whole space.

Let $\mathcal{H}$ be a Hilbert space and let $L : \mathcal{D}(L) \subseteq \mathcal{H} \to \mathcal{H}$ be a densely defined linear operator. For each complex number z consider the operator $L - zI$.

DEFINITION 2.16. The complex number z is said to be in the *spectrum* of L if $(L - zI)^{-1}$ does not exist as a bounded linear operator. Otherwise z is said to be in the *resolvent set* of L, and $(L - zI)^{-1}$ is called the *resolvent* of L at z.

Thus, when z is in the resolvent set, $(L - zI)^{-1}$ exists and is bounded. The equation $(L - zI)^{-1}f = \mu f$ is then equivalent to $f = (L - zI)(\mu f)$, and hence also to $Lf = (z + \frac{1}{\mu})f$. Thus to find the eigenvalues of L we can study the resolvent $(L - zI)^{-1}$. If L is Hermitian and we choose a real k in the resolvent set for L, then $(L - kI)^{-1}$ is Hermitian. For L as in the Sturm-Liouville set-up, the resolvent is a *compact* operator. In general, an unbounded operator L on a Hilbert space has *compact resolvent* if there is a z for which $(L - zI)^{-1}$ is compact. A generalization of Theorem 2.13 holds when L is self-adjoint and has compact resolvent.

In order to prove Theorem 2.13, we need to know that the resolvent $(L - kI)^{-1}$ is compact. We will use Green's functions.

11.1. The Green's function. In this subsection we construct the Green's function G in a fashion often used in Physics and Engineering. It will follow that a complete orthonormal system exists in the Sturm-Liouville setting. Let L be the operator defined in (50).

First we find a solution u to $Lu = 0$ that satisfies the boundary condition at a. Then we find a solution v to $Lv = 0$ that satisfies the boundary condition at b. We put

$$c = p(x)W(x) = p(x)(u(x)v'(x) - u'(x)v(x)). \tag{56}$$

By Lemma 2.4, when u and v are linearly independent, c is a non-zero constant.

We then define the Green's function as follows. Put $G(x,t) = \frac{1}{c}u(t)v(x)$ for $t < x$ and $G(x,t) = \frac{1}{c}u(x)v(t)$ for $t > x$. Then G extends to be continuous when $x = t$. Thus $Lu = 0$ and $Lv = 0$. The following important theorem and its proof illustrate the importance of the Green's function.

THEOREM 2.14. *Consider the Sturm-Liouville equation (SL). Let L be the Hermitian operator defined by (50). Let u be a solution to $Lu = 0$ satisfying boundary condition (SL.1) and v a solution to $Lv = 0$ with boundary condition (SL.2). Assume u and v are linearly independent, and define c by (56). Given f continuous, define y by*

$$y(x) = \frac{1}{c}\int_a^x u(x)(vfw)(t)dt + \frac{1}{c}\int_x^b v(x)(ufw)(t)dt = \int_a^b G(x,t)f(t)\ dt. \tag{57}$$

Then y is twice differentiable and $Ly = f$.

PROOF. We start with (57) and the formula (58) for L:

$$Ly = \frac{-p}{w}y'' - \frac{p'}{w}y' - \frac{q}{w}y. \tag{58}$$

We apply L to (57) using the fundamental theorem of calculus and compute. The collection of terms obtained where we differentiate past the integral must vanish because u, v satisfy $Lu = Lv = 0$. The remaining terms arise because of the fundamental theorem of calculus. The first time we differentiate we get

$$\frac{1}{c}(uvp)(x) - \frac{1}{c}(uvp)(x) = 0.$$

The minus sign arises because the second integral goes from x to b, rather than from b to x.

The next time we differentiate we obtain the term

$$\frac{p}{c}(u_x v - u v_x)fw,$$

with all terms evaluated at x. The term in parentheses is minus the Wronskian. By Lemma 2.4, the entire expression simplifies to $-(fw)(x)$. When we multiply by $\frac{-1}{w}$, from formula (58) of L, this expression becomes $f(x)$. We conclude, as desired, that $Ly = f$. Since u, v are twice differentiable, p is continuously differentiable, and w, f are continuous, it follows that y is twice differentiable.   □

Things break down when we cannot find linearly independent u and v, and the Green's function need not exist. In that case we must replace L by $L - kI$ for a suitable constant k. The following example illustrates several crucial points.

EXAMPLE 2.4. Consider the equation $Ly = y'' = 0$ with $y'(0) = y'(1) = 0$. The only solutions to $Lu = 0$ are constants, and hence linearly dependent. If c satisfies (56), then $c = 0$. We cannot solve $Ly = f$ for general f. Suppose that $y'(0) = y'(1) = 0$, and that $y'' = f$. Integrating twice we then must have

$$y(x) = y(0) + \int_0^x \int_0^t f(s)ds dt.$$

By the fundamental theorem of calculus, $y'(0) = 0$ and $y'(1) = \int_0^1 f(s)ds$. If $\int_0^1 f$ is not 0, then we cannot solve the equation $Ly = f$. In this case, 0 is an eigenvalue for L and hence L^{-1} does not exist. The condition $\int_0^1 f = 0$ means that the function f must be orthogonal to the constants.

To finish the proof of the Sturm-Liouville theorem, we need to show that there is a real k such that $(L - kI)^{-1}$ exists as a bounded operator. This statement holds for all k sufficiently negative, but we omit the proof. Assuming this point, we can find linearly independent u and v satisfying the equation, with u satisfying the boundary condition at a and v satisfying it at b. We construct the Green's function for $L - kI$ as above. We write $(L - kI)^{-1}f(x) = \int_a^b f(t)G(x, t)dt$. Since G is continuous on the rectangle $[a, b] \times [a, b]$, $(L - kI)^{-1}$ is compact, by Proposition 2.12. Theorem 2.11 then yields the desired conclusions.

We can express things in terms of orthonormal expansion. Let L be the operator defined in (50). Given f, we wish to solve the equation $Lg = f$. Let $\{\phi_j\}$ be the complete orthonormal system of eigenfunctions for $(L - kI)^{-1}$. This system exists because $(L - kI)^{-1}$ is compact and Hermitian. We expand g in an orthonormal series as in (49), obtaining

$$g(x) = \sum_{j=1}^{\infty} \int_a^b g(t)\overline{\phi_j(t)}w(t)dt \; \phi_j(x).$$

Differentiating term by term yields

$$(Lg)(x) = f(x) = \sum_{j=1}^{\infty} \left(\int_a^b g(t)\overline{\phi_j(t)}w(t)dt \right) \lambda_j \phi_j(x).$$

The function f also has an orthonormal expansion:

$$f(x) = \sum_{j=1}^{\infty} \left(\int_a^b f(t)\overline{\phi_j(t)}w(t)dt \right) \phi_j(x).$$

We equate coefficients to obtain

$$g(x) = \int_a^b \sum_{j=1}^{\infty} \frac{\phi_j(x)\overline{\phi_j(t)}}{\lambda_j} w(t)f(t) \, dt = \int_a^b G(x,t)f(t)w(t) \, dt. \qquad (59)$$

We summarize the story. Assume that $(L - kI)^{-1}$ has a continuous Green's function. Then $(L - kI)^{-1}$ is compact and Hermitian, and a complete orthonormal system of eigenfunctions exists. Decompose the Hilbert space into eigenspaces E_{λ_j}. If $h \in E_{\lambda_j}$ we have $(L - kI)h = \lambda_j h$. Note that no λ_j equals 0. Thus, restricted to E_{λ_j}, we can invert $L - kI$ by

$$(L - kI)^{-1}(h) = \frac{1}{\lambda_j}h.$$

We invert in general by inverting on each eigenspace and adding up the results. Things are essentially the same as in Section 4 of Chapter 1, where we solved a linear system when there was an orthonormal basis of eigenvectors. In this setting we see that the Green's function is given by

$$G(x,t) = \sum_{j=1}^{\infty} \frac{\phi_j(x)\overline{\phi_j(t)}}{\lambda_j}.$$

We consider the simple special case where $Ly = -y''$ on the interval $[0,1]$ with boundary conditions $y(0) = y(1) = 0$. For each positive integer m, there is an eigenvalue $\pi^2 m^2$, corresponding to the normalized eigenfunction $\sqrt{2}\sin(m\pi x)$. In this case $G(x,t)$ has the following expression:

$$G(x,t) = \begin{cases} x(1-t) & x < t \\ t(1-x) & x > t \end{cases}. \qquad (60)$$

We can check this formula directly by differentiating twice the relation

$$y(x) = (1-x)\int_0^x tf(t) \, dt + x \int_x^1 (1-t)f(t) \, dt.$$

Of course, we discovered this formula by the prescription from Theorem 2.14. The function x is the solution vanishing at 0. The function $1-x$ is the solution vanishing at 1. See Figure 5. Using orthonormal expansion, we have another expression for $G(x,t)$:

$$G(x,t) = 2\sum_{m=1}^{\infty} \frac{\sin(m\pi x)\sin(m\pi t)}{\pi^2 m^2}.$$

See [F2] and [G] for many computational exercises involving Green's functions for Sturm-Liouville equations and generalizations. See also [GS] for excellent intuitive discussion concerning the construction of the Green's function and its connections with the Dirac delta function.

EXERCISE 2.47. Assume $0 \le x < \frac{1}{2}$. Put $L = -(\frac{d}{dx})^2$ on $[0,1]$ with boundary conditions $y(0) = y(1) = 0$. Equate the two expressions for the Green's function to establish the identity

$$x = \frac{4}{\pi^2} \sum_{r=0}^{\infty} \frac{(-1)^r \sin((2r+1)\pi x)}{(2r+1)^2}.$$

Prove that this identity remains true at $x = \frac{1}{2}$.

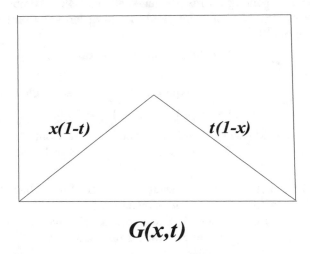

$$G(x,t)$$

FIGURE 5. Green's function for the second derivative

EXERCISE 2.48. Consider the equation $y'' + \lambda y = 0$ with boundary conditions $y(0) - y(1) = 0$ and $y'(0) + y'(1) = 0$. Show that every λ is an eigenvalue. Why doesn't this example contradict Theorem 2.13? Hint: Look carefully at (SL.1) and (SL.2).

EXERCISE 2.49. Suppose $L \in \mathcal{L}(\mathcal{H})$ is Hermitian. Find $\lim_{n \to \infty} ||L^n||^{\frac{1}{n}}$. Suggestion: If $L = L^*$, then $||L^2|| = ||L||^2$.

EXERCISE 2.50. Put the Bessel equation $x^2 y'' + x y' + (\lambda^2 x^2 - \nu^2) y = 0$ into Sturm-Liouville form.

EXERCISE 2.51. Find the Green's function for the equation $Ly = x^2 y'' - 2xy' + 2y = f$ on the interval $[1, 2]$ with $y(1) = y(2) = 0$. (First put the equation in Sturm-Liouville form.) How does the answer change if the boundary condition is replaced by $y'(1) = y'(2) = 0$?

11.2. Exercises on Legendre polynomials. The next several exercises involve the Legendre polynomials. These polynomials arise throughout pure and applied mathematics. We will return to them in Section 13.

We first remind the reader of a method for finding solutions to linear ordinary differential equations, called *reduction of order*. Consider a linear differential operator L of order m. Suppose we know one solution f to $Ly = g$. We then seek a solution of the form $y = uf$ for some unknown function u. The function u' will then satisfy a homogeneous linear differential equation of order $m - 1$. We used a similar idea in Subsection 4.1 of Chapter 1, where we replaced a constant c with a function $c(x)$ when solving an inhomogeneous equation. We note, when $m = 2$, that the method of reduction of order yields a first order equation for u' which can often be solved explicitly.

EXERCISE 2.52. Verify that the method of reduction of order works as described above.

EXERCISE 2.53. The Legendre equation (in Sturm-Liouville form) is

$$((1 - x^2)y')' + n(n+1)y = 0. \tag{61}$$

Find all solutions to (61) when $n = 0$ and when $n = 1$. Comment: When $n = 1$, finding one solution is easy. The method of reduction of order can be used to find an independent solution.

EXERCISE 2.54. Let n be a non-negative integer. Show that there is a polynomial solution P_n to (61) of degree n. Normalize to make $P_n(1) = 1$. This P_n is called the n-th Legendre polynomial. Show that an alternative definition of P_n is given for $|x| \leq 1$ and $|t| < 1$ by the generating function

$$\frac{1}{\sqrt{1 - 2xt + t^2}} = \sum_{n=0}^{\infty} P_n(x)t^n.$$

Show that the collection of these polynomials forms a complete orthogonal system for $L^2([-1,1], dx)$. Show that $||P_n||^2 = \frac{2}{2n+1}$. If needed, look ahead to the next section for one method to compute these norms.

EXERCISE 2.55. Obtain the first few Legendre polynomials by applying the Gram-Schmidt process to the monomials $1, x, x^2, x^3, x^4$.

EXAMPLE 2.5. The first few Legendre polynomials:
- $P_0(x) = 1$.
- $P_1(x) = x$.
- $P_2(x) = \frac{3x^2 - 1}{2}$
- $P_3(x) = \frac{5x^3 - 3x}{2}$.
- $P_4(x) = \frac{35x^4 - 30x^2 + 3}{8}$.

EXERCISE 2.56. Let P_n be the n-th Legendre polynomial. Show that

$$(n + 1)P_{n+1}(x) - (2n + 1)xP_n(x) + nP_{n-1}(x) = 0.$$

Use the method of difference equations to find constants a_k such that

$$P_n(x) = \sum_{k=0}^{n} a_k(1 + x)^k(1 - x)^{n-k}.$$

EXERCISE 2.57. Here is an alternative proof that the Legendre polynomials are orthogonal. First show that $P_n = c_n(\frac{d}{dx})^n(x^2 - 1)^n$. Then integrate by parts to show that

$$\langle P_n, f \rangle = c_n(-1)^n \langle (x^2 - 1)^n, (\frac{d}{dx})^n f \rangle.$$

In other words, f is orthogonal to P_n if f is a polynomial of degree less than n.

EXERCISE 2.58. Let P_l denote a Legendre polynomial. Define the *associated Legendre functions* with parameters l and m by

$$P_l^m(x) = (1 - x^2)^{\frac{m}{2}}(\frac{d}{dx})^m P_l(x).$$

- Show when m is even that P_l^m is a polynomial.
- Obtain a differential equation satisfied by P_l^m by differentiating m-times the Sturm-Liouville equation (61) defining P_l.
- Show that $P_l^m(x)$ is a constant times a power of $(1-x^2)$ times a derivative of a power of $(1-x^2)$.

The associated Legendre functions arise in Section 13 on spherical harmonics.

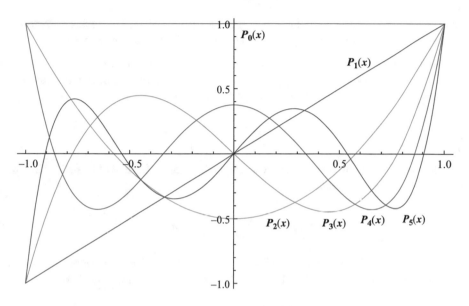

FIGURE 6. Legendre polynomials

12. Generating functions and orthonormal systems

Many of the complete orthonormal systems used in physics and engineering are defined via the *Gram-Schmidt process*. Consider an interval I in $\mathbf{R}$ and the Hilbert space $L^2(I, w(x)dx)$ of square integrable functions with respect to some weight function w. Starting with a nice class of functions, such as the monomials, and then orthonormalizing them, one obtains various *special functions*. The Gram-Schmidt process often leads to tedious computation.

Following the method of Exercise 2.54, we use *generating functions* to investigate orthonormal systems. In addition to the Legendre polynomials, we give two examples of importance in physics, the Laguerre polynomials and the Hermite polynomials. We return to the Hermite polynomials in Chapter 3, where we relate them to eigenfunctions of the Fourier transform.

We will use a simple proposition relating orthonormal systems and generating functions. We then show how the technique works for the Laguerre and Hermite polynomials.

Before stating and proving this proposition, we discuss vector-valued convergent power series. Let $\mathbf{B}$ denote the open unit disk in $\mathbf{C}$. Let $\mathcal{H}$ be a Hilbert space; it is often useful to consider complex analytic functions $f : \mathbf{B} \to \mathcal{H}$.

Consider a power series $A(z) = \sum A_n z^n$, where the coefficients A_n lie in $\mathcal{H}$. This series converges at the complex number z if its partial sums there form a Cauchy sequence in $\mathcal{H}$. We define a function $A : \mathbf{B} \to \mathcal{H}$ to be complex analytic if there is a sequence $\{A_n\}$ in $\mathcal{H}$ such that the series

$$\sum_{n=0}^{\infty} A_n z^n$$

converges to $A(z)$ for all z in $\mathbf{B}$. On compact subsets of $\mathbf{B}$, the series converges in norm, and we may therefore rearrange the order of summation at will.

PROPOSITION 2.17. *Let $\mathcal{H}$ be a Hilbert space, and suppose $A : \mathbf{B} \to \mathcal{H}$ is complex analytic with $A(t) = \sum_{n=0}^{\infty} A_n t^n$. Then the collection of vectors $\{A_n\}$ forms an orthonormal system in $\mathcal{H}$ if and only if, for all $t \in \mathbf{B}$,*

$$\|A(t)\|^2 = \frac{1}{1 - |t|^2}.$$

PROOF. Using the absolute convergence on compact subsets to order the summation as we wish, we obtain

$$\|A(t)\|^2 = \sum_{m,n=0}^{\infty} \langle A_n, A_m \rangle t^n \overline{t}^m. \tag{62}$$

Comparison with the geometric series yields the result: the right-hand side of (62) equals $\frac{1}{1-|t|^2}$ if and only if $\langle A_n, A_m \rangle$ equals 0 for $n \neq m$ and equals 1 for $n = m$. $\square$

DEFINITION 2.17. The formal series

$$\sum_{n=0}^{\infty} L_n t^n$$

is the *ordinary generating function* for the sequence $\{L_n\}$. The formal series

$$\sum_{n=0}^{\infty} L_n \frac{t^n}{n!}$$

is the *exponential generating function* for the sequence $\{L_n\}$.

Explicit formulas for these generating functions often provide powerful insight as well as simple proofs of orthogonality relations.

EXAMPLE 2.6 (Laguerre polynomials). Let $\mathcal{H} = L^2([0,\infty), e^{-x}dx)$ be the Hilbert space of square integrable functions on $[0,\infty)$ with respect to the measure $e^{-x}dx$. Consider functions L_n defined via their generating function by

$$A(x,t) = \sum_{n=0}^{\infty} L_n(x)t^n = (1-t)^{-1}\exp\left(\frac{-xt}{1-t}\right).$$

Note that $x \geq 0$ and $|t| < 1$. In order to study the inner products $\langle L_n, L_m \rangle$, we compute $\|A(x,t)\|^2$. We will find an explicit formula for this squared norm; Proposition 2.17 implies that the L_n form an orthonormal system.

We have

$$|A(x,t)|^2 = (1-t)^{-1} \exp\left(\frac{-xt}{1-t}\right)(1-\bar{t})^{-1}\exp\left(\frac{-x\bar{t}}{1-\bar{t}}\right).$$

Multiplying by the weight function e^{-x} and integrating we obtain

$$\|A(x,t)\|^2 = (1-t)^{-1}(1-\bar{t})^{-1}\int_0^\infty \exp\left(-x(1+\frac{t}{1-t}+\frac{\bar{t}}{1-\bar{t}})\right)dx.$$

Computing the integral on the right-hand side and simplifying shows that

$$\|A(x,t)\|^2 = \frac{1}{(1-t)(1-\bar{t})}\frac{1}{1+\frac{t}{1-t}+\frac{\bar{t}}{1-\bar{t}}} = \frac{1}{1-|t|^2}.$$

From Proposition 2.17 we see that $\{L_n\}$ forms an orthonormal system in $\mathcal{H}$.

The series defining the generating function converges for $|t| < 1$, and each L_n is real-valued. In Exercise 2.60 we ask the reader to show that the functions L_n satisfy the Rodrigues formula

$$L_n(x) = \frac{e^x}{n!}\left(\frac{d}{dx}\right)^n(x^n e^{-x}) \tag{63}$$

and hence are polynomials of degree n. They are called the Laguerre polynomials, and they form a *complete* orthonormal system for $L^2([0,\infty), e^{-x}dx)$. Laguerre polynomials arise in solving the Schrödinger equation for a hydrogen atom.

A similar technique works for the Hermite polynomials, which arise in many problems in physics, such as the quantum harmonic oscillator. See pages 120–122 in [GS]. We discuss these polynomials at the end of Chapter 3. One way to define the Hermite polynomials is via the exponential generating function

$$\exp(2xt-t^2) = \sum H_n(x)\frac{t^n}{n!}. \tag{64}$$

The functions H_n are polynomials and form an orthogonal set for $\mathcal{H}=L^2(\mathbf{R}, e^{-x^2}dx)$. With this normalization the norms are not equal to unity. In Exercise 2.62 the reader is asked to study the Hermite polynomials by mimicking the computations for the Laguerre polynomials. Other normalizations of these polynomials are also common. Sometimes the weight function used is $e^{\frac{-x^2}{2}}$. The advantage of our normalization is Theorem 3.9.

The technique of generating functions can also be used to find normalizing coefficients. Suppose, such as in the Sturm-Liouville setting, that the collection $\{f_n\}$ for $n \geq 0$ forms a complete orthogonal system. We wish to find $\|f_n\|_{L^2}$. Assume that we have found the generating function

$$B(x,t) = \sum_{n=0}^\infty f_n(x)t^n$$

explicitly. We may assume t is real. Taking L^2 norms (in x) we discover that $\|f_n\|^2$ must be the coefficient of t^{2n} in the series expansion of $\|B(x,t)\|_{L^2}^2$.

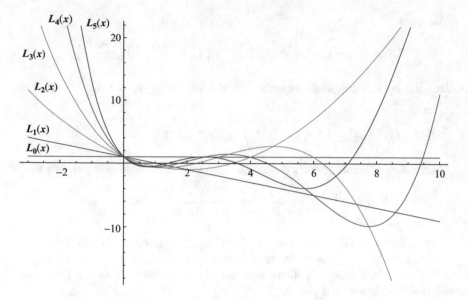

FIGURE 7. Laguerre polynomials

We illustrate this result by solving part of Exercise 2.54. The generating function for the Legendre polynomials is known to be

$$B(x,t) = \frac{1}{\sqrt{1 - 2xt + t^2}}.$$

By elementary calculus, its L^2 norm on $[-1, 1]$ is found to satisfy

$$\|B(x,t)\|_{L^2}^2 = \frac{1}{t}\left(\log(1+t) - \log(1-t)\right).$$

Expanding $\log(1 \pm t)$ in a Taylor series shows that

$$\|B(x,t)\|_{L^2}^2 = 2\sum_{n=0}^{\infty} \frac{t^{2n}}{2n+1}.$$

Hence $\|P_n\|_{L^2}^2 = \frac{2}{2n+1}$.

EXERCISE 2.59. Fill in the details from the previous paragraph.

EXERCISE 2.60. (1) With L_n as in Example 2.6, verify the Rodrigues formula (63). Suggestion: Write the power series of the exponential on the right-hand side of (63) and interchange the order of summation.

(2) Show that each L_n is a polynomial in x. Hint: The easiest way is to use (1).

(3) Prove that $\{L_n\}$ forms a *complete* system in $L^2\left([0,\infty), e^{-x}dx\right)$.

EXERCISE 2.61. For $x > 0$ verify that

$$\sum_{n=0}^{\infty} \frac{L_n(x)}{n+1} = \int_0^{\infty} \frac{e^{-xt}}{t+1} dt.$$

Suggestion: Integrate the relation

$$\sum_{n=0}^{\infty} L_n(x)s^n = (1-s)^{-1} \exp(\frac{-xs}{1-s})$$

over the interval $[0, 1]$ and then change variables in the integral.

EXERCISE 2.62 (Hermite polynomials). Here H_n is defined by (64).
(1) Use (64) to find a simple expression for

$$\sum_{n=0}^{\infty} H_n(x)t^n \sum_{m=0}^{\infty} H_m(x)s^m.$$

(2) Integrate the result in (1) over $\mathbf{R}$ with respect to the measure $e^{-x^2} dx$.
(3) Use (2) to show that the Hermite polynomials form an orthogonal system with

$$||H_n||^2 = 2^n n! \sqrt{\pi}.$$

(4) Prove that the system of Hermite polynomials is complete in $L^2(\mathbf{R}, e^{-x^2} dx)$.

Comment: Sometimes the functions defined by $H_n(x)e^{\frac{-x^2}{2}}$ are called *Hermite functions*. Thus the Hermite functions form a complete orthogonal system for $L^2(\mathbf{R})$. Look ahead to Theorem 3.9 to see a remarkable property of these functions.

EXERCISE 2.63. Replace the generating function used for the Legendre polynomials by $(1 - 2xt + t^2)^{-\lambda}$ for $\lambda > -\frac{1}{2}$ and carry out the same steps. The resulting polynomials are the *ultraspherical* or *Gegenbauer* polynomials. Note that the Legendre polynomials are the special case when $\lambda = \frac{1}{2}$. See how many properties of the Legendre polynomials you can generalize.

13. Spherical harmonics

We close this chapter by discussing spherical harmonics. This topic provides one method to generalize Fourier series on the unit circle to orthonormal expansions on the unit sphere. One approach to spherical harmonics follows a thread of history, based on the work of Legendre. This approach relates the exercises from Section 11 on Legendre polynomials to elementary physics, and relies on spherical coordinates from calculus. Perhaps the most elegant approach, given in Theorems 2.15 and 2.16, uses spaces of homogeneous polynomials. We discuss both approaches.

Let S^2 denote the unit sphere in real Euclidean space $\mathbf{R}^3$. Let Δ denote the Laplace operator $\sum_{j=1}^{3} \frac{\partial^2}{\partial x_j^2}$. We would like to find a complete orthonormal system for $L^2(S^2)$ whose properties are analogous to those of the exponentials e^{inx} on the unit circle. Doing so is not simple.

Recall that Newton's law of gravitation and Coulomb's law of electric charge both begin with a potential function. Imagine a mass or charge placed at a single point p in real Euclidean space $\mathbf{R}^3$. The potential at $\mathbf{x}$ due to this mass or charge is then a constant times the reciprocal of the distance from $\mathbf{x}$ to p. Let us suppose that the mass or charge is located at the point $(0, 0, 1)$. The potential at the point $\mathbf{x} = (x_1, x_2, x_3)$ is then

$$\frac{c}{||\mathbf{x} - p||} = \frac{c}{\sqrt{(x_1^2 + x_2^2 + (x_3 - 1)^2)}}. \tag{65}$$

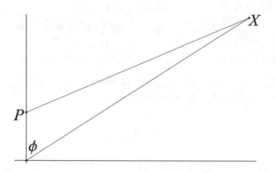

FIGURE 8. The co-latitude ϕ

We wish to express (65) in spherical coordinates. We write

$$\mathbf{x} = (x_1, x_2, x_3) = (\rho\cos(\theta)\sin(\phi), \rho\sin(\theta)\sin(\phi), \rho\cos(\phi))$$

where ρ is the distance to the origin, θ is the usual polar coordinate angle in the (x_1, x_2) plane measuring longitude, and ϕ is the co-latitude. Thus $0 \leq \theta \leq 2\pi$ whereas $0 \leq \phi \leq \pi$. These conventions are common in calculus books, but the physics literature often interchanges θ and ϕ. Also, sometimes r is used instead of ρ. In many sources, however, r is reserved for its role in cylindrical coordinates, and thus $r^2 = x^2 + y^2$.

Writing (65) in spherical coordinates we obtain

$$\frac{c}{||\mathbf{x} - p||} = \frac{c}{\sqrt{1 + \rho^2 - 2\rho\cos(\phi)}}. \tag{66}$$

The denominator in (66) is the same expression as in the generating function for the Legendre polynomials P_n from Exercise 2.54, with t replaced by ρ and x replaced by $\cos(\phi)$. Therefore we can rewrite (66) as follows:

$$\frac{c}{||\mathbf{x} - p||} = c \sum_{n=0}^{\infty} P_n(\cos(\phi))||\mathbf{x}||^n. \tag{67}$$

The potential function from (65) is harmonic away from p. We leave the computation to Exercise 2.64. We write the Laplace operator in spherical coordinates:

$$\Delta(f) = \frac{1}{\rho^2} \frac{\partial}{\partial \rho}(\rho^2 f_\rho) + \frac{1}{\rho^2 \sin(\phi)} \frac{\partial}{\partial \phi}(\sin(\phi) f_\phi) + \frac{1}{\rho^2 \sin^2(\phi)} f_{\theta\theta}. \tag{68}$$

We attempt to solve the Laplace equation $\Delta(f) = 0$ using separation of variables, generalizing Exercise 1.60. Thus we assume that

$$f(\rho, \theta, \phi) = A(\rho)B(\phi)C(\theta). \tag{69}$$

Plugging (69) into the Laplace equation yields the equation

$$0 = (\rho^2 A' BC)_\rho + \frac{1}{\sin(\phi)}(\sin(\phi) AB'C)_\phi + \frac{1}{\sin^2(\phi)} ABC''. \tag{70}$$

After dividing by ABC, we obtain

$$0 = \left(\frac{\rho^2 A'' + 2\rho A'}{A}\right) + \frac{B''}{B} + \cot(\phi)\frac{B'}{B} + \frac{1}{\sin^2(\phi)}\frac{C''}{C}. \tag{71}$$

The first fraction in (71) depends on ρ; the other terms do not. Hence there is a constant λ such that

$$\rho^2 A'' + 2\rho A' = \lambda A. \tag{72}$$

Furthermore we also have

$$\left(\frac{B''}{B} + \cot(\phi)\frac{B'}{B}\right)\sin^2(\phi) + \frac{C''}{C} = -\lambda\sin^2(\phi). \tag{73}$$

The only solutions to the equation (72) for A that are continuous at zero are $A(\rho) = c\rho^l$ for non-negative integers l. It follows that $\lambda = l(l+1)$.

Now we look at equation (73). Again by grouping the θ and ϕ terms separately we obtain two equations:

$$\frac{C''}{C} = -\mu \tag{74}$$

$$\sin^2(\phi)\left(\frac{B''}{B} + \cot(\phi)\frac{B'}{B} + \lambda\right) = \mu. \tag{75}$$

Now (74) must be periodic in θ. Hence μ is the square of an integer k. We see that $C(\theta) = ce^{ik\theta}$. Also (75) becomes

$$\sin^2(\phi)\left(\frac{B''}{B} + \cot(\phi)\frac{B'}{B} + \lambda\right) = k^2. \tag{76}$$

Simplifying (76) leads to the equation

$$B'' + \cot(\phi)B' + \left(l(l+1) - \frac{k^2}{\sin^2(\phi)}\right)B = 0. \tag{77}$$

Equation (77) evokes the differential equation defining the Legendre polynomials. In fact, if we make the substitution $x = \cos(\phi)$, then (77) is precisely equivalent (See Exercise 2.66) to the equation

$$(1 - x^2)B_{xx} - 2xB_x + \left(l(l+1) - \frac{k^2}{(1-x^2)}\right)B = 0. \tag{78}$$

The solutions P_l^k to (78) are the associated Legendre functions from Exercise 2.58 when $k \geq 0$, and related expressions when $k < 0$. The function $e^{ik\theta}P_l^k(\cos(\phi))$ is the spherical harmonic $Y_l^k(\theta, \phi)$. The integer parameter k varies from $-l$ to l, yielding $2l+1$ independent functions. The functions $\rho^l e^{ik\theta}P_l^k(\cos(\phi))$ are harmonic. The functions Y_l^k are not themselves harmonic in general; on the sphere each Y_l^k is an eigenfunction of the Laplacian with eigenvalue $-l(l+1)$.

A Wikipedia page called *Table of spherical harmonics* lists these Y_l^k, including the normalizing constants, for $0 \leq l \leq 10$ and all corresponding k. The functions Y_l^k and Y_b^a are orthogonal, on $L^2(S^2)$, unless $k = a$ and $l = b$. These functions form a complete orthogonal system for $L^2(S^2)$. Remarkable additional properties whose discussion is beyond the scope of this book hold as well.

We next approach spherical harmonics via homogeneous polynomials. Things are simpler this way, but perhaps less useful in applied mathematics.

We will work in $\mathbf{R}^n$, although we will write some formulas explicitly when $n = 3$. Let $\mathbf{x} = (x_1, ..., x_n)$ denote the variables. A polynomial $p(\mathbf{x})$ is homogeneous of degree k if $p(t\mathbf{x}) = t^k p(\mathbf{x})$. Homogeneous polynomials are therefore determined by their values on the unit sphere. It is often useful to identify a homogeneous polynomial $p(\mathbf{x})$ with the function

$$P(\mathbf{x}) = \frac{p(\mathbf{x})}{||\mathbf{x}||^k},$$

which is defined in the complement of the origin, agrees with p on the sphere, and is homogeneous of degree 0. See Proposition 2.18. For each m, we write $\mathbf{H}_m$ for the vector space of homogeneous harmonic polynomials of degree m. In Theorem 2.16, we will compute the dimension of $\mathbf{H}_m$. When $n = 3$, its dimension turns out be $2m + 1$. We obtain spherical harmonics by restricting harmonic homogeneous polynomials to the unit sphere.

EXAMPLE 2.7. Put $n = 3$. When $m = 1$, the harmonic polynomials x, y, z form a basis for $\mathbf{H}_1$. For $m = 2$, the following five polynomials form a basis for $\mathbf{H}_2$:

- xy
- xz
- yz
- $x^2 + y^2 - 2z^2$
- $x^2 - 2y^2 + z^2$.

Note that the harmonic polynomial $-2x^2 + y^2 + z^2$ is linearly dependent on the last two items in the list.

It will be as easy to work in $\mathbf{R}^n$ as it is in $\mathbf{R}^3$. We write $v \cdot w$ for the usual inner product of v, w in $\mathbf{R}^n$. We assume $n \geq 2$.

Let V_m denote the vector space of homogeneous polynomials of degree m in the variable $\mathbf{x}$ in $\mathbf{R}^n$. We regard $\mathbf{H}_m$ as a subspace of V_m. The dimension of V_m is the binomial coefficient $\binom{m+n-1}{n-1}$. We have a map $M : V_m \to V_{m+2}$ given by multiplication by $||\mathbf{x}||^2$. The Laplace operator Δ maps the other direction. These operators turn out to be adjoints. See Theorem 2.16.

We begin with a remarkable formula involving the Laplacian on harmonic, homogeneous polynomials on $\mathbf{R}^n$. The function P in Proposition 2.18 below is homogeneous of degree 0, and hence its Laplacian is homogeneous of degree -2. This observation explains why we must divide by $||\mathbf{x}||^2$ in (79).

PROPOSITION 2.18. *Let p be a harmonic, homogeneous polynomial of degree l on $\mathbf{R}^n$. Outside the origin, consider the function P defined by*

$$P(\mathbf{x}) = \frac{p(\mathbf{x})}{||\mathbf{x}||^l}.$$

Then we have

$$\Delta(P) = -l(l + n - 2)\frac{P(\mathbf{x})}{||\mathbf{x}||^2}. \tag{79}$$

Restricted to the sphere, P defines an eigenfunction of the Laplacian with eigenvalue $-l(l + n - 2)$. When $n = 3$, P is therefore a linear combination of the spherical harmonics Y_l^k with $-l \leq k \leq l$.

PROOF. See Exercise 2.72 for the computation yielding (79). The second statement follows from (79) by putting $||\mathbf{x}||^2$ equal to 1. The last statement follows from the discussion just after (78). □

Consider the Hilbert space $L^2(S^{n-1})$, where S^{n-1} is the unit sphere in n-dimensions, and $n \geq 2$. In order to integrate over the unit sphere, we use n-dimensional spherical coordinates. We put $\mathbf{x} = \rho \mathbf{v}$, where $\rho = ||\mathbf{x}||$ and $\mathbf{v}$ lies on the unit sphere. We then can write the volume form dV on $\mathbf{R}^n$ as

$$dV(\mathbf{x}) = \rho^{n-1} d\rho\, d\sigma(\mathbf{v}).$$

Let f be a function on $\mathbf{R}^n$. Away from $\mathbf{0}$, we define a function F by

$$F(\mathbf{x}) = f(\frac{\mathbf{x}}{||\mathbf{x}||}) = f(\mathbf{v}).$$

The function F satisfies $F(t\mathbf{x}) = F(\mathbf{x})$ when $t > 0$. Such a function is called positive homogeneous of degree 0. We note a special case of Euler's formula for such functions, when F is differentiable. See Exercise 2.71 for a more general statement.

PROPOSITION 2.19. *Assume F is differentiable and $F(t\mathbf{x}) = F(\mathbf{x})$ for $t > 0$ and all $\mathbf{x}$. Then $dF(\mathbf{x}) \cdot \mathbf{x} = 0$.*

PROOF. Apply $\frac{d}{dt}$ to the equation $F(t\mathbf{x}) = F(\mathbf{x})$ and set $t = 1$. □

Let χ be a smooth function on $\mathbf{R}$ with the following properties:

(1) $\chi(0) = 0$.
(2) $\chi(t)$ tends to 0 as t tends to infinity.
(3) $\int_0^\infty \chi(t^2) t^{n-1} dt = 1$. (Here n is the dimension.)

Given a smooth function w, we wish to compute $\int_{S^{n-1}} w d\sigma$. Because of property (3) of χ, the integration formula (80) holds. It allows us to express integrals over the sphere as integrals over Euclidean space:

$$\int_{\mathbf{R}^n} \chi(||\mathbf{x}||^2) w(\frac{\mathbf{x}}{||\mathbf{x}||}) dV = \int_{S^{n-1}} \int_0^\infty \chi(\rho^2) \rho^{n-1} d\rho\, w(\mathbf{v}) d\sigma(\mathbf{v}) = \int_{S^{n-1}} w d\sigma. \quad (80)$$

The other two properties of χ will be useful in an integration by parts.

THEOREM 2.15. *For $k \neq l$, the subspaces $\mathbf{H}_k$ and $\mathbf{H}_l$ are orthogonal in $L^2(S^{n-1})$.*

PROOF. Given harmonic homogeneous polynomials f of degree k and g of degree l, let F and G be the corresponding homogeneous functions of degree 0 defined above. By Proposition 2.18, these functions are eigenfunctions of the Laplacian on the sphere, with distinct eigenvalues. We claim that the Laplacian is Hermitian:

$$\int_{S^{n-1}} \Delta F\, \overline{G}\, d\sigma = \int_{S^{n-1}} F\, \overline{\Delta G}\, d\sigma. \quad (81)$$

Given the claim, eigenfunctions corresponding to distinct eigenvalues are orthogonal. Thus harmonic, homogeneous polynomials of different degrees are orthogonal on the unit sphere.

It remains to prove (81). We may assume that G is real. Let

$$W = \Delta F\, G - F\, \Delta G = \sum_{j=1}^n (\frac{\partial}{dx_j})(F_{x_j} G - F G_{x_j}).$$

We integrate by parts in (80), moving each $\frac{\partial}{\partial x_j}$. Note that $\frac{\partial}{\partial x_j}(||\mathbf{x}||^2) = 2x_j$.

$$\int_{S^{n-1}} W\, d\sigma = \int_{\mathbf{R}^n} \chi(||\mathbf{x}||^2) W(\frac{\mathbf{x}}{||\mathbf{x}||})\, dV$$

$$= -\int_{\mathbf{R}^n} \sum_{j=1}^{n} (F_{x_j} G - F G_{x_j}) \chi'(||\mathbf{x}||^2) 2x_j \, dV. \tag{82}$$

The last term in (82) is zero by Proposition 2.19, because F and G are positive homogeneous of degree 0. Thus Δ is Hermitian. $\qquad\square$

It is convenient to define particular inner products on the spaces V_m, which differ from the usual inner product given by integration. By linearity, to define the inner product on V_m it suffices to define the inner product of monomials. We illustrate for $n = 3$. Put

$$\langle x^a y^b z^c, x^A y^B z^C \rangle_{V_m} = 0 \tag{83}$$

unless $a = A$, $b = B$, and $c = C$. In this case, we put $||x^a y^b z^c||^2_{V_m} = a!b!c!$. The generalization to other dimensions is evident:

$$|| \prod_{j=1}^{n} x_j^{a_j} ||^2_{V_m} = \prod_{j=1}^{n} a_j!.$$

Thus distinct monomials are decreed to be orthogonal.

THEOREM 2.16. *The mapping $M : V_m \to V_{m+2}$ is the adjoint of the mapping $\Delta : V_{m+2} \to V_m$. In other words,*

$$\langle Mf, g \rangle_{V_{m+2}} = \langle f, \Delta g \rangle_{V_m}. \tag{84}$$

Hence the image of M is orthogonal to the harmonic space $\mathbf{H}_{m+2}$ and

$$V_{m+2} = M(V_m) \oplus \mathbf{H}_{m+2}.$$

Furthermore, $\mathbf{H}_m$ is of dimension $\binom{m+n-1}{n-1} - \binom{m+n-3}{n-1}$. When $n = 3$, this dimension is $2m + 1$.

PROOF. To be concrete, we write out the proof when $n = 3$. By linearity, it suffices to check (84) on monomials $f = x^a y^b z^c$ and $g = x^A y^B z^C$, where it follows by computing both sides of (84) in terms of factorials. There are three possible circumstances in which the inner product is not zero:

- $(a, b, c) = (A - 2, B, C)$
- $(a, b, c) = (A, B - 2, C)$
- $(a, b, c) = (A, B, C - 2)$.

In the first case, we must check that $(a + 2)!b!c! = A(A - 1)(A - 2)!B!C!$, which holds. The other two cases are similarly easy, and hence (84) holds.

Next, suppose that h is in the image of M and that g is in the nullspace of Δ. Then (84) gives

$$\langle h, g \rangle_{V_{m+2}} = \langle Mf, g \rangle_{V_{m+2}} = \langle f, \Delta g \rangle_{V_m} = 0.$$

The desired orthogonality thus holds and the direct sum decomposition follows. Finally, the dimension of V_m is $\binom{m+n-1}{n-1}$. Since M is injective, the dimension of the image of M is the dimension of V_m. The dimension of $\mathbf{H}_{m+2}$ is therefore

$$\binom{m+n+1}{n-1} - \binom{m+n-1}{n-1}.$$

When $n = 3$, the dimension of $\mathbf{H}_{m+2}$ therefore is

$$\frac{(m+4)(m+3)}{2} - \frac{(m+2)(m+1)}{2} = 2m + 5,$$

and hence the dimension of $\mathbf{H}_m$ is $2m + 1$. $\qquad\square$

REMARK 2.3. The formula in Theorem 2.16 for the dimension of $\mathbf{H}_m$ defines a polynomial of degree $n - 2$ in m. See Exercise 2.75.

COROLLARY 2.9. *On the sphere we have* $V_m = \mathbf{H}_m \oplus \mathbf{H}_{m-2} \oplus \ldots$.

PROOF. The formula follows by iterating the equality $V_m = M(V_{m-2}) \oplus \mathbf{H}_m$ and noting that $||\mathbf{x}||^2 = 1$ on the sphere. $\qquad\square$

COROLLARY 2.10. *Suppose f is continuous on the unit sphere. Then there is a sequence of harmonic polynomials converging uniformly to f.*

PROOF. This proof assumes the Stone-Weierstrass theorem to the effect that a continuous function on a compact subset S of $\mathbf{R}^n$ is the uniform limit on S of a sequence of polynomials. We proved this result in Corollary 1.8 when S is the circle. Given this theorem, the result follows from Corollary 2.9, because each polynomial can be decomposed on the sphere in terms of harmonic polynomials. $\quad\square$

COROLLARY 2.11. *The spherical harmonics form a complete orthogonal system for $L^2(S^2)$.*

We illustrate Corollary 2.11 for $m = 0$ and $m = 1$, when $n = 3$. Of course V_0 is the span of the constant 1. Its image under M is the span of $x^2 + y^2 + z^2$. The space $\mathbf{H}_2$ is spanned by the five functions $xy, xz, yz, x^2 + y^2 - 2z^2, x^2 - 2y^2 + z^2$. Each of these is orthogonal to $x^2 + y^2 + z^2$, which spans the orthogonal complement of $\mathbf{H}_2$. Next, V_1 is spanned by x, y, z. Its image under M is the span of $x(x^2 + y^2 + z^2), y(x^2 + y^2 + z^2), z(x^2 + y^2 + z^2)$. The space V_3 has dimension ten. The seven-dimensional space $\mathbf{H}_3$ is the orthogonal complement of the span of $M(V_1)$.

EXERCISE 2.64. Show that (65) defines a harmonic function away from $(0, 0, 1)$. Use both Euclidean coordinates and spherical coordinates.

EXERCISE 2.65. Verify formula (68).

EXERCISE 2.66. Use the chain rule (and some computation) to show that (77) and (78) are equivalent. Suggestion: First show that

$$B_{\phi\phi} = B_{xx}x_\phi^2 + B_x x_{\phi\phi}.$$

EXERCISE 2.67. For $n = 3$, express the harmonic polynomials of degree two using spherical coordinates.

EXERCISE 2.68. For $n = 3$, find seven linearly independent harmonic polynomials of degree three.

EXERCISE 2.69. (Difficult) Analyze (78) fully in terms of Legendre polynomials.

EXERCISE 2.70. Verify (79) if $p(x, y, z) = x^2 - y^2$.

EXERCISE 2.71. Verify Euler's identity: if f is differentiable and homogeneous of degree k on $\mathbf{R}^n$, then

$$df(\mathbf{x}) \cdot \mathbf{x} = kf(\mathbf{x}).$$

Proposition 2.19 was the case $k = 0$. What is the geometric interpretation of the result in this case?

EXERCISE 2.72. Verify (79). Euler's identity is useful.

EXERCISE 2.73. Take $n = 2$, and identify $\mathbf{R}^2$ with $\mathbf{C}$. Consider the harmonic polynomial $\mathrm{Re}(z^{2m})$. Give a much simpler proof of the analogue of formula (79) using the formula $\Delta(u) = 4u_{z\bar{z}}$ from Section 11 of Chapter 1.

EXERCISE 2.74. Again identify $\mathbf{R}^2$ with $\mathbf{C}$. Write down a basis for the homogeneous harmonic polynomials of degree m in terms of z and $\bar{z}$. Comment: The answer is obvious!

EXERCISE 2.75. For $n \geq 2$, simplify the formula in Theorem 2.16 to show that $\dim(\mathbf{H}_m)$ is a polynomial of degree $n - 2$ in m.

CHAPTER 3

Fourier transform on R

1. Introduction

We define and study the Fourier transform in this chapter. Rather than working with functions defined on the circle, we consider functions defined on the real line **R**. Among many books, the reader can consult [E], [G], and [GS] for applications of Fourier transforms to applied mathematics, physics, and engineering. See [F1] for an advanced mathematical treatment.

When $|f|$ is integrable on **R**, we will define the Fourier transform of f by

$$\mathcal{F}(f)(\xi) = \hat{f}(\xi) = \frac{1}{\sqrt{2\pi}} \int_{-\infty}^{\infty} f(x)e^{-ix\xi}dx. \tag{1}$$

In (1), the variable ξ is real. Thus $\hat{f}$ will be another function defined on the real line. We will then extend the definition of the Fourier transform by using methods of functional analysis.

Along the way we will develop a deeper understanding of approximate identities and the Dirac delta function. We will define *distributions* or *generalized functions* and thereby place the Dirac function on firm theoretical ground. For nice functions f we have the Fourier inversion formula

$$f(x) = \frac{1}{\sqrt{2\pi}} \int_{-\infty}^{\infty} \hat{f}(\xi)e^{ix\xi}d\xi. \tag{2}$$

Our abstract approach leads to a definition of $\hat{f}$ for $f \in L^2(\mathbf{R})$ or even when it is a distribution. We prove the fundamental result (Plancherel theorem) that the Fourier transform is unitary on $L^2(\mathbf{R})$ and hence

$$||f||_{L^2}^2 = ||\hat{f}||_{L^2}^2. \tag{3}$$

We combine the Plancherel theorem and the Cauchy-Schwarz inequality to establish the famous inequality which yields the Heisenberg uncertainty principle from quantum mechanics. We include a brief introduction to pseudo-differential operators which includes the Sobolev lemma in one dimension. We close this chapter with a section on inequalities.

For functions defined on the circle, we observed that the more differentiable the function, the faster its Fourier coefficients decay at infinity. An analogous phenomenon happens for functions on **R**. It therefore makes sense to begin our study of the Fourier transform by restricting to smooth functions of rapid decay at infinity.

© Springer Nature Switzerland AG 2019
J. P. D'Angelo, *Hermitian Analysis*, Cornerstones,
https://doi.org/10.1007/978-3-030-16514-7_3

2. The Fourier transform on the Schwartz space

The Schwartz space $\mathcal{S}$ consists of the smooth functions of rapid decay at infinity. This space is named for Laurent Schwartz, a different person from Hermann Schwarz, whose name is associated with the Cauchy-Schwarz inequality. Here is the precise definition:

DEFINITION 3.1. The Schwartz space $\mathcal{S}$ consists of those infinitely differentiable complex-valued functions f on **R** such that, for all non-negative integers a, b,

$$\lim_{|x| \to \infty} |x|^a \left(\frac{d}{dx}\right)^b f(x) = 0.$$

Functions in the Schwartz space decay so rapidly at infinity that, even after differentiating or multiplying by x an arbitrary (finite) number of times, the resulting functions still decay at infinity. For any $\epsilon > 0$, the Gaussian $e^{-\epsilon x^2}$ is in $\mathcal{S}$. Smooth functions of compact support provide additional examples. For convenience we recall the existence of such functions.

EXAMPLE 3.1. First we define a function h on **R** by $h(t) = 0$ for $t \leq 0$ and by $h(t) = \exp(\frac{-1}{t})$ for $t > 0$. This function is infinitely differentiable on all of **R**, and all of its derivatives vanish at 0. Put $g(t) = h(t)h(1 - t)$. Then g is also infinitely differentiable. Furthermore, $g(t) > 0$ for $0 < t < 1$ and $g(t) = 0$ otherwise. Thus g is a smooth function with compact support.

The technique from Example 3.1 can be extended to prove the stronger result stated in Theorem 3.1. Exercise 3.19 suggests the standard proof using convolution integrals.

THEOREM 3.1. *Let I denote any closed bounded interval on **R** and let J denote any open interval containing I. Then there is an infinitely differentiable function $\chi : \mathbf{R} \to [0, 1]$ such that $\chi = 1$ on I and $\chi = 0$ off J.*

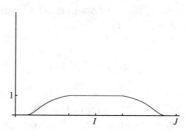

FIGURE 1. Cutoff Function

PROPOSITION 3.1. *The Schwartz space $\mathcal{S}$ is a complex vector space. It is closed under differentiation and under multiplication by x.*

PROOF. Left to the reader.   □

DEFINITION 3.2. We define the Fourier transform for $f \in \mathcal{S}$ as an integral:

$$\mathcal{F}(f)(\xi) = \hat{f}(\xi) = \frac{1}{\sqrt{2\pi}} \int_{-\infty}^{\infty} f(x) e^{-ix\xi} dx. \tag{4}$$

Sometimes the Fourier transform is defined without the factor $\frac{1}{\sqrt{2\pi}}$, and sometimes it is defined with a factor of 2π in the exponent. The Wolfram Mathematica
9 web page on the Fourier transform allows the user to set two parameters, thereby
obtaining whichever convention the user wishes to employ.

The definition of $\mathcal{S}$ regards differentiation and multiplication on an equal footing. Let $D = \frac{d}{dx}$ denote differentiation and let $M = M_{i\xi}$ denote multiplication
by $i\xi$. Working on $\mathcal{S}$ is convenient for several reasons; in particular, the Fourier
transform exchanges these operations. Furthermore, as we will show in Theorem
3.2, the Fourier transform maps $\mathcal{S}$ to itself bijectively. We can interpret the last
two items of the following Proposition as saying that $\mathcal{F}$ *diagonalizes* differentiation.

PROPOSITION 3.2. *The following elementary properties hold for Fourier trans-*
forms defined on $\mathcal{S}$.

   (1) $\mathcal{F}$ *is linear.*
   (2) $\|\hat{f}\|_{L^\infty} \le \frac{1}{\sqrt{2\pi}}\|f\|_{L^1}$.
   (3) $\overline{\hat{f}}(\xi) = \hat{\bar{f}}(-\xi)$.
   (4) *Put* $f_h(x) = f(x+h)$. *Then* $\hat{f}_h(\xi) = e^{ih\xi}\hat{f}(\xi)$.
   (5) $\frac{d}{d\xi}\hat{f}(\xi) = -i\mathcal{F}(M_x f)$. *That is,* $D_{i\xi}\mathcal{F} = \mathcal{F}M_x$.
   (6) $\hat{f}'(\xi) = i\xi\hat{f}(\xi)$. *That is,* $\mathcal{F}D_x = M_{i\xi}\mathcal{F}$.
   (7) $D = \mathcal{F}^{-1}M\mathcal{F}$ *and* $M = \mathcal{F}D\mathcal{F}^{-1}$.

PROOF. The first six items are left to the reader. The last item follows from
the penultimate item and Theorem 3.2 below.   □

The reader should compare Proposition 3.2 with Lemma 1.7.

PROPOSITION 3.3. *The Fourier transform maps $\mathcal{S}$ to itself.*

PROOF. Differentiating equation (4) from Definition 3.2 under the integral sign,
justified by the rapid decay of f at infinity, shows that $\hat{f}$ is infinitely differentiable.
Since $\mathcal{F}D = M\mathcal{F}$ it follows for each positive integer k and integration by parts that

$$\mathcal{F}D^k = M_{i\xi}^k\mathcal{F}.$$

All boundary terms vanish since we are working in $\mathcal{S}$. It follows that f decays
rapidly if and only if $\hat{f}$ does. Hence $\mathcal{F}: \mathcal{S} \to \mathcal{S}$.   □

PROPOSITION 3.4. *Let G_σ denote the Gaussian:* $G_\sigma(x) = \frac{1}{\sigma\sqrt{2\pi}}e^{\frac{-x^2}{2\sigma^2}}$. *Then:*

   (1) $\int_{-\infty}^{\infty} G_\sigma(x)dx = 1$ *for* $\sigma > 0$. *Thus G_σ is a probability density.*
   (2) *The Fourier transform (up to a factor of $\frac{1}{\sigma}$) is another Gaussian:*

$$\mathcal{F}(G_\sigma)(\xi) = \frac{1}{\sqrt{2\pi}}e^{\frac{-\sigma^2\xi^2}{2}} = \frac{1}{\sigma}G_{\frac{1}{\sigma}}(\xi).$$

PROOF. We will give two proofs of (2). First we prove (1). We must evaluate

$$I = \int_{-\infty}^{\infty} e^{\frac{-x^2}{2}}\,dx.$$

The computation, a standard example in two-variable calculus, illustrates the power
of polar coordinates. Consider I^2. We have

$$I^2 = \int_{-\infty}^{\infty}\int_{-\infty}^{\infty} e^{\frac{-x^2}{2}}e^{\frac{-y^2}{2}}\,dxdy = \int_0^{2\pi}\int_0^{\infty} e^{\frac{-r^2}{2}}r\,drd\theta = 2\pi.$$

Thus $I = \sqrt{2\pi}$. To prove (1), we have

$$\int_{-\infty}^{\infty} G_\sigma(x)dx = \frac{1}{\sigma\sqrt{2\pi}} \int_{-\infty}^{\infty} e^{\frac{-x^2}{2\sigma^2}}\, dx.$$

Changing variables by putting $x = \sigma u$ gives

$$\int_{-\infty}^{\infty} G_\sigma(x)dx = \frac{1}{\sigma\sqrt{2\pi}}\, \sigma I = 1.$$

The first method used in proving (2) is completing the square in the exponential. After changing variables with $x = t\sigma$, we must compute $\frac{1}{\sqrt{2\pi}} \int_{-\infty}^{\infty} e^{\frac{-t^2}{2}} e^{-it\sigma\xi} dt$. By completing the square, we see that

$$e^{\frac{-t^2}{2}} e^{-it\sigma\xi} = e^{-\frac{1}{2}(t+i\sigma\xi)^2} e^{\frac{-\sigma^2\xi^2}{2}}.$$

The second factor is independent of t, and we are integrating with respect to t.

By Exercise 3.3, for x, y real, $\int_{-\infty}^{\infty} e^{\frac{-(x+iy)^2}{2}}\, dx$ is independent of y. A careful proof requires some knowledge of complex analytic function theory. Using this equality, we obtain the result.

To avoid this use of complex analysis, we provide a second proof of (2). Let $\phi(\xi)$ denote the Fourier transform of G_σ. Because of the rapid decay of e^{-x^2}, it is valid to differentiate under the integral sign to find a formula for $\phi'(\xi)$. Doing so introduces a factor of $-ix$ in the integrand, which we group with $e^{\frac{-x^2}{2\sigma^2}}$ in order to integrate by parts. The boundary term vanishes and we obtain

$$\phi'(\xi) = -\xi\sigma^2\phi(\xi).$$

Solving this first order differential equation shows, for a positive constant k, that

$$\phi(\xi) = k e^{\frac{-\sigma^2\xi^2}{2}}.$$

Thus ϕ is a multiple of a Gaussian with σ replaced by its reciprocal. Since $k = \phi(0)$, which in turn equals $\frac{1}{\sqrt{2\pi}}$, we set $\xi = 0$ and obtain (2). $\qquad\square$

One cannot show that $\int_{-\infty}^{\infty} e^{\frac{-(x+iy)^2}{2}}\, dx$ is independent of y by the naive change of variables $u = x + iy$. This change of variables alters the path of integration; the integral remains the same because of the Cauchy theorem applied to the complex analytic function $z \to e^{\frac{-z^2}{2}}$. See Exercise 3.3.

We next prove several lemmas which get used in establishing the fundamental properties of the Fourier transform. The proofs of Lemma 3.2 and Theorem 1.5 use the same basic idea. When the function g in Lemma 3.2 is non-negative, it is the density function of a random variable with finite expectation. First we establish a fundamental symmetry property of the Fourier transform, leading to Theorem 3.5.

LEMMA 3.1. *For $f, g \in \mathcal{S}$ we have*

$$\int_{-\infty}^{\infty} f(x)\hat{g}(x)dx = \int_{-\infty}^{\infty} \hat{f}(\xi)g(\xi)d\xi \tag{5}$$

PROOF. Because of the rapid decay of f and g, we can write either side of (5) as a double integral, and integrate in either order. Then each side of (5) equals the double integral

$$\frac{1}{\sqrt{2\pi}} \int_{-\infty}^{\infty} \int_{-\infty}^{\infty} f(x)g(\xi)e^{-ix\xi} dx d\xi. \qquad\square$$

LEMMA 3.2. *Assume f is differentiable on $\mathbf{R}$ and that f' is bounded. Let g satisfy the following:*

- $\int_{-\infty}^{\infty} g(y)dy = 1$
- $\int_{-\infty}^{\infty} |y|\,|g(y)|dy < \infty$

Then

$$\lim_{\epsilon \to 0} \int_{-\infty}^{\infty} f(x + \epsilon y)g(y)\, dy = f(x). \tag{6.1}$$

PROOF. Since f is differentiable and f' is bounded, the mean-value theorem of calculus implies the following inequality:

$$|f(b) - f(a)| \leq \sup_t |f'(t)|\,|b - a| = M|b - a|. \tag{mvt}$$

Since g integrates to 1, we can write $f(x) = \int_{-\infty}^{\infty} f(x)g(y)dy$. Using (mvt) we get

$$\left| \int_{-\infty}^{\infty} f(x + \epsilon y)g(y)\, dy - f(x) \right| = \left| \int_{-\infty}^{\infty} (f(x + \epsilon y) - f(x))\, g(y)\, dy \right|$$

$$\leq M \int_{-\infty}^{\infty} |\epsilon y||g(y)|dy. \tag{6.2}$$

Since $|yg(y)|$ is integrable, the expression in (6.2) is bounded by a constant times ϵ. The desired conclusion (6.1) then follows by the definition of a limit. $\square$

LEMMA 3.3. *Suppose $h \in \mathcal{S}$. Then*

$$\lim_{\epsilon \to 0} \int_{-\infty}^{\infty} h(t)e^{\frac{-\epsilon^2 t^2}{2}}\, dt = \int_{-\infty}^{\infty} h(t)dt$$

PROOF. Given $\epsilon' > 0$ we must show that

$$\left| \int_{-\infty}^{\infty} h(t)(1 - e^{\frac{-\epsilon^2 t^2}{2}})dt \right| < \epsilon'$$

for sufficiently small ϵ. Since h decays rapidly at ∞, there is an R such that

$$\left| \int_{|t| \geq R} h(t)(1 - e^{\frac{-\epsilon^2 t^2}{2}})dt \right| \leq \int_{|t| \geq R} |h(t)|dt < \frac{\epsilon'}{2}.$$

Once this R is determined, we can choose ϵ sufficiently small such that

$$\left| \int_{-R}^{R} h(t)(1 - e^{\frac{-\epsilon^2 t^2}{2}})dt \right| \leq 2R\sup(|h|)(1 - e^{\frac{-\epsilon^2 R^2}{2}}) < \frac{\epsilon'}{2}.$$

The needed inequality follows. $\square$

REMARK 3.1. It is tempting to plug $\epsilon = 0$ into the left-hand side of (6.1) or into the limit in Lemma 3.3. Doing so is not valid without some assumptions; the limit of an integral is not necessarily the integral of the limit. The reason is that an integral is itself a limit, and one cannot in general interchange the order of limits. See Exercise 3.8. This simple issue recurs throughout analysis; one needs appropriate hypotheses before one can interchange the order when taking limits.

REMARK 3.2. Even if a continuous (or smooth) function is integrable on $\mathbf{R}$, it need not vanish at infinity. See Exercise 3.10.

EXERCISE 3.1. Prove that $e^{-x^2} \in \mathcal{S}$.

EXERCISE 3.2. Prove Propositions 3.1 and 3.2.

EXERCISE 3.3. Fill in the details of the computations in Proposition 3.4. In particular, prove that $\int_{-\infty}^{\infty} e^{\frac{-(x+iy)^2}{2}} dx$ is independent of y. Use the Cauchy integral theorem on a rectangle with vertices at $\pm R$, $\pm R + iy$; then let R tend to infinity. Verify also the details in the second proof given for part (2).

EXERCISE 3.4. Compute the Fourier transform of $e^{-a(x-\mu)^2}$ for $a > 0$ and $\mu \in \mathbf{R}$. Comment: This result is of fundamental importance in probability theory. We use it in the proofs of Theorem 3.2 and Theorem 3.9.

EXERCISE 3.5. Compute the Fourier transform of $\frac{1}{1+x^2}$ using residues. Note that the residue calculation depends on the sign of ξ.

EXERCISE 3.6. Verify the assertions in Example 3.1.

EXERCISE 3.7. Put $g(\eta) = \frac{1}{\sqrt{2\pi}} e^{\frac{-\eta^2}{2}}$ for $\eta \in \mathbf{R}$. Show that g satisfies the hypotheses of Lemma 3.2.

EXERCISE 3.8. Put $f(x,y) = |x|^{|y|}$ for $(x,y) \neq (0,0)$. Show that

$$\lim_{x \to 0} \lim_{y \to 0} f(x,y) \neq \lim_{y \to 0} \lim_{x \to 0} f(x,y).$$

EXERCISE 3.9. It is not necessary that h be in $\mathcal{S}$ for the proof of Lemma 3.3 to work. Give a weaker condition on h for which Lemma 3.3 remains valid.

EXERCISE 3.10. Show that there is continuous, nonnegative function f on **R** such that $f(n) = n$ for all positive integers n and $\int_{\mathbf{R}} f(x)dx = 1$. One can even make f infinitely differentiable.

We have now understood the Schwartz space and developed some computational facility with Fourier transforms. We are therefore in a position to prove the Fourier inversion formula; this theorem is one of the most important results in pure and applied analysis. Nearly all the rest of this chapter depends upon the inversion formula.

THEOREM 3.2. $\mathcal{F}$ is a bijective map of $\mathcal{S}$ to itself. Furthermore, for $f \in \mathcal{S}$, we have the Fourier inversion formula

$$f(x) = \frac{1}{\sqrt{2\pi}} \int_{-\infty}^{\infty} \hat{f}(\xi) e^{ix\xi} d\xi. \tag{7}$$

PROOF. We use the Gaussian (with $\sigma = 1$) as an approximate identity and apply Lemma 3.2. Put

$$g(\eta) = \frac{1}{\sqrt{2\pi}} e^{\frac{-\eta^2}{2}}.$$

By Exercise 3.7, g satisfies the hypotheses of Lemma 3.2, and we obtain

$$f(x) = \lim_{\epsilon \to 0} \int_{-\infty}^{\infty} f(x + \epsilon y) g(y) dy = \lim_{\epsilon \to 0} \frac{1}{\sqrt{2\pi}} \int_{-\infty}^{\infty} f(x + \epsilon y) e^{\frac{-y^2}{2}} dy. \tag{8}$$

By Proposition 2.4, the Gaussian is its own Fourier transform. We exploit this fact in (8) to obtain

$$f(x) = \lim_{\epsilon \to 0} \int_{-\infty}^{\infty} \frac{1}{\sqrt{2\pi}} f(x + \epsilon y) \frac{1}{\sqrt{2\pi}} \int_{-\infty}^{\infty} e^{\frac{-\eta^2}{2}} e^{-iy\eta} d\eta \, dy. \tag{9.1}$$

In (9.1) we make the change of variables $t = x + \epsilon y$, obtaining

$$f(x) = \lim_{\epsilon \to 0} \int_{-\infty}^{\infty} \int_{-\infty}^{\infty} \frac{1}{\sqrt{2\pi}} f(t) e^{\frac{-\eta^2}{2}} e^{-i(t-x)\frac{\eta}{\epsilon}} \frac{1}{\epsilon\sqrt{2\pi}} d\eta \, dt. \tag{9.2}$$

Now we change variables by putting $\eta = \epsilon\xi$; doing so introduces the factor $e^{\frac{-\epsilon^2\xi^2}{2}}$ and enables us to interchange the order of integration. The result gives

$$f(x) = \lim_{\epsilon \to 0} \int_{-\infty}^{\infty} \int_{-\infty}^{\infty} \frac{1}{\sqrt{2\pi}} f(t) e^{-it\xi} dt \, e^{\frac{-\epsilon^2\eta^2}{2}} e^{ix\xi} \frac{d\xi}{\sqrt{2\pi}}. \tag{9.3}$$

The inner integral is simply $\hat{f}(\xi)$. Hence we obtain

$$f(x) = \lim_{\epsilon \to 0} \frac{1}{\sqrt{2\pi}} \int_{-\infty}^{\infty} \hat{f}(\xi) e^{ix\xi} e^{\frac{-\epsilon^2\xi^2}{2}} d\xi. \tag{10}$$

To finally obtain the inversion formula, we use Lemma 3.3 to interchange the limit and integral in (10). □

The reader should note the extent to which the proof of Theorem 3.2 resembles the proofs of results such as Theorem 1.5.

The inversion formula has the following consequence: For $f \in \mathcal{S}$, we have $(\mathcal{F}^2 f)(x) = f(-x)$. Hence $\mathcal{F}^4$ is the identity operator.

EXERCISE 3.11. Compute the Fourier transform of χ, if $\chi(x) = 1$ for $|x| \leq 1$ and $\chi(x) = 0$ otherwise. (Note that χ is not smooth, but that it is integrable.)

The next exercise is a bit akin to opening Pandora's box. Taking functions of operations such as differentiation is natural (see for example Section 6 of Chapter 1) but somewhat hard to justify. Doing so without great care can lead to delicate logical issues.

EXERCISE 3.12. Use the property $D = \mathcal{F}^{-1} M \mathcal{F}$ to give a plausible definition of the α-th derivative of a nice function, where $0 < \alpha < 1$. Check that $D^{\alpha+\beta} = D^\alpha D^\beta$. More generally, try to define $g(D)$ for various functions g. What technical difficulties arise?

3. The dual space

The Schwartz space $\mathcal{S}$ is not a normed space, but we nonetheless require a notion of convergence. This notion is achieved via *semi-norms*. We define measurements of a function f in $\mathcal{S}$ as follows:

DEFINITION 3.3. Let a, b be non-negative integers. We define $||f||_{a,b}$ by

$$||f||_{a,b} = \sup \left(|x|^a \left| (\frac{d}{dx})^b f(x) \right| \right).$$

These measurements are not norms because $||f||_{a,b}$ can be zero without f being 0. If $||f||_{a,0}$ vanishes for some a, however, then f is the zero function. Note that we could replace supremum by maximum in the definition of the semi-norm, because functions in $\mathcal{S}$ are continuous and decay rapidly at infinity. The number of semi-norms is countable, and hence we can make $\mathcal{S}$ into a metric space. The distance between two functions is given by the formula

$$\text{dist}(f, g) = \sum_{a,b} c_{ab} \frac{||f - g||_{a,b}}{1 + ||f - g||_{a,b}},$$

where $c_{ab} > 0$ and is chosen to make the sum converge. For example, $c_{ab} = 2^{-a-b}$ is often used. With this distance function, $\mathcal{S}$ is a complete metric space. See Exercise 3.16. It is adequate to state the notion of convergence in terms of the semi-norms, rather than in an equivalent manner using this distance function.

DEFINITION 3.4. A sequence $\{f_n\}$ converges to f in $\mathcal{S}$ if, for all a, b,

$$||f_n - f||_{a,b} \to 0.$$

Since $\mathcal{S}$ is a vector space, it would have sufficed to give the notion of convergence to 0. To say that a sequence $\{f_n\}$ converges to 0 means that, any derivative of any polynomial multiple of f_n tends to 0 uniformly.

DEFINITION 3.5. Let $L : \mathcal{S} \to \mathbf{C}$ be a linear functional. Then L is called *continuous* if, whenever f_n converges to f in $\mathcal{S}$, then $L(f_n)$ converges to $L(f)$ in $\mathbf{C}$.

DEFINITION 3.6. The *dual space* $\mathcal{S}'$ is the vector space consisting of all continuous linear functionals on $\mathcal{S}$. Elements of $\mathcal{S}'$ are called *tempered distributions*.

It is often convenient to write the action of a linear functional using inner product notation:

$$\phi(f) = \langle f, \phi \rangle.$$

There is no complex conjugate used here.

Each element g of $\mathcal{S}$ can be regarded as a distribution by the formula

$$g(f) = \langle f, g \rangle = \int_{-\infty}^{\infty} f(x)g(x)dx. \tag{11}$$

The integral in (11) defines a distribution more generally. For example, when g is bounded and continuous, (11) makes sense and defines g as an element of $\mathcal{S}'$. When g is any function such that (11) makes sense for all $f \in \mathcal{S}$, we regard g as the element of $\mathcal{S}'$ defined by (11). Distributions are more general than functions.

EXAMPLE 3.2 (Distributions). The most famous example of a distribution is the Dirac delta function, henceforth called the Dirac delta distribution. We define $\delta \in \mathcal{S}'$ by

$$\delta(f) = \langle f, \delta \rangle = f(0).$$

Another example is given by its derivative:

$$\delta'(f) = \langle f, \delta' \rangle = \langle -f', \delta \rangle = -f'(0).$$

More generally, if ϕ is a tempered distribution, we define its k-th derivative $\phi^{(k)}$ by

$$\phi^{(k)}(f) = \langle f, \phi^{(k)} \rangle = (-1)^k \langle f^{(k)}, \phi \rangle. \tag{12}$$

By Exercise 3.13, (12) defines a continuous linear functional on $\mathcal{S}$, and hence $\phi^{(k)} \in \mathcal{S}'$. Formula (12) is the natural definition of *distribution derivative*. If ϕ were itself k times differentiable, then (12) would hold; we integrate by parts k times and all boundary terms vanish.

Let us clarify these definitions. Let V be a topological vector space with dual space V'. As above, we use the notation $\langle f, \phi \rangle$ for the action of $\phi \in V'$ on $f \in V$. When $L : V \to V$ is linear, we define its transpose L^t, mapping V' to V', by

$$\langle \phi, Lf \rangle = \langle L^t \phi, f \rangle.$$

It is consistent with standard practice not to introduce a complex conjugation when using the inner product notation for the action of a distribution on a function.

Let D denote differentiation. By integration by parts, and because all boundary terms vanish when we are working on the Schwartz space, $(D)^t = -D$. We extend differentiation to the dual space by preserving this property. It follows that $(D^k)^t$, the transpose of differentiating k times, is $(-1)^k D^k$.

By Lemma 3.1, the transpose of the Fourier transform, when acting on functions in $\mathcal{S}$, is itself. In Definition 3.7, we will *define* the Fourier transform of a distribution by setting $\mathcal{F}^t = \mathcal{F}$.

Let us give another example of a distribution and its derivative. Define a function u by $u(x) = x$ for $x \geq 0$ and $u(x) = 0$ for $x < 0$. This function is sometimes called the *ramp function*. Then u', which is not defined at 0 as a function, nonetheless defines a distribution. We have

$$u'(f) = -u(f') = -\int_0^\infty x \, f'(x) \, dx = \int_0^\infty f(x) \, dx. \tag{13}$$

In (13) the first equality is the definition of distribution derivative, the next equality holds because u is a function, and the last equality holds via integration by parts. We also can compute the second derivative of u:

$$u''(g) = -u'(g') = -\int_0^\infty g'(t) \, dt = g(0).$$

Thus $u'' = \delta$. The Dirac delta distribution is thus the second distribution derivative of the ramp function u. The distribution $H = u'$ is known as the *Heaviside* function. It is named after Oliver Heaviside, rather than for the following reason. Note that $H = 1$ on the positive axis and $H = 0$ on the negative axis. Thus H is "heavy" on the right side. See [H] for a fascinating discussion of Heaviside's life.

EXERCISE 3.13. Verify that $\delta \in \mathcal{S}'$. If $\phi \in \mathcal{S}'$, show that ϕ', as defined by (12), also is in $\mathcal{S}'$.

If f is a continuous function, and ϕ is a distribution, then we naturally define $f \cdot \phi$ by $(f \cdot \phi)(g) = \phi(fg)$. It is not possible to define the product of distributions in general. See [SR] and its references for discussion of this issue.

EXERCISE 3.14. Let f be continuous and let δ be the Dirac delta distribution. Find the distribution derivative $(f \cdot \delta)'$. Assuming f is differentiable, find $f' \cdot \delta + f \cdot \delta'$.

DEFINITION 3.7 (The generalized Fourier transform). Let $\phi \in \mathcal{S}'$. We define its Fourier transform $\mathcal{F}(\phi)$ by duality as follows. For each $f \in \mathcal{S}$ we decree that

$$\langle \mathcal{F}(\phi), f \rangle = \langle \phi, \mathcal{F}(f) \rangle.$$

Definition 3.7 is justified by Lemma 3.1. In our current notation, this Lemma states that $\langle \mathcal{F}(f), g \rangle = \langle f, \mathcal{F}(g) \rangle$ for functions f, g in $\mathcal{S}$.

The Fourier transform $\mathcal{F}(\phi)$ is itself a distribution. It is obviously linear. We verify continuity. If f_n converges to 0 in $\mathcal{S}$, then $\hat{f}_n$ also converges to 0 in $\mathcal{S}$. Hence $\langle \phi, \mathcal{F}(f_n) \rangle$ converges to 0 in $\mathbf{C}$.

EXAMPLE 3.3. What is the Fourier transform of the Dirac delta? We have

$$\langle \hat{\delta}, f \rangle = \langle \delta, \hat{f} \rangle = \hat{f}(0) = \frac{1}{\sqrt{2\pi}} \int_{-\infty}^{\infty} f(x)dx = \langle \frac{1}{\sqrt{2\pi}}, f \rangle.$$

Thus $\hat{\delta}$ is the constant function $\frac{1}{\sqrt{2\pi}}$.

EXERCISE 3.15. Compute the Fourier transforms $\hat{H}$ and $\hat{u}$.

EXERCISE 3.16. Fill in the discussion between Definitions 3.3 and 3.4 as follows. Verify that the given alleged distance function is in fact a distance function. Then use the Arzela-Ascoli theorem to show that $\mathcal{S}$ is a complete metric space.

EXERCISE 3.17. Let ϕ be a linear functional on $\mathcal{S}$. Show that $\phi \in \mathcal{S}'$ if and only if there is a constant M and an integer N such that

$$|\phi(f)| \leq M \max\{||f||_{a,b} : a + b \leq N\}.$$

4. Convolutions

We have already observed the power of convolution in understanding Fourier series. We extend the notion of convolution to **R** and obtain similarly powerful results.

DEFINITION 3.8. Suppose f and g are integrable functions on **R**. We define $f * g$ by

$$(f * g)(x) = \int_{-\infty}^{\infty} f(x - y)g(y)dy = \int_{-\infty}^{\infty} f(y)g(x - y)dy. \tag{14}$$

The equality of the two integrals follows by change of variables and also implies that $f * g = g * f$. We also can easily check that if $f \in L^1$ and $g \in L^2$, then $f * g \in L^2$.

THEOREM 3.3. If f and g are in L^1, then $(f * g)^\wedge = \sqrt{2\pi} \hat{f} \hat{g}$.

PROOF. We write out $(f * g)^\wedge(\xi)$ as a double integral and interchange the order of integration, obtaining $\sqrt{2\pi} \hat{f} \hat{g}$. ☐

We wish to extend our work on approximate identities to this setting. First let χ denote any integrable smooth function such that $\int_{-\infty}^{\infty} \chi(x)dx = 1$. For $\epsilon > 0$ we then define χ_ϵ by

$$\chi_\epsilon(x) = \frac{\chi(\frac{x}{\epsilon})}{\epsilon}. \tag{15}$$

Then, by change of variables, $\int_{-\infty}^{\infty} \chi_\epsilon(x)dx = 1$ also.

DEFINITION 3.9. For χ_ϵ as in (15), put $J_\epsilon(f) = \chi_\epsilon * f$. We call J_ϵ a *mollifier*.

THEOREM 3.4. If $f \in L^1$, then $J_\epsilon f$ converges to f in L^1 as ϵ tends to 0. If $f \in L^2$, then $J_\epsilon f$ converges to f in L^2 as ϵ tends to 0. If f is uniformly continuous near x, then $J_\epsilon f$ converges uniformly to f near x. If f is integrable and χ is infinitely differentiable, then $J_\epsilon f$ is infinitely differentiable.

PROOF. We refer the reader to Chapter 8 of [F2]. ☐

EXERCISE 3.18. Show that the function f defined by $f(x) = e^{\frac{-1}{x}}$ for $x > 0$ and by $f(x) = 0$ otherwise is infinitely differentiable. Sketch the graph of the function $x \to f(x)f(1 - x)$.

EXERCISE 3.19. Prove Theorem 3.1. Suggestion: First find a continuous function that is 1 on I and 0 off J. Mollify it, using a function χ as constructed in the previous exercise.

EXERCISE 3.20. The *support* of a function f is the smallest closed set outside of which f is identically zero. Suppose f is supported in $[a, b]$ and g is supported in $[c, d]$. What can you say about the support of $f * g$?

5. Plancherel theorem

The Parseval formula equates the l^2-norm of the Fourier coefficients of a function with the L^2 norm of the function. Its polarized form, Corollary 2.7, states that

$$\langle \hat{f}, \hat{g} \rangle_2 = \langle f, g \rangle_{L^2}.$$

The Plancherel theorem (which holds both in higher dimensions and in more abstract settings) extends the Parseval result by establishing that the Fourier transform is a unitary operator on $L^2(\mathbf{R})$.

Recall that the Fourier transform is defined on L^2 in a subtle manner; the integral in (1) need not converge for $f \in L^2$. We define $\mathcal{F}$ on the Schwartz space via integration as in (1), and then we extend $\mathcal{F}$ to $\mathcal{S}'$ by duality. We then regard an element of L^2 as an element of $\mathcal{S}'$. It would also be possible to define $\mathcal{F}$ on $L^1 \cap L^2$ by the integral (1) and proceed by limiting arguments.

THEOREM 3.5. *The Fourier transform* $\mathcal{F} : L^2(\mathbf{R}) \to L^2(\mathbf{R})$ *is unitary.*

PROOF. By Proposition 2.6, it suffices to check that $\|\mathcal{F}f\|_{L^2}^2 = \|f\|_{L^2}^2$ for all f in L^2. The norm is continuous, and hence it suffices to check this equality on the dense set $\mathcal{S}$. Put $\hat{g} = \overline{f}$ in Lemma 3.1. Then $\overline{g} = \hat{f}$ and Lemma 3.1 gives $\|f\|_{L^2}^2 = \|\hat{f}\|_{L^2}^2$. $\qquad\square$

EXERCISE 3.21. Note that $\mathcal{F}^4 = I$. Use this fact and Proposition 3.4 to find all eigenvalues of $\mathcal{F}$. Harder: Find all eigenfunctions. Suggestion: Apply $\frac{d}{dx} - x$ to $e^{\frac{-x^2}{2}}$ and use formula (64) from Chapter 2. Or, look ahead to Theorem 3.9.

EXERCISE 3.22. Put $\chi(x) = 1$ for $|x| \le 1$ and $\chi(x) = 0$ otherwise. Find $\hat{\chi}$.

EXERCISE 3.23. Use Exercise 3.22 and the Plancherel Theorem to find $\int_{-\infty}^{\infty} \frac{\sin^2(x)}{x^2} dx$. Also use contour integration to check your answer.

EXERCISE 3.24. Assume $b \ge 0$. Compute the integrals:

$$\int_0^{\infty} e^{-u(1+x^2)} du.$$

$$\int_{-\infty}^{\infty} \frac{e^{-ibx}}{1+x^2} dx.$$

$$\int_0^{\infty} \frac{e^{-t}}{\sqrt{t}} e^{\frac{-b^2}{4t}} dt.$$

Suggestion: Use the first, the second, and the Fourier transform of a Gaussian to compute the third. The answer to the third is $\sqrt{\pi} \exp(-b)$.

EXERCISE 3.25. Put $f(x) = e^{-x} x^{a-1}$ for $x \ge 0$ and $f(x) = 0$ otherwise. Find the condition on a for which $f \in L^1(\mathbf{R})$. Under this condition, find $\hat{f}$. Comment: One can use contour integrals from complex analysis here.

6. Heisenberg uncertainty principle

The famous Heisenberg uncertainty principle from quantum mechanics states something to the effect that it is not possible to determine precisely both the position and momentum of a moving particle. This principle can be formulated as an inequality involving the Fourier transform. After giving an intuitive explanation, we state and prove this inequality.

Let $|f|^2$ denote the probability density on **R** determined by the position of a moving particle. By definition, the probability that this particle is found in the interval $[a, b]$ is the integral $\int_a^b |f(x)|^2 dx$ and of course $\int_{-\infty}^{\infty} |f(x)|^2 dx = 1$ (the particle is somewhere).

The *mean* (expected value) of position is by definition the integral

$$\mu = \int_{-\infty}^{\infty} x|f(x)|^2 dx,$$

when this integral converges (which we assume in this section). After a translation we may assume without loss of generality that $\mu = 0$.

The picture for momentum looks the same, except that we use $|\hat{f}|^2$ to define the density for momentum. By the Plancherel theorem, $|\hat{f}|^2$ also defines a density. Again, without loss of generality, we may assume (after multiplying f by a function e^{iax} of modulus 1) that the mean of momentum is 0. See Exercise 3.26.

The *variance* of the position of a particle of mean 0 equals

$$\int_{-\infty}^{\infty} x^2|f(x)|^2 dx = ||xf(x)||_{L^2}^2,$$

and the variance of its momentum is

$$\int_{-\infty}^{\infty} \xi^2|\hat{f}(\xi)|^2 d\xi = ||\xi\hat{f}(\xi)||_{L^2}^2.$$

Again we assume these integrals converge.

The following famous inequality gives a positive lower bound on the product of the two variances.

THEOREM 3.6 (Heisenberg's inequality). *Assume both f and f' are square-integrable on* **R**. *Then*

$$||xf(x)||_{L^2}^2 \, ||\xi\hat{f}(\xi)||_{L^2}^2 \geq \frac{1}{4}||f||_{L^2}^4. \tag{16}$$

PROOF. We assume that $f \in \mathcal{S}$. The general case follows because $\mathcal{S}$ is dense in L^2. Consider the integral

$$I = \int_{-\infty}^{\infty} x\left(f(x)\overline{f}'(x) + f'(x)\overline{f}(x)\right) dx.$$

Using integration by parts, we obtain

$$I = x|f(x)|^2\big|_{-\infty}^{\infty} - \int_{-\infty}^{\infty} |f(x)|^2 dx = -\int_{-\infty}^{\infty} |f(x)|^2 dx,$$

because the boundary terms are zero. By the Cauchy-Schwarz inequality and the Plancherel theorem, we also have

$$|I| \leq 2||xf(x)||_{L^2} \, ||f'||_{L^2} = 2||xf(x)||_{L^2} \, ||\xi\hat{f}(\xi)||_{L^2}. \tag{17}$$

Putting the two formulas together gives

$$\frac{1}{2}||f||_{L^2}^2 \leq ||xf(x)||_{L^2} \,||\xi\hat{f}(\xi)||_{L^2},\tag{18}$$

which yields (16) upon squaring both sides.   □

COROLLARY 3.1. *The following inequality holds:*

$$||f||_{L^2}^2 \leq ||xf(x)||_{L^2}^2 + ||\xi\hat{f}(\xi)||_{L^2}^2 = ||xf(x)||_{L^2}^2 + ||f'(x)||_{L^2}^2.$$

PROOF. For non-negative real numbers s, t we always have $st \leq \frac{s^2+t^2}{2}$. (The arithmetic-geometric mean inequality, or the Cauchy-Schwarz inequality!). Applying this simple fact to the product on the right-hand side of (18) yields the inequality. The equality follows by Proposition 3.2.   □

EXERCISE 3.26. What is the effect on the mean of position if we replace $f(x)$ by $e^{ixh}f(x)$? What is the effect on the mean of momentum in doing so?

EXERCISE 3.27. When does equality hold in (16)?

A less precise form of the Heisenberg uncertainty formula says the following. Unless f is identically 0, then f and $\hat{f}$ cannot both vanish outside of a bounded set. We prove this result next.

THEOREM 3.7. *Suppose f is integrable on $[-r, r]$ and $f(x) = 0$ for $|x| > r$. If there is an R such that $\hat{f}(\xi) = 0$ for $|\xi| > R$, then f is identically 0.*

PROOF. We start with

$$\hat{f}(\xi) = \frac{1}{\sqrt{2\pi}} \int_{-r}^{r} f(x)e^{-ix\xi}dx.\tag{19}$$

In (19), we let ξ be a complex variable. Since we may differentiate under the integral as often as we wish (introducing factors of $-ix$) and the integral still converges, $\xi \to \hat{f}(\xi)$ is an entire complex analytic function. If an R exists as in the theorem, then this complex analytic function vanishes on an interval in $\mathbf{R}$. By basic complex variable theory, a complex analytic function vanishing on an interval of the real axis must be identically 0. Thus both $\hat{f}$ and f vanish identically.

One can also prove the theorem as follows. Expand $e^{-ix\xi}$ in a Taylor series about any point and interchange the order of integration and summation. We see that $\hat{f}$ is an entire analytic function with a zero of infinite order, and hence it is identically 0.   □

7. Differential equations

The Fourier transform plays a major role in partial differential equations. Although most of this material is beyond the scope of this book, we can glimpse some of the ideas in simple settings. The key point is that the Fourier transformation *diagonalizes* differentiation. When we have diagonalized an operation, we can take functions of the operation.

We first consider diagonalization is a simple but interesting context. Consider an affine function $z \mapsto mz + w = f(z)$ on $\mathbf{C}$. We write f^{*n} for the iteration of f a total of n times. Thus $f^{*1} = f$ and $f^{*(n+1)} = f \circ f^{*n}$. Using diagonalization, we can compute f^{*n} easily.

There are two cases. When $f(z) = z + w$ (thus f is a translation), we obtain $f^{*n}(z) = z + nw$. When $f(z) = mz + w$ for $m \neq 1$, we can write

$$f(z) = mz + w = m(z + \zeta) - \zeta,$$

where $m\zeta - \zeta = w$. We see that $f = T^{-1}MT$, where T is translation by ζ and M is multiplication by m. Thus $f^{*n} = T^{-1}M^nT$, or

$$f^{*n}(z) = m^n(z + \zeta) - \zeta.$$

The simplicity of these formulas is evident. Furthermore, the formulas make sense when n is replaced by an arbitrary real number α as long as we are careful in our definition of the multi-valued function $m \to m^\alpha$. The crucial point is that we can take functions of operators. Often we simply want to find f^{-1}, but we can do more, such as composing f with itself α times, where α is not necessarily an integer.

This technique of diagonalization applies to differential equations via the Fourier transform. The starting point is the inversion formula (3). When differentiating underneath the integral sign is valid, we obtain

$$f^{(k)}(x) = \frac{1}{\sqrt{2\pi}} \int_{-\infty}^{\infty} (i\xi)^k \hat{f}(\xi) e^{ix\xi} d\xi.$$

Thus the process of differentiating k times can be expressed as follows: first take the Fourier transform. Then multiply by $(i\xi)^k$. Then take the inverse Fourier transform. The reader should compare this process with the discussion in Section 4 of Chapter 1 as well as with the above paragraphs.

Let p be a monic polynomial of degree k. Consider a linear differential equation of the form

$$p(D)y = y^{(k)} - \sum_{j=1}^{k-1} c_j y^{(j)} = f. \tag{20}$$

Taking Fourier transforms of both sides, we obtain a relation

$$(i\xi)^k \hat{y}(\xi) - \sum_{j=1}^{k-1} c_j (i\xi)^j \hat{y}(\xi) = p(i\xi)\hat{y}(\xi) = \hat{f}(\xi), \tag{21}$$

which is an *algebraic* equation for $\hat{y}$. Thus $\hat{y}(\xi) = \frac{\hat{f}(\xi)}{p(i\xi)}$ for the polynomial p which defines the differential equation. To solve (20), we take the Fourier transform of f, divide by this polynomial in $i\xi$, and recover y by finally taking the inverse Fourier transform. The problem of integrating a differential equation gets replaced by the problem of dividing by a polynomial and taking the inverse Fourier transform of the result.

EXAMPLE 3.4. Consider $y'' - y = f$. We obtain $\hat{y}(\xi) = \frac{-\hat{f}(\xi)}{1+\xi^2}$. Using the Fourier inversion formula, we get

$$y(x) = \frac{-1}{2\pi} \int \int f(t) \frac{e^{i\xi(x-t)}}{1+\xi^2} dt \, d\xi.$$

For nice f we can invert the order of integration and obtain a formula $y(x) = \int G(x,t)f(t)dt$.

EXAMPLE 3.5. We return to the wave equation $u_{xx} = u_{tt}$. Assume $u(x,0) = f(x)$ and $u_t(x,0) = g(x)$. We apply the Fourier transform in x, regarding t as a parameter. The wave equation becomes a second order ODE with two initial conditions:

$$\hat{u}_{tt}(\xi, t) = -\xi^2 \hat{u}(\xi, t)$$
$$\hat{u}(\xi, 0) = \hat{f}(\xi)$$
$$\hat{u}_t(\xi, 0) = \hat{g}(\xi).$$

Solving this second order ODE is easy, as it has constant coefficients when t is regarded as the variable. We get

$$\hat{u}(\xi, t) = \hat{f}(\xi)\cos(|\xi|t) + \hat{g}(\xi)\frac{\sin(|\xi|t)}{|\xi|}. \tag{22}$$

Exercise 3.29 asks for the simple details. Now we can find u by applying the Fourier inversion formula, obtaining the same result as in Theorem 1.4.

EXERCISE 3.28. Suppose the f in Example 3.4 lies in the Schwartz space. Find $G(x,t)$. Be careful about the sign of $x - t$.

EXERCISE 3.29. Fill in the details in Example 3.5, including the use of the inversion formula.

EXERCISE 3.30. Let V be a complex vector space. Put $f(z) = Mz + b$, for $M \in \mathcal{L}(V,V)$. Under what condition does $f = T^{-1}MT$ for some translation T?

8. Pseudo-differential operators

This section indicates how the Fourier transform has been used in modern analysis. Let us repeat the idea from the previous section, by differentiating the Fourier inversion formula for a Schwartz function u:

$$(\frac{d}{dx})^k u(x) = \frac{1}{\sqrt{2\pi}} \int_{-\infty}^{\infty} e^{ix\xi}(i\xi)^k \hat{u}(\xi)d\xi. \tag{23}$$

Let $p_k(x)$ be a smooth function of x and let $(Lu)(x) = \sum_{k=0}^{m} p_k(x)u^{(k)}(x)$ denote a differential operator of order m. By (23) we have

$$Lu(x) = \sum_{k=0}^{m} p_k(x)u^{(k)}(x) = \frac{1}{\sqrt{2\pi}} \int_{-\infty}^{\infty} e^{ix\xi} \sum_{k=0}^{m} (i\xi)^k p_k(x)\hat{u}(\xi)d\xi.$$

As before, Lu is computed by a three-step process. We find $\hat{u}$; we multiply by a polynomial in ξ with coefficients in x, namely $\sum_k p_k(x)(i\xi)^k$; finally we take the inverse Fourier transform. To invert L we proceed in a similar fashion, with multiplication replaced by division.

A pseudo-differential operator P is a linear operator, defined on a space of functions or distributions, obtained in the following fashion. Given u, we compute Pu as follows. We find $\hat{u}(\xi)$; we multiply by a smooth function $p(x,\xi)$ satisfying appropriate smoothness conditions in (x,ξ) and growth restrictions in ξ; finally we take the inverse Fourier transform. The function $p(x,\xi)$ is called the *symbol* of P.

One of the most useful pseudo-differential operators is written Λ^s. Its symbol is $(1+\xi^2)^{\frac{s}{2}}$. The operator Λ^2 is the same as $1 - (\frac{d}{dx})^2$. Note that Λ^{-2} is its inverse. Hence we can solve the differential equation $(1 - (\frac{d}{dx})^2)u = f$ by writing $f = \Lambda^{-2}u$.

The operator Λ^{-2} is certainly not a differential operator, although it is the inverse of one.

Pseudo-differential operators extend this idea to both ordinary and partial linear differential equations. The key idea is to perform algebraic operations (multiplication and division) on the symbols, rather than to differentiate and integrate.

What functions are allowed as symbols? Perhaps the most basic class of symbols, but not the only one, is defined as follows. For $m \in \mathbf{R}$, we let $\mathbf{S}^m$ denote the space of infinitely differentiable functions u in (x, ξ) such that, for all a, b there is a constant C_{ab} such that

$$|(\frac{d}{dx})^a(\frac{d}{d\xi})^b u(x, \xi)| \le C_{ab}(1 + \xi^2)^{\frac{m-b}{2}}. \tag{24}$$

Elements of $\mathbf{S}^m$ are called *symbols of order m*. A pseudo-differential operator has *order m* if its symbol has order m. In particular, a differential operator gives an example of a pseudo-differential operator of the same order. For each real number s, the operator Λ^s has order s.

We obviously have $\mathbf{S}^l \subseteq \mathbf{S}^m$ if $l \le m$. We therefore naturally define $\mathbf{S}^\infty = \cup \mathbf{S}^m$ and $\mathbf{S}^{-\infty} = \cap \mathbf{S}^m$. The reader might wonder what an operator of order $-\infty$ might be. Mollifiers as we defined earlier provide nice examples. These operators smooth things out, as we indicate below.

These ideas are part of a sophisticated theory including *Sobolev spaces*. See [SR]. For us the key point to mention extends our earlier remark to the effect that the smoothness of a function on the circle is related to the rate of decay of its Fourier coefficients at ∞. Sobolev spaces measure the rate of decay at ∞ of the Fourier transform of a function. We give the following definition of the Sobolev space $W^s(\mathbf{R})$. Henceforth we drop the $\mathbf{R}$ from the notation.

DEFINITION 3.10. Assume $u \in \mathcal{S}'$. Then $u \in W^s$ if and only if $\Lambda^s u \in L^2(\mathbf{R})$.

This definition is equivalent to demanding that $\hat{u}$ be a function for which

$$||u||_{W^s}^2 = \int_{-\infty}^{\infty} (1 + \xi^2)^s |\hat{u}(\xi)|^2 d\xi < \infty. \tag{24}$$

Note that $W^s \subseteq W^t$ if $s \ge t$. We naturally put $W^{-\infty} = \cup_s W^s$ and $W^\infty = \cap_s W^s$. We then have the obvious additional containments, each of which is strict:

$$\mathcal{S} \subseteq W^\infty \subseteq W^{-\infty} \subseteq \mathcal{S}'.$$

EXERCISE 3.31. Show that $\frac{1}{1+x^2}$ is in W^∞ but not in $\mathcal{S}$.

LEMMA 3.4. *Suppose $|g|$ is integrable on* **R**. *Then both the Fourier transform of g and the inverse Fourier transform of g are continuous.*

PROOF. See Exercise 3.32.   □

The significance of Lemma 3.4 arises from the following crucial idea. To prove that a function or distribution u has k continuous derivatives, we take Fourier transforms and prove that $\xi^k \hat{u}(\xi)$ is integrable. This method clarifies the relationship between smoothness of a function and the behavior of its Fourier transform at infinity.

EXERCISE 3.32. Prove that the Fourier transform of an L^1 function is continuous. Be careful in your proof, as the real line is not compact.

THEOREM 3.8. *Assume $u \in L^2(\mathbf{R})$ and let k be a nonnegative integer. Then:*
- *$u \in W^k$ if and only if $u^{(k)} \in L^2(\mathbf{R})$.*
- *(Sobolev lemma, special case) If $u \in W^s$ for $s > k + \frac{1}{2}$, then u has k continuous derivatives.*

Hence $u \in W^\infty$ if and only if all derivatives of u are in $L^2(\mathbf{R})$.

PROOF. First we note two obvious estimates.

$$|\xi|^{2k} \leq (1 + |\xi|^2)^k \leq 2^k \text{ if } |\xi| \leq 1 \tag{25.1}$$

$$|\xi|^{2k} \leq (1 + |\xi|^2)^k \leq 2^k |\xi|^{2k} \text{ if } 1 \leq |\xi|. \tag{25.2}$$

The inequalities in (25.1) and (25.2) show that $u \in W^k$ is equivalent to $u^{(k)}$ being square-integrable, and the first statement holds. To prove the second statement, we estimate using the Cauchy-Schwarz inequality:

$$\int |\xi|^k |\hat{u}(\xi)| = \int \frac{|\xi|^k}{(1 + |\xi|^2)^{\frac{s}{2}}} |\hat{u}(\xi)|(1 + |\xi|^2)^{\frac{s}{2}} \leq ||u||_{W^s} \left(\int \frac{|\xi|^{2k}}{(1 + |\xi|^2)^s} \right)^{\frac{1}{2}}. \tag{26}$$

The integral on the far right in (26) is convergent if and only if $2k - 2s < -1$. Hence, if $u \in W^s$ and $s > k + \frac{1}{2}$, then the expression $|\xi|^k \hat{u}(\xi)$ is integrable. Since we recover the k-th derivative of u by taking the inverse Fourier transform, the second statement now follows from Lemma 3.4. $\square$

COROLLARY 3.2. *Suppose $u \in \mathcal{S}'$ and $u^{(k)} \in L^2$ for all k. Then u agrees with an infinitely differentiable function almost everywhere.*

EXERCISE 3.33. For what s is the Dirac delta distribution in W^s?

EXERCISE 3.34. Consider the analogue of the integral in (26) in n dimensions:

$$\int_{\mathbf{R}^n} \frac{||\xi||^{2k}}{(1 + ||\xi||^2)^s} dV.$$

What is the condition on k, s, n such that this integral converges?

9. Hermite polynomials

We saw in Chapter 2 that the Hermite polynomials $H_n(x)$ have the exponential generating function $\exp(2xt - t^2)$. In other words,

$$\exp(2xt - t^2) = \sum_{n=0}^{\infty} H_n(x) \frac{t^n}{n!}. \tag{27}$$

These polynomials are closely related to the Fourier transform. As mentioned in Chapter 2, the functions in Theorem 3.9 are sometimes called the Hermite functions. See the comment after Exercise 2.62.

THEOREM 3.9. *For each non-negative integer n, the function $e^{\frac{-x^2}{2}} H_n(x)$ is an eigenfunction of the Fourier transform with eigenvalue $(-i)^n$.*

PROOF. We start with (27) and multiply both sides by $e^{\frac{-x^2}{2}}$. We then take Fourier transforms. Doing so yields

$$\mathcal{F}\left(\exp(-\frac{x^2}{2} + 2xt - t^2)\right)(\xi) = \sum_n \mathcal{F}\left(e^{\frac{-x^2}{2}} H_n(x)\right)(\xi)\frac{t^n}{n!}. \qquad (28)$$

We will simplify the left-hand side of (28) and use (27) again to obtain the result. Note that

$$\exp(\frac{-x^2}{2} + 2xt - t^2) = \exp(\frac{-1}{2}(x - 2t)^2)\exp(t^2). \qquad (29)$$

Plugging (29) into (28) replaces the left-hand side with

$$\exp(t^2)\mathcal{F}\left(\exp(\frac{-1}{2}(x - 2t)^2)\right)(\xi). \qquad (30)$$

The second factor in (30) is the Fourier transform of a Gaussian with mean $2t$ and variance 1. By Exercise 3.4, the Fourier transform of the Gaussian with mean μ and variance 1 is $\exp(-i\mu\xi)$ times itself. The expression in (30) therefore becomes

$$\exp(t^2 - 2i\xi t)\exp(\frac{-\xi^2}{2}). \qquad (31)$$

The first factor in (31) is the left-hand side of (27) with t replaced by $(-it)$ and x replaced by ξ. Using the generating function expansion from (27) with t replaced by $-it$, and equating coefficients of t^n, yields the conclusion:

$$\mathcal{F}\left(\exp(\frac{-x^2}{2})H_n(x)\right)(\xi) = (-i)^n \exp(\frac{-\xi^2}{2})H_n(\xi).$$

$\square$

EXAMPLE 3.6. The first few Hermite polynomials:

- $H_0(x) = 1$
- $H_1(x) = 2x$
- $H_2(x) = -2 + 4x^2$
- $H_3(x) = -12x + 8x^3$
- $H_4(x) = 12 - 48x^2 + 16x^4$
- $H_5(x) = 120x - 160x^3 + 32x^5$
- $H_6(x) = -120 + 720x^2 - 480x^4 + 64x^6$
- $H_7(x) = -1680x + 3360x^3 - 1344x^5 + 128x^7$.

Figure 2 shows the polynomials $H_n(x)/2^n$ for $0 \le n \le 5$. We divide by 2^n to make the leading coefficient equal to 1, and the graphs become nicer.

EXERCISE 3.35. Write simple code to get Mathematica or something similar to print out the first twenty Hermite polynomials. Observe some patterns and then prove them.

EXERCISE 3.36. For each n find $(x - \frac{d}{dx})^n \exp\left(\frac{-x^2}{2}\right)$.

EXERCISE 3.37. Prove that $H_n'(x) = 2xH_n(x) - H_{n+1}(x)$.

EXERCISE 3.38. Prove for each n that $H_n(x)$ has integer coefficients.

EXERCISE 3.39. Let M denote multiplication by $\exp(\frac{-x^2}{2})$ and let T denote $x - \frac{d}{dx}$. Express $M^{-1}T^n M$ in terms of Hermite polynomials.

EXERCISE 3.40. We saw in Chapter 2 that the Hermite polynomials form a complete orthogonal system for $\mathcal{H} = L^2(\mathbf{R}, \exp(-x^2))$. Show that $||H_n||^2 = 2^n n! \sqrt{\pi}$.

EXERCISE 3.41. Find a combinatorial interpretation of the sequence of (absolute values of) coefficients $0, 0, 0, 2, 12, 48, 160, 480, \ldots$ of the second highest power of x in $H_n(x)$.

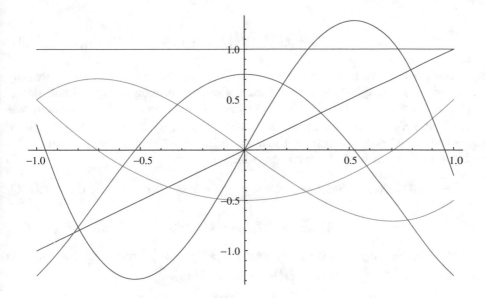

FIGURE 2. Scaled Hermite polynomials

10. More on Sobolev spaces

We begin by proving the following theorem. Its proof is quite similar to several proofs from Chapter 1. The analogues in higher dimensions of this result and the subsequent proposition are important tools in partial differential equations. See [F1] and [Ra].

THEOREM 3.10. *Assume* $s < t$. *Let* $\{f_n\}$ *be a sequence of functions such that:*

- *There is a constant* C *such that* $||f_n||_{W^t} \leq C$ *for all* n.
- $\hat{f}_n(\xi)$ *converges uniformly to* 0 *on compact subsets of* $\mathbf{R}$.

Then $||f_n||_{W^s}$ *converges to* 0.

PROOF. We start with

$$||f_n||^2_{W^s} = \int (1 + |\xi|^2)^s |\hat{f}_n(\xi)|^2 \, d\xi. \tag{32}$$

As in many proofs in this book, we estimate the integral by breaking it into two parts. First consider the set where $|\xi| > R$. On this set, and remembering that $s < t$, we have

$$(1 + |\xi|^2)^s = (1 + |\xi|^2)^{s-t}(1 + |\xi|^2)^t \leq (1 + R^2)^{s-t}(1 + |\xi|^2)^t. \tag{33}$$

Using (33) in (32) we obtain

$$\|f_n\|_{W^s}^2 \leq \int_{|\xi| \leq R} (1 + |\xi|^2)^s |\hat{f}_n(\xi)|^2 \, d\xi + (1 + R^2)^{s-t}\|f_n\|_{W^t}^2. \tag{34}$$

Suppose $\epsilon > 0$ is given. Since the terms $\|f_n\|_{W^t}^2$ are bounded, and $s < t$, we may choose R large enough to guarantee that the second term on the right-hand side of (34) is at most $\frac{\epsilon}{2}$. Fix this R. Now consider the first term on the right-hand side of (34). It is bounded by a constant (independent of n) times $\sup |\hat{f}_n(\xi)|^2$. Since the set where $|\xi| \leq R$ is compact, the assumption of uniform convergence allows us to bound this term by $\frac{\epsilon}{2}$, by choosing n large enough. □

PROPOSITION 3.5. *Suppose $s < t < u$ and $\epsilon > 0$. Then there is a constant C_ϵ such that*

$$\|f\|_{W^t}^2 \leq \epsilon\|f\|_{W^u}^2 + C_\epsilon\|f\|_{W^s}^2. \tag{35}$$

PROOF. For any positive x we have $1 \leq x^{u-t} + x^{s-t}$ since one of x and $\frac{1}{x}$ is already at least 1. Plug in $(1 + |\xi|^2)\epsilon^{\frac{1}{u-t}}$ for x. We get

$$1 \leq \epsilon(1 + |\xi|^2)^{u-t} + \epsilon^{\frac{s-t}{u-t}}(1 + |\xi|^2)^{u-t}.$$

Then multiply this inequality by $(1 + |\xi|^2)^t|\hat{f}(\xi)|^2$ and integrate. The result is that

$$\|f\|_{W^t}^2 \leq \epsilon\|f\|_{W^u}^2 + C_\epsilon\|f\|_{W^s}^2,$$

where $C_\epsilon = \epsilon^{\frac{s-t}{u-t}}$. Note that C_ϵ is a negative power of ϵ. □

One can also write this proof using the operators Λ^r where $r = u - t$ and $r = s - t$. Equivalently, put $g = \Lambda^t f$ in (35). We obtain an estimate of the L^2 norm of g in terms of a small constant times a Sobolev norm with a positive index and a large constant times a Sobolev norm with negative index.

These results are closely related to the Rellich Lemma. See [F1]. Often one considers Sobolev spaces $W^s(\Omega)$ on a bounded domain or compact manifold Ω. The Rellich Lemma then states that the inclusion map of $W^t(\Omega)$ into $W^s(\Omega)$ is a compact operator when $s < t$. If we work with Sobolev spaces on **R**, we may state the Rellich Lemma as follows:

THEOREM 3.11 (Rellich lemma). *Assume that $\{g_n\}$ is a bounded sequence in W^t and each g_n vanishes outside a fixed compact set K. Then, whenever $s < t$, there is a subsequence $\{g_{n_k}\}$ converging in W^s.*

The proof is similar to that of Theorem 3.10. Under the hypotheses of the Rellich lemma, and using the Arzela-Ascoli theorem, one can find a subsequence $\{\hat{g}_{n_k}\}$ which converges uniformly on compact sets. Then one applies Theorem 3.10 to show that g_{n_k} is Cauchy in W^s.

11. Inequalities

We close this chapter with several standard inequalities for functions defined on subsets of the real line. These inequalities hold more generally. Again see [F1]. We begin with a result from elementary calculus.

LEMMA 3.5. *Let f be continuous and increasing on $[0, a]$, with $f(0) = 0$. Then*

$$af(a) = \int_0^a f(x)dx + \int_0^{f(a)} f^{-1}(y)dy. \tag{36}$$

Suppose $0 < b < f(a)$. Then

$$ab \leq \int_0^a f(x)dx + \int_0^b f^{-1}(y)dy. \tag{37}$$

PROOF. Both sides of (36) represent the area of the rectangle with vertices at $(0, 0), (a, 0), (a, f(a)), (0, f(a))$, and hence (36) holds. If $0 < b < f(a)$, then the right-hand side of (37) represents the area of a set strictly containing the rectangle with vertices $(0, 0), (a, 0), (a, b), (0, b)$. See Figure 3. □

REMARK 3.3. Lemma 3.5 has an amusing corollary. Assume f is a monotone, elementary function whose indefinite integral is also an elementary function. Then the indefinite integral of f^{-1} is also an elementary function. Changing variables by putting $y = f(x)$ in the right-hand integral in (36) shows that one can find the indefinite integral of f^{-1} by using integration by parts and Lemma 3.5.

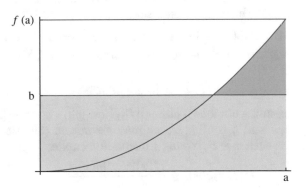

FIGURE 3. Lemma 3.5

PROPOSITION 3.6 (Young's inequality). *For $1 < p$, set $q = \frac{p}{p-1}$. For positive numbers a, b we then have*

$$ab \leq \frac{a^p}{p} + \frac{b^q}{q}. \tag{38}$$

PROOF Take $f(x) = x^{p-1}$ in Lemma 3.5. We obtain

$$ab \leq \int_0^a x^{p-1}dx + \int_0^b y^{\frac{1}{p-1}}dy = \frac{a^p}{p} + \frac{b^{\frac{p}{p-1}}}{\frac{p}{p-1}} = \frac{a^p}{p} + \frac{b^q}{q}. \qquad\qquad □$$

The condition that $q = \frac{p}{p-1}$ is often written $\frac{1}{p} + \frac{1}{q} = 1$. These numbers are called *conjugate exponents*. Proposition 3.6 still holds when $p = 1$, if we set $q = \infty$. See Exercise 3.42.

REMARK 3.4. Inequality (37), used to prove Proposition 3.6, is sometimes also known as Young's inequality.

To state and prove several additional inequalities, we use the notation of Lebesgue integration. Let S be a (measurable) subset of **R**. Assume $|f|$ is integrable on S. For $1 \le p < \infty$ we write (not including S in the notation)

$$\|f\|_{L^p}^p = \int_S |f(x)|^p dx.$$

In order to regard $L^p(S)$ as a normed vector space, we must first remember that elements of L^p are equivalence classes of functions. The set of these equivalence classes forms a vector space. The zero vector is the equivalence class consisting of all functions that vanish except on a set of measure 0. We add equivalence classes by selecting representatives and adding as usual, and we multiply an equivalence class by a scalar by selecting a representative and multiplying it by the scalar. As usual, one needs to verify the trivial assertions that the results are independent of the representatives chosen. We omit this pedantic point.

We must also verify that $\| \cdot \|_{L^p}$ defines a norm. Property (2) from Definition 2.1 is clear; property (1) of a norm is clear once we realize we are working with equivalence classes. Property (3), the triangle inequality, is Theorem 3.12 below. It is known as Minkowski's inequality and is a subtle point. In order to prove it, we first prove Hölder's inequality. See [S] both for additional versions of this inequality and for some interesting historical discussion.

PROPOSITION 3.7 (Hölder's inequality). *Again assume $1 < p$ and $q = \frac{p}{p-1}$. Assume $|f|^p$ is integrable on S and $|g|^q$ is integrable on S. Then $|fg|$ is integrable on S and*

$$\|fg\|_{L^1} \le \|f\|_{L^p}\|g\|_{L^q}. \tag{39}$$

PROOF. The result is obvious if either $\|f\|_{L^p}$ or $\|g\|_{L^q}$ is zero. Otherwise, after dividing f by $\|f\|_{L^p}$ and g by $\|g\|_{L^q}$, we may assume in (39) that each of these norms equals 1. For each $x \in S$, Young's inequality implies

$$|f(x)g(x)| \le \frac{|f(x)|^p}{p} + \frac{|g(x)|^q}{q}. \tag{40}$$

Integrating (40) shows that

$$\|fg\|_{L^1} \le \frac{\|f\|_{L^p}^p}{p} + \frac{\|g\|_{L^q}^q}{q} = \frac{1}{p} + \frac{1}{q} = 1.$$

$\square$

Hölder's inequality remains true when $p = 1$, in which case it is obvious. (Exercise 3.42).

EXERCISE 3.42. Verify Young's inequality and Hölder's inequality when $p = 1$.

EXERCISE 3.43. Verify the statements from Remark 3.3.

EXERCISE 3.44. Suppose $q = \frac{p}{p-1}$. Show that $\|h^{p-1}\|_{L^q} = \|h\|_{L^p}^{p-1}$.

THEOREM 3.12 (Minkowski). *Assume* $1 \leq p < \infty$. *The triangle inequality holds in L^p. In other words, if $|f|^p$ and $|g|^p$ are integrable, then $|f + g|^p$ is integrable and*

$$||f + g||_{L^p} \leq ||f||_{L^p} + ||g||_{L^p}. \tag{41}$$

PROOF. First we note that the statement is immediate if $f + g = 0$ and is easy when $p = 1$. We thus assume $p > 1$ and that $f + g$ is not the zero function (equivalence class). The following string of statements is elementary:

$$||f + g||_{L^p}^p \leq \int (|f| + |g|) |f + g|^{p-1} = \int |f| |f + g|^{p-1} + \int |g||f + g|^{p-1}. \tag{42}$$

Now use Hölder's inequality in (42) to get

$$||f + g||_{L^p}^p \leq (||f||_{L^p} + ||g||_{L^p}) || |f + g|^{p-1}||_{L^q}. \tag{43}$$

Since $q = \frac{p}{p-1}$, the last term in (43) becomes $|| |f + g| ||_{L^p}^{p-1}$. (See Exercise 3.44) Dividing both sides by this term gives the triangle inequality (41). □

REMARK 3.5. The L^p spaces are complete; the proof requires results from the theory of the Lebesgue integral.

By Theorem 3.12, the L^p norm of a sum is at most the sum of the L^p norms. That result suggests that the L^p norm of an integral should be at most the integral of the L^p norms. Such a result holds; it is often called Minkowski's inequality for integrals. See [F1].

Next we use Hölder's inequality to establish the integral versions of Hilbert's inequality and Hardy's inequality, formulas (57.1) and (57.2) of Chapter 1. Note that (44) is obvious when $p = 1$, as the right-hand side is infinite.

THEOREM 3.13 (Hilbert's inequality revisited). *Let p, q be conjugate exponents with $p > 1$. Assume that $f \in L^p([0, \infty))$ and that $g \in L^q([0, \infty))$. Then*

$$\int_0^\infty \int_0^\infty \frac{f(x)g(y)}{x + y} dx dy \leq \frac{\pi}{\sin(\frac{\pi}{p})}||f||_{L^p}||g||_{L^q}. \tag{44}$$

PROOF. Change variables in the integral in (44) by replacing x by yt. We get

$$I = \int_0^\infty \int_0^\infty \frac{f(x)g(y)}{x + y} dx dy = \int_0^\infty \int_0^\infty \frac{f(yt)g(y)}{1 + t} dt dy. \tag{45}$$

Let F_t denote the function given by $F_t(y) = f(yt)$. Interchange the order of integration in (45). Then apply Hölder's inequality to the inner integral, obtaining

$$I \leq \int_0^\infty ||F_t||_{L^p} \frac{dt}{1 + t}||g||_{L^q}.$$

By changing variables, note that $||F_t||_{L^p} = \frac{||f||_{L^p}}{t^{\frac{1}{p}}}$. Plugging in this result gives

$$I \leq \int_0^\infty \frac{dt}{(1 + t)t^{\frac{1}{p}}} ||f||_{L^p}||g||_{L^q}. \tag{46}$$

The integral on the right-hand side of (46) equals $\frac{\pi}{\sin(\frac{\pi}{p})}$, which gives (44). We discuss the evaluation of the integral after the proof. □

For $p > 1$, put $C_p = \int_0^\infty \frac{dt}{(t+1)t^{\frac{1}{p}}}$. This integral can be evaluated by using contour integrals and complex analysis. See page 133 in [D2] or pages 157–8 in [A]. The contour used is pictured in Figure 4, where the positive real axis is a *branch cut*.

One can also reduce the computation of C_p (Exercise 3.45) to the Euler Beta function discussed in Exercise 4.63, and the formula (Exercise 4.67)

$$\Gamma(z)\Gamma(1 - z) = \frac{\pi}{\sin(\pi z)}. \tag{47}$$

The best way to establish (47), however, uses contour integrals to evaluate C_p.

EXERCISE 3.45. For $p > 1$, put $C_p = \int_0^\infty \frac{dt}{(t+1)t^{\frac{1}{p}}}$. Verify that

$$C_p = \int_0^1 s^{\frac{1}{p}-1}(1 - s)^{\frac{-1}{p}} ds. \tag{48}$$

If you are familiar with the Gamma and Beta functions (See Chapter 4), show that

$$C_p = \Gamma(\frac{1}{p})\Gamma(1 - \frac{1}{p}) = \frac{\pi}{\sin(\frac{\pi}{p})}.$$

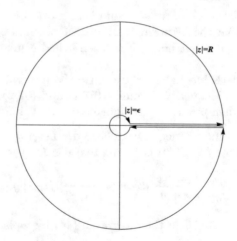

FIGURE 4. Contour used to evaluate C_p

The next exercise provides another generalization of Hilbert's inequality. For simplicity we work in L^2. To solve it, mimic the proof of Theorem 3.13.

EXERCISE 3.46. Consider a continuous function $K : (0, \infty) \times (0, \infty) \to [0, \infty)$ such that $K(\lambda x, \lambda y) = \frac{K(x,y)}{\lambda}$. Show that

$$\int_0^\infty \int_0^\infty K(x, y)f(x)g(y)dxdy \leq C\|f\|_{L^2}\|g\|_{L^2},$$

where C satisfies

$$C = \int_0^\infty \frac{K(1, y)}{\sqrt{y}}dy.$$

The following beautiful exercise also fits into this general discussion, providing a bound from the other direction.

EXERCISE 3.47. Let $\{x_j\}$ be a sequence of distinct positive numbers. Consider the infinite matrix A whose entries are $\frac{1}{x_j+x_k}$. Prove that

$$0 \leq \sum_{j,k} \frac{z_j \overline{z_k}}{x_j + x_k}$$

for all sequences $\{z_j\}$, and equality holds only if z is the zero sequence. Thus A is non-negative definite. Suggestion: Write

$$\frac{1}{x_j + x_k} = \int_0^\infty e^{-(x_j+x_k)t} dt.$$

EXERCISE 3.48. Suppose $f = \hat{g}$, where $g \geq 0$. Show that the matrix with entries $f(x_j - x_k)$ is non-negative definite. Comment: The converse assertion is a special case of a general result in harmonic analysis known as Bochner's theorem. On the real line, Bochner's theorem states that a continuous function f satisfies this matrix non-negativity condition if and only if it is the Fourier transform of a positive measure. The one-dimensional result can be derived from Herglotz's theorem, mentioned in Section 13 of Chapter 1. See [K].

EXERCISE 3.49. The Laplace transform of a function f is defined by

$$(\mathcal{L}f)(x) = \int_0^\infty e^{-xt} f(t) dt,$$

whenever the improper integral converges. Show that

$$(\mathcal{L}^2 f)(x) = \int_0^\infty \frac{f(t)}{x + t} dt.$$

CHAPTER 4

Geometric considerations

The purpose of this chapter is to discuss various geometric problems which are informed by orthogonality and related considerations. We begin with Hurwitz's proof of the isoperimetric inequality using Fourier series. We prove Wirtinger's inequality, both by Fourier series and by compact operators. We continue with a theorem comparing areas of the images of the unit disk under complex analytic mappings. We again give two proofs, one using power series and one using Green's (Stokes') theorem. The maps $z \mapsto z^d$ from the circle to itself play a prominent part in our story. We naturally seek the higher dimensional versions of some of these results. It turns out, not surprisingly, that one can develop the ideas in many directions. We limit ourselves here to a small number of possible paths, focusing on the unit sphere in $\mathbf{C}^n$, and we travel only a small distance along each of them.

Complex analytic mappings sending the unit sphere in $\mathbf{C}^n$ to the unit sphere in some $\mathbf{C}^N$ play a major role in this chapter. For example, we study polynomial mappings that are also invariant under finite subgroups of the unitary group, and we discover a surprising connection to Chebyshev polynomials. We also compute many explicit integrals. The author's technique of *orthogonal homogenization* is introduced and is used to prove a sharp inequality about volumes (with multiplicity accounted for) of complex analytic images of the unit ball. To prove this inequality we develop needed information about differential forms and complex vector fields.

We do not consider the Fourier transform in higher dimensions. Many books on partial differential equations and harmonic analysis tell that story well.

1. The isoperimetric inequality

Geometric inequalities range from easy observations to deep assertions. One of the easiest such inequalities is that the rectangle of a given perimeter with maximum area is a square. The proof follows from $(x+y)(x-y) = x^2 - y^2 \leq x^2$, with equality when $y = 0$. One of the most famous inequalities solves the *isoperimetric problem*; given a closed curve in the plane of length L, the area A enclosed satisfies $A \leq \frac{L^2}{4\pi}$. Equality happens only if the curve is a circle. We use Fourier series to prove this *isoperimetric inequality*, assuming that the curve is smooth.

Recall from calculus that a smooth planar curve is a smooth function $\gamma : [a, b] \to \mathbf{R}^2$ for which $\gamma'(t)$ does not vanish. Officially speaking, the curve is the function, but it is natural to think also of the curve as the image of the function, traced out in some order. The curve is called *closed* if $\gamma(a) = \gamma(b)$ and *simple* if $\gamma(t_1) \neq \gamma(t_2)$ for $t_1 \neq t_2$ unless $t_1 = a, t_2 = b$ or $t_1 = b, t_2 = a$. This complicated sounding condition is clear in geometric terms; if one thinks of the curve as its image, then the curve is simple if it neither crosses itself nor covers itself multiple times. Note, for example, that the curve $\gamma : [0, 2\pi] \to \mathbf{C}$ given by $\gamma(t) = e^{2it}$ is closed but not simple, because it covers the circle twice.

© Springer Nature Switzerland AG 2019
J. P. D'Angelo, *Hermitian Analysis*, Cornerstones,
https://doi.org/10.1007/978-3-030-16514-7_4

The length of γ is the integral $\int_\gamma ds$, where ds is the arc length form. In terms of the function $t \mapsto \gamma(t)$, we have the equivalent formula $L = \int_a^b ||\gamma'(t)||dt$; this value is unchanged if we reparametrize the curve. It is often convenient to parametrize using arc length; in this case $||\gamma'(s)|| = ||\gamma'(s)||^2 = 1$.

We can integrate 1-forms along nice curves γ. We give a precise definition of 1-form in Section 5. For now we assume the reader knows the meaning of the line integral $\int_\gamma Pdx + Qdy$, assuming P and Q are continuous functions on γ. This integral measures the work done in moving along γ against a force given by (P, Q). We also assume Green's theorem from calculus. In Green's theorem, the curve γ is assumed to be *positively oriented*. Intuitively, this condition means the (image of the) curve is traversed counter-clockwise as the parameter t increases from a to b.

PROPOSITION 4.1 (Green's theorem). *Let γ be a piecewise smooth, positively oriented, simple closed curve in $\mathbf{R}^2$, bounding a region Ω. Assume that P and Q are continuously differentiable on Ω and continuous on $\Omega \cup \gamma$. Then*

$$\int_\gamma Pdx + Qdy = \int_\Omega \left(\frac{\partial Q}{\partial x} - \frac{\partial P}{\partial y} \right) dxdy.$$

The area A enclosed by γ is of course given by a double integral. Assume that γ is positively oriented. Using Green's theorem, we see that A is also given by a line integral:

$$A = \int_\Omega dxdy = \frac{1}{2} \int_\gamma xdy - ydx = \frac{1}{2} \int_a^b (x(t)y'(t) - x'(t)y(t))\, dt. \qquad (1)$$

Notice the appearance of the Wronskian.

EXERCISE 4.1. Graph the set of points where $x^3 + y^3 = 3xy$. Use a line integral to find the area enclosed by the loop. Solve the same problem when the defining equation is $x^{2k+1} + y^{2k+1} = (2k + 1)x^k y^k$. Comment: Set $y = tx$ to parametrize the curve. Then $xdy - ydx = x(tdx + xdt) - txdx = x^2 dt$.

EXERCISE 4.2. Verify Green's theorem when Ω is a rectangle. Explain how to extend Green's theorem to a region whose boundary consists of finitely many sides, each parallel to one of the coordinate axes.

THEOREM 4.1 (Isoperimetric inequality, smooth version). *Let γ be a smooth simple closed curve in $\mathbf{R}^2$ of length L and enclosing a region of area A. Then $A \le \frac{L^2}{4\pi}$ and equality holds only when γ defines a circle.*

PROOF. This proof goes back to Hurwitz in 1901. After a change of scale we may assume that the length L of γ is 2π. After mapping $[a, b]$ to $[0, 2\pi]$ we parametrize by arc length s and thus assume $\gamma : [0, 2\pi] \to \mathbf{R}^2$ and $||\gamma'(s)|| = 1$.

Since the curve is closed, γ may be thought of as periodic of period 2π. In terms of Fourier series we may therefore write:

$$\gamma(s) = (x(s), y(s)) = \left(\sum_{-\infty}^\infty a_n e^{ins}, \sum_{-\infty}^\infty b_n e^{ins} \right) \qquad (2)$$

$$\gamma'(s) = (x'(s), y'(s)) = \left(\sum_{-\infty}^\infty in a_n e^{ins}, \sum_{-\infty}^\infty in b_n e^{ins} \right). \qquad (3)$$

Since $(x'(s), y'(s))$ is a unit vector, we have $2\pi = \int_0^{2\pi} (x'(s))^2 + (y'(s))^2 ds$. The only term that matters in computing the integral of a trigonometric series is the constant term. Constant terms in $x'(s)^2$ and $y'(s)^2$ arise precisely when the term with index n is multiplied by the term with index $-n$. It therefore follows that

$$\sum_{-\infty}^{\infty} n^2(|a_n|^2 + |b_n|^2) = 1. \tag{4}$$

We do a similar computation of $xy' - yx'$ to find the area A. We have

$$A = \frac{1}{2}\left| \int_0^{2\pi} (x(s)y'(s) - x'(s)y(s))\, ds \right| = \frac{1}{2}2\pi\left| \sum in(a_n\overline{b_n} - b_n\overline{a_n}) \right|$$

$$= \pi\left| \sum in(a_n\overline{b_n} - b_n\overline{a_n}) \right| \le 2\pi \sum n|a_n||b_n|. \tag{5}$$

Next we use $|n| \le n^2$ and

$$|a_n b_n| \le \frac{1}{2}(|a_n|^2 + |b_n|^2) \tag{6}$$

in the last term in (5). These inequalities and (4) yield

$$A \le \pi \sum n^2(|a_n|^2 + |b_n|^2) = \pi = \frac{L^2}{4\pi},$$

where we have also used the value $L = 2\pi$.

We check when equality holds in the inequality. It must be that the only non-zero terms are those with $|n| = n^2$, that is $n = 0, \pm 1$. We must also have equality in (6), and hence $|a_n| = |b_n|$. By (4) we then must have $|a_1| = |b_1| = \frac{1}{2}$. Put $a_1 = \frac{1}{2}e^{i\mu}$ and $b_1 = \frac{1}{2}e^{i\nu}$. Since $x(s)$ and $y(s)$ are real, $a_{-1} = \overline{a_1}$ and $b_{-1} = \overline{b_1}$. In other words we must have

$$(x(s), y(s)) = \left(a_0 + a_1 e^{is} + \overline{a_1}e^{-is}, b_0 + b_1 e^{is} + \overline{b_1}e^{-is} \right).$$

Under these conditions we get $(x - a_0, y - b_0) = (\cos(s + \mu), \cos(s + \nu))$. Finally, remembering that $(x')^2 + (y')^2 = 1$, we conclude that $\cos(s + \nu) = \pm\sin(s + \mu)$. Hence γ defines a circle of radius 1. $\qquad \square$

EXERCISE 4.3. Given an ellipse E, create a family E_t of ellipses such that the following all hold:

(1) $E = E_0$.
(2) Each E_t has the same perimeter.
(3) The area enclosed by E_t is nondecreasing as a function of t.
(4) E_1 is a circle.

EXERCISE 4.4. A region Ω in the plane is *convex* if, whenever $p, q \in \Omega$, the line segment connecting p and q also lies in Ω. Assume that Ω is bounded, has a nice boundary, but is not convex. Find, by a simple geometric construction, a region Ω' with the same perimeter as Ω but with a larger area. (Reflect a *dent* across a line segment. See Figure 1.)

REMARK 4.1. The isoperimetric inequality holds in higher dimensions. For example, of all simple closed surfaces in $\mathbf{R}^3$ with a given surface area, the sphere encloses the maximum volume.

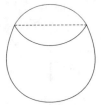

FIGURE 1. Convexity and the isoperimetric inequality

2. Elementary L^2 inequalities

In this section we prove several inequalities relating L^2 norms of functions and their derivatives. The setting for the first example is an interval on the real line, whereas the setting for the second example is the unit disk in $\mathbf{C}$.

We begin with the Wirtinger inequality, an easy one-dimensional version of various higher dimensional inequalities relating functions and their derivatives. We give two proofs to help unify topics in this book.

THEOREM 4.2 (Wirtinger inequality). *Assume f is continuously differentiable on $[0,1]$ and $f(0) = f(1) = 0$. The following inequality holds and is sharp:*

$$\|f\|_{L^2}^2 \leq \frac{1}{\pi^2}\|f'\|_{L^2}^2.$$

PROOF. First we show that there is a function for which equality occurs. Put $f(x) = \sin(\pi x)$. Then $\|f'\|_{L^2}^2 = \pi^2\|f\|_{L^2}^2$ because

$$\|f\|_{L^2}^2 = \int_0^1 \sin^2(\pi x)dx - \frac{1}{2}$$

$$\|f'\|_{L^2}^2 = \int_0^1 \pi^2\cos^2(\pi x)dx = \frac{\pi^2}{2}.$$

Next we use Fourier series to prove the inequality. By putting $f(-x) = -f(x)$, we extend f to be an odd function (still called f) on the interval $[-1,1]$. The extended f is still continuously differentiable, even at 0. Then f equals its Fourier series, which involves only the functions $\sin(n\pi x)$. Put $f(x) = \sum c_n\sin(n\pi x)$. Since f is odd, $c_0 = \hat{f}(0) = 0$. Let L^2 continue to denote $L^2([0,1])$. By either the Parseval formula or by orthonormality we get

$$\|f\|_{L^2}^2 = \frac{1}{2}\sum_{-\infty}^{\infty}|c_n|^2 = \sum_{n=1}^{\infty}|c_n|^2$$

$$\|f'\|_{L^2}^2 = \sum_{n=1}^{\infty}n^2\pi^2|c_n|^2 = \pi^2\sum_{n=1}^{\infty}n^2|c_n|^2.$$

Since $1 \leq n^2$ for all $n \geq 1$ we obtain

$$\pi^2\|f\|_{L^2}^2 \leq \|f'\|_{L^2}^2.$$

The proof also tells us when equality occurs! Put $c_n = 0$ unless $n = 1$; that is, put $f(x) = \sin(\pi x)$.   □

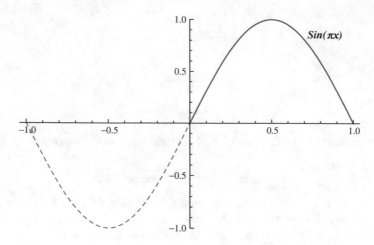

FIGURE 2. Wirtinger inequality

PROOF. We sketch a different proof using compact operators, noting that this sketch had an error in the first edition. Define a linear operator T on the continuous functions in $L^2([0,1])$ by $Tf(x) = \int_0^x f(t)dt$. Computation (see Exercise 4.5) shows that $T^*f(x) = \int_x^1 f(u)du$. The operator T^*T is compact and self-adjoint. It is easy to check that each eigenvalue of T^*T is non-negative. By the first part of the proof of the Spectral theorem, the maximal eigenvalue λ_M of T^*T satisfies $\lambda_M = ||T^*T|| = ||T||^2$. We find all eigenvalues.

Set $T^*Tf = \lambda f$ to get

$$\int_x^1 \int_0^t f(u)dudt = \lambda f(x).$$

Differentiating once, using the fundamental theorem of calculus, and evaluating at 0 gives $f'(0) = 0$. Differentiating again gives

$$-f(x) = \lambda f''(x).$$

Since $f'(0) = f(1) = 0$, we conclude that $f(x) = c\cos(\frac{(2n+1)\pi x}{2})$. Here

$$\lambda = \frac{4}{(2n+1)^2\pi^2}.$$

The maximum happens when $n = 0$. Thus the norm of T is $\frac{2}{\pi}$. To arrange things such that $f(0) = 0$, we change variables by setting $x = 2t - 1$. This step is analogous to the extension to $[-1, 1]$ in the first proof. Then

$$\cos\left(\frac{\pi(2t-1)}{2}\right) = \sin(\pi t).$$

This function vanishes at both 0 and 1. It follows by symmetry that the norm of the operator whose domain consists of functions g with these boundary conditions is $\frac{1}{\pi}$. Hence $||Tg||_{L^2} \le \frac{1}{\pi}||g||_{L^2}$ for all g. Setting $g = f'$ gives the desired conclusion. $\square$

COROLLARY 4.1. *Assume f is continuously differentiable with $f(a) = f(b) = 0$. Then*

$$\int_a^b |f(x)|^2 dx \leq \left(\frac{b-a}{\pi}\right)^2 \int_a^b |f'(x)|^2 dx.$$

PROOF. The result follows from changing variables (Exercise 4.6).   □

EXERCISE 4.5. Put $Tf(x) = \int_a^b K(x,y)f(y)dy$. Express T^* as an integral operator. Check your answer when T is as in the second proof of Theorem 4.2.

EXERCISE 4.6. Prove Corollary 4.1.

Higher dimensional analogues of the Wirtinger inequality are called Poincaré inequalities. Given a region Ω in $\mathbf{R}^n$, a Poincaré inequality is an estimate of the form (for some constant C)

$$||u||_{L^2}^2 \leq C^2 \left(\left|\int_\Omega u\right|^2 + ||\nabla u||_{L^2}^2\right). \tag{P}$$

Let A denote the volume of Ω and let $u_0 = \frac{1}{A}\int_\Omega u$ denote the average value of u on Ω. Since u_0 is a constant, applying (P) to $u - u_0$ gives

$$||u - u_0||_{L^2}^2 \leq C^2 ||\nabla u||_{L^2}^2. \tag{P.1}$$

By expanding the squared norm on the left-hand side of (P.1) and doing some simple manipulations, the left-hand-side of (P.1) becomes $||u||_{L^2}^2 - A|u_0|^2$. We can therefore also rewrite (P) in the form

$$||u||_{L^2}^2 \leq \frac{1}{A}\left|\int u\right|^2 + C^2 ||\nabla u||_{L^2}^2. \tag{P.2}$$

The technique of subtracting the average value and expanding the squared norm appears, in various guises, many times in this book. This reasoning is standard in elementary probability, as used in Proposition 6.4 from the appendix. Observe also, for f, f_0 in a Hilbert space, that

$$||f - f_0||^2 = ||f||^2 - ||f_0||^2$$

whenever $(f - f_0) \perp f_0$. This version of the Pythagorean theorem was used in the proof of Bessel's inequality, where f_0 was the orthogonal projection of f onto the subspace spanned by a finite orthonormal system.

Poincaré estimates do not hold for all domains. When such an inequality does hold, the smallest value of C that works is called the Poincaré constant for the domain.

We make one additional observation. In our proof of the Wirtinger inequality, we assumed that f vanished at both endpoints. We could have assumed that f vanished at only one endpoint, or instead that the average value of f was 0, and in each case proved similar results. The condition that the average value of f vanishes means that f is orthogonal to the one-dimensional subspace of constant functions. The condition that f vanish at the endpoints means that f lies in a subspace of *codimension* two.

EXERCISE 4.7. Find the Poincaré constant for the interval $[-A, A]$. (The function $\sin(\frac{\pi x}{2A})$ achieves the bound. The answer is $\frac{2A}{\pi}$.)

REMARK 4.2. The Wirtinger inequality provides a bound on the L^2 norm of a function in terms of the L^2 norm of its derivative. Various inequalities that bound the maximum of the derivative p' of a *polynomial* in terms of the maximum of p (thus going in the other direction) and its degree are called Bernstein inequalities and Markov inequalities. We do not consider such results in this book.

We next prove a simple geometric inequality in one complex dimension. It motivates a more difficult higher dimensional analogue which we prove in Section 9. The orthogonality of the functions $e^{in\theta}$ again features prominently.

Let f be a complex analytic function on the unit disk B_1. Let A_f denote the area of the image, with multiplicity counted. For example, if $f(z) = z^m$, then f covers the disk m times and $A_f = m\pi$. The formula for A_f involves the L^2 norm of the derivative. We make the concept of *counting multiplicity* precise by defining A_f as follows:

DEFINITION 4.1. Let Ω be open in $\mathbf{C}$. Assume $f : \Omega \to \mathbf{C}$ is complex analytic. The area, written $A_f(\Omega)$ or A_f, of the image of f, with multiplicity counted, is defined by

$$A_f = ||f'||^2_{L^2(\Omega)}. \tag{7}$$

We next note that this concept agrees with what we expect when f is injective.

LEMMA 4.1. *Let $f : \Omega \to \mathbf{C}$ be complex analytic and injective. Then the area of $f(\Omega)$ equals $||f'||^2_{L^2(\Omega)}$.*

PROOF. Let $A(f)$ denote the area of the image of f. Write $f = u + iv$ and define $F(x,y) = (u(x,y), v(x,y))$. The Cauchy-Riemann equations and the definition of f' imply $\det(F') = u_x v_y - u_y v_x = u_x^2 + u_y^2 = |f'|^2$. Since F is injective, the change of variables formula for double integrals applies and gives

$$A(f) = \int_{F(\Omega)} du\,dv = \int_\Omega |\det(F')|dx\,dy = \int_\Omega |f'(z)|^2 dx\,dy = ||f'||^2_{L^2}. \qquad \square$$

Versions of the change of variables formula hold more generally. Suppose that f is m-to-one for some fixed m. The change of variables formula gives

$$m \int_{F(\Omega)} du\,dv = \int_\Omega |\det(F')|dx\,dy = \int_\Omega |f'(z)|^2 dx\,dy = ||f'||^2_{L^2}.$$

In general, the multiplicity varies from point to point. For complex analytic functions, things are nonetheless quite nice. See [A] for the following result. Suppose that f is complex analytic near z_0 and the function $z \mapsto f(z) - f(z_0)$ has a zero of order m. Then, for w sufficiently close to $f(z_0)$, there is a (deleted) neighborhood of z_0 on which the equation $f(z) = w$ has precisely m solutions. By breaking Ω into sets on which f has constant multiplicity, we justify the definition of A_f.

We return to the unit disk. The natural Hilbert space here is the set $\mathcal{A}^2$ of square-integrable complex analytic functions f on the unit disk. The inner product on $\mathcal{A}^2$ is given by

$$\langle f, g \rangle = \int_{B_1} f(z)\overline{g(z)}dx\,dy.$$

The subspace $\mathcal{A}^2$ is closed in L^2 and hence is itself a Hilbert space. See, for example, pages 70–71 in [D1] for a proof. The main point of the proof is that, on any compact subset K of the disk, we can estimate (the L^∞ norm) $\sup_K |f|$ by a constant times (the L^2 norm) $||f||$. Hence, if $\{f_n\}$ is Cauchy in L^2, then $\{f_n\}$ converges uniformly on compact subsets. By a standard fact in complex analysis (see [A]), the limit function is also complex analytic.

We are also concerned with the subspace of $\mathcal{A}^2$ consisting of those f for which f' is square integrable.

LEMMA 4.2. *The functions z^n for $n = 0, 1, 2, \ldots$ form a complete orthogonal system for $\mathcal{A}^2$.*

PROOF. Using polar coordinates we have

$$\langle z^n, z^m \rangle = \int_0^1 \int_0^{2\pi} r^{n+m+1} e^{i(n-m)\theta} \, d\theta \, dr. \tag{8}$$

By (8), the inner product vanishes unless $m = n$. To check completeness, we observe that a complex analytic function in the unit disk has a power series based at 0 that converges in the open unit disk. If f is orthogonal to each monomial, then each Taylor coefficient of f vanishes and f is identically 0. $\qquad\square$

Lemma 4.2 illustrates a beautiful aspect of Hilbert spaces of complex analytic functions. Let f be complex analytic in the unit disk, with power series $\sum a_n z^n$. By basic analysis, the partial sums S_N of this series converge uniformly to f on compact subsets of the unit disk. By Lemma 4.2, the partial sum S_N can also be interpreted as the orthogonal projection of f onto the subspace of polynomials of degree at most N. Hence the partial sums also converge to f in the Hilbert space sense.

In Proposition 4.2 we relate $||f||^2_{L^2}$ to the l^2 norm of the Taylor coefficients of f. By (9) below we can identify elements of $\mathcal{A}^2$ with sequences $\{b_n\}$ such that $\sum \frac{|b_n|^2}{n+1}$ converges.

Consider the effect on the area of the image if we multiply f by z. Since $|z| < 1$, the inequality $|zf(z)| \leq |f(z)|$ is immediate. But the area of the image under zf exceeds the area of the image under f, unless f is identically 0. In fact we can explain and determine precisely how the area grows.

PROPOSITION 4.2. *Let $f(z) = \sum_{n=0}^\infty b_n z^n$ be a complex analytic function on the unit disk B_1. We assume that both f and f' are in $L^2(B_1)$. Then*

$$||f||^2_{L^2} = \pi \sum_{n=0}^\infty \frac{|b_n|^2}{n+1} \tag{9}$$

$$||f'||^2_{L^2} = \pi \sum_{n=0}^\infty (n+1)|b_{n+1}|^2 \tag{10}$$

$$||(zf)'||^2_{L^2} = ||f'||^2_{L^2} + \pi \sum_{n=0}^\infty |b_n|^2. \tag{11.1}$$

Thus $A_{zf} \geq A_f$ and equality occurs only when f vanishes identically.

PROOF. The proof of (9) is an easy calculation in polar coordinates, using the orthogonality of $e^{in\theta}$. Namely, we have

$$\|f\|_{L^2}^2 = \int_{\mathbb{B}_1} |\sum b_n z^n|^2 dx dy = \int_0^1 \int_0^{2\pi} \sum b_n \bar{b}_m r^{m+n} e^{i\theta(m-n)} r dr d\theta.$$

The only terms that matter are those for which $m = n$ and we see that

$$\|f\|_{L^2}^2 = 2\pi \sum |b_n|^2 \int_0^1 r^{2n+1} dr = \pi \sum_{n=0}^{\infty} \frac{|b_n|^2}{n+1}.$$

Formula (10) follows immediately from (9). To prove (11.1) observe that $(zf)'(z) = \sum_{n=0}^{\infty}(n+1)b_n z^n$. By (10) we have

$$\|(zf)'\|_{L^2}^2 = \pi \sum_{n=0}^{\infty}(n+1)|b_n|^2 = \pi \sum_{n=0}^{\infty} n|b_n|^2 + \pi \sum_{n=0}^{\infty} |b_n|^2$$

$$= \|f'\|_{L^2}^2 + \pi \sum_{n=0}^{\infty} |b_n|^2.$$

$\square$

We express (11.1) in operator-theoretic language. Let $D = \frac{d}{dz}$ with domain $\{f \in \mathcal{A}^2 : f' \in \mathcal{A}^2\}$. Then D is an unbounded linear operator. Let M denote the bounded operator of multiplication by z. When f extends continuously to the circle, we write Sf for its restriction to the circle, that is, its boundary values. Thus $\|Sf\|^2 = \frac{1}{2\pi} \int_0^{2\pi} |f|^2$. The excess area has a simple geometric interpretation:

COROLLARY 4.2. *Suppose Mf is in the domain of D. Then Sf is square-integrable on the circle and*

$$\|DMf\|_{L^2}^2 - \|Df\|_{L^2}^2 = \frac{1}{2} \int_0^{2\pi} |f(e^{i\theta})|^2 d\theta = \pi \|Sf\|^2. \tag{11.2}$$

PROOF. The result is immediate from (11.1). $\square$

Corollary 4.2 suggests an alternate way to view (11.1) and (11.2). We can use Green's theorem to relate the integral over the unit disk to the integral over the circle. The computation uses the notation of differential forms. We discuss differential forms in detail in Sections 5 and 6. For now we need to know less. In particular $dz = dx + idy$ and $d\bar{z} = dx - idy$. We can differentiate in these directions. See Section 5.1 for detailed discussion. For any differentiable function f, we write ∂f for $\frac{\partial f}{\partial z} dz$ and $\bar{\partial} f$ for $\frac{\partial f}{\partial \bar{z}} d\bar{z}$. If f is complex analytic, then $\bar{\partial} f = 0$ (the Cauchy-Riemann equations), and we have

$$df = (\partial + \bar{\partial})f = \partial f = f'(z)dz.$$

The area form in the plane becomes

$$dx \wedge dy = \frac{-1}{2i} dz \wedge d\bar{z} = \frac{i}{2} dz \wedge d\bar{z}.$$

Finally, we use Green's theorem, expressed in complex notation, in formula (12) of the geometric proof below. We generalize this proof in Section 9.

EXERCISE 4.8. Express Green's theorem in complex notation: express the line integral of $Adz + Bd\bar{z}$ around γ as an area integral over the region bounded by γ.

EXERCISE 4.9. Use Exercise 4.8 to show that $\int_\gamma f(z)dz = 0$ when f is complex analytic and γ is a closed curve as in Green's theorem. (This result is an easy form of the Cauchy integral theorem.)

Here is a beautiful geometric proof of (11.2), assuming f' extends continuously to the circle:

PROOF. For any complex analytic f we have

$$A_f = ||f'||_{L^2}^2 = \frac{i}{2}\int_{B_1} \partial f \wedge \overline{\partial f} = \frac{i}{2}\int_{B_1} d(f\overline{\partial f}).$$

We apply this formula also to $(zf)'$. The difference in areas satisfies

$$A_{zf} - A_f = ||(zf)'||_{L^2}^2 - ||f'||_{L^2}^2 = \frac{i}{2}\int_{B_1} d\left(zf\overline{\partial(zf)} - f\overline{\partial f}\right).$$

Assuming f' extends continuously to the circle, we may use Green's theorem to rewrite this integral as an integral over the circle:

$$A_{zf} - A_f = \frac{i}{2}\int_{S^1} zf\overline{\partial(zf)} - (f\overline{\partial f}). \tag{12}$$

By the product rule, $\partial(zf) = fdz + z\partial f$. We plug this formula into (12) and simplify, getting

$$A_{zf} - A_f = \frac{i}{2}\int_{S^1} (|z|^2 - 1)f\overline{\partial f} + \frac{i}{2}\int_{S^1} z|f(z)|^2 d\overline{z}.$$

The first integral vanishes because $|z|^2 = 1$ on the circle. We rewrite the second integral by putting $z = e^{i\theta}$ to obtain

$$\frac{i}{2}\int_{S^1} e^{i\theta}|f(e^{i\theta})|^2(-i)e^{-i\theta}d\theta = \frac{1}{2}\int_{S^1} |f(e^{i\theta})|^2 d\theta = \pi\frac{1}{2\pi}\int_{S^1} |f(e^{i\theta})|^2 d\theta = \pi||Sf||^2.$$

$\square$

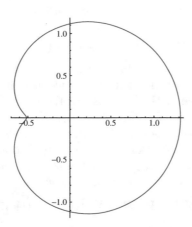

FIGURE 3. Injective image of unit disk

In the next several exercises, the operators D and M are defined as in the paragraph preceding Corollary 4.2. In particular, the Hilbert space is $\mathcal{A}^2$, the square-integrable complex analytic functions on the unit disk.

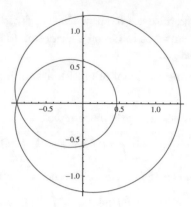

FIGURE 4. Overlapping image of unit disk

EXERCISE 4.10. Show that Corollary 4.2 can be stated as $M^*D^*DM - D^*D = \pi S^*S$.

EXERCISE 4.11. What are the eigenfunctions and eigenvalues of DM and of MD? Show that the commutator $[D, M] = DM - MD$ is the identity. This example arises in quantum mechanics. See Section 12 for more information.

EXERCISE 4.12. Find a closed formula for $\sum_{j=0}^{\infty} \frac{|z|^{2j}}{c_j}$, where $c_j = ||z^j||^2$ is the squared norm in $\mathcal{A}^2$. The answer is the Bergman kernel function of the unit disk.

EXERCISE 4.13. For $0 \leq a \leq 1$ and for $|z| < 1$, put $f_a(z) = \sqrt{1 - a^2}z + az^2$. Find $||f_a'||_{L^2}^2$ in terms of a. For several values of a, graph the image of the unit disk under f. For what values of a is f injective? See Figures 3 and 4.

EXERCISE 4.14. Put $f(z) = z + z^2 + z^3$. Describe or graph the image of the set $|z| = r$ under f for several values of r. Suggestion: Use polar coordinates.

3. Unitary groups

We now begin studying geometric problems in several complex variables. Recall that $\langle z, w \rangle$ denotes the Hermitian inner product of points in complex Euclidean space $\mathbf{C}^n$. The unitary group $\mathbf{U}(n)$ consists of the linear transformations T which preserve the inner product; $\langle Tz, Tw \rangle = \langle z, w \rangle$. Setting $z = w$ shows that such transformations also preserve norms. The converse is also true: if $||Tz||^2 = ||z||^2$ for all z, then $\langle Tz, Tw \rangle = \langle z, w \rangle$ for all z and w, by Proposition 2.6.

The group law in $\mathbf{U}(n)$ is composition. Let U, V be unitary transformations on $\mathbf{C}^N$. Then the composition UV is also unitary, because

$$(UV)^* = V^*U^* = V^{-1}U^{-1} = (UV)^{-1}.$$

It follows that the collection $\mathbf{U}(N)$ of unitary transformations on $\mathbf{C}^N$ is a subgroup of the group of invertible linear transformations.

We will often deal with complex Euclidean spaces of different dimensions. It is convenient to omit the dimension in the notation for the inner products and norms. When doing so, we must be careful. Suppose $L : \mathbf{C}^n \to \mathbf{C}^{n+1}$ is given by $L(z) = (z, 0)$. Then L is linear and $||L(z)|| = ||z||$, but L is not unitary. Distance preserving maps are called *isometries*. In this setting, when $N > n$, we often identify $\mathbf{C}^n$ with the subspace $\mathbf{C}^n \oplus 0 \subseteq \mathbf{C}^N$.

Our first result (which holds much more generally than we state here) provides a polarization technique and gets used several times in the sequel. We use it several times in the special case when f and g are polynomial mappings.

THEOREM 4.3. *Let B be a ball centered at 0 in $\mathbf{C}^n$. Suppose $f : B \to \mathbf{C}^{N_1}$ and $g : B \to \mathbf{C}^{N_2}$ are complex analytic mappings and $||f(z)||^2 = ||g(z)||^2$ for all $z \in B$. Assume that the image of g lies in no lower dimensional subspace, and that $N_1 \geq N_2$. Then there is an isometry $U : \mathbf{C}^{N_2} \to \mathbf{C}^{N_1}$ such that $f(z) = Ug(z)$ for all z. When f and g are as above and also $N_2 = N_1$, then U is unitary.*

PROOF. We expand f and g as convergent power series about 0, writing $f(z) = \sum_\alpha A_\alpha z^\alpha$ and $g(z) = \sum_\alpha B_\alpha z^\alpha$; the coefficients A_α lie in $\mathbf{C}^{N_1}$ and the B_α lie in $\mathbf{C}^{N_2}$. Equating the Taylor coefficients in $||f(z)||^2 = ||g(z)||^2$ leads, for each pair α and β of multi-indices, to

$$\langle A_\alpha, A_\beta \rangle = \langle B_\alpha, B_\beta \rangle. \tag{13}$$

It follows from (13) that $A_{\alpha_1}, \ldots, A_{\alpha_K}$ is a linearly independent set if and only if $B_{\alpha_1}, \ldots, B_{\alpha_K}$ is a linearly independent set. We then define U by $U(B_\alpha) = A_\alpha$ for a maximal linearly independent set of the B_α. If B_μ is a linear combination of some B_α, then we define $U(B_\mu)$ as the same linear combination of the A_α. The relations (13) guarantee that U is well-defined. Furthermore, these relationships imply that U preserves inner products. Hence U is an isometry on the span of the B_α. When $N_1 = N_2$, an isometry U must be unitary. $\square$

EXAMPLE 4.1. The parallelogram law provides an example of Theorem 4.3. Suppose $g(z_1, z_2) = (\sqrt{2}z_1, \sqrt{2}z_2)$ and $f(z_1, z_2) = (z_1 + z_2, z_1 - z_2)$. Then

$$||g(z)||^2 = 2|z_1|^2 + 2|z_2|^2 = |z_1 + z_2|^2 + |z_1 - z_2|^2 = ||f(z)||^2.$$

In this case $f = Ug$, where U is given by

$$U = \begin{pmatrix} \frac{1}{\sqrt{2}} & \frac{1}{\sqrt{2}} \\ \frac{1}{\sqrt{2}} & \frac{-1}{\sqrt{2}} \end{pmatrix}.$$

Our next example illustrates the situation when $N_1 > N_2$ in Theorem 4.3.

EXAMPLE 4.2. Put $f(z) = (z_1^2, z_1 z_2, z_1 z_2, z_2^2)$ and $g(z) = (z_1^2, \sqrt{2}z_1 z_2, z_2^2)$. Here $f : \mathbf{C}^2 \to \mathbf{C}^4$ and $g : \mathbf{C}^2 \to \mathbf{C}^3$. Also,

$$||f(z)||^2 = |z_1|^4 + 2|z_1|^2 |z_2|^2 + |z_2|^4 = (|z_1|^2 + |z_2|^2)^2 = ||g(z)||^2.$$

The map $U : \mathbf{C}^3 \to \mathbf{C}^4$ for which $f = Ug$ is given by the matrix (with respect to the usual bases)

$$U = \begin{pmatrix} 1 & 0 & 0 \\ 0 & \frac{1}{\sqrt{2}} & 0 \\ 0 & \frac{1}{\sqrt{2}} & 0 \\ 0 & 0 & 1 \end{pmatrix}.$$

If $\zeta = (\zeta_1, \zeta_2, \zeta_3)$, then $||U\zeta||^2 = |\zeta_1|^2 + |\zeta_2|^2 + |\zeta_3|^2 = ||\zeta||^2$. Thus U is an isometry, but U is not unitary.

Observe that the maps f and g from Example 4.2 each send the unit sphere in the domain to the unit sphere in the target. We will now consider such mappings in detail.

We begin with some examples of symmetries of the unit sphere. If $e^{i\theta}$ lies on the unit circle S^1, and z lies on the unit sphere S^{2n-1}, the scalar multiple $e^{i\theta}z$ lies on S^{2n-1} as well. Thus S^1 acts on S^{2n-1}. We can replace S^1 with the n-torus $S^1 \times \dots \times S^1$. In this case we map $z = (z_1, z_2, \dots, z_n)$ to $(e^{i\theta_1}z_1, e^{i\theta_2}z_2, \dots, e^{i\theta_n}z_n)$. Furthermore, for $z \in S^{2n-1}$ and $U \in \mathbf{U}(n)$, we have of course that $Uz \in S^{2n-1}$.

The next example of a symmetry is a bit more complicated. Choose a point a in the open unit ball $\mathbb{B}_n$. First define a linear mapping $L : \mathbf{C}^n \to \mathbf{C}^n$ by

$$L(z) = sz + \frac{1}{s+1}\langle z, a\rangle a,$$

where $s = \sqrt{1 - ||a||^2}$. Then define ϕ_a by

$$\phi_a(z) = \frac{a - L_a(z)}{1 - \langle z, a\rangle}.$$

The mapping ϕ_a is a complex analytic automorphism of the unit ball, and it maps the unit sphere to itself. See Exercise 4.15, Exercise 4.16, and the discussion in Section 4 for more information.

EXERCISE 4.15. Verify the following properties of the mapping ϕ_a.

- $\phi_a(0) = a$.
- $\phi_a(a) = 0$.
- $\phi_a : S^{2n-1} \to S^{2n-1}$.
- $\phi_a \circ \phi_a$ is the identity.

EXERCISE 4.16. Carefully compute $\phi_b \circ \phi_a$. The result is **not** of the form ϕ_c for any c with $||c|| < 1$. Show, however, that the result can be written $U\phi_c$ for some unitary U. Suggestion: first do the computation when $n = 1$.

REMARK 4.3. In complex analysis or harmonic analysis, it is natural to consider the group of all complex analytic automorphisms preserving the sphere. Each element of this group can be written $U \circ \phi_a$ for some unitary U and some ϕ_a. We will consider the full group in the next chapter; for now we focus on the unitary group $\mathbf{U}(n)$ and its finite subgroups. Various interesting combinatorial and number-theoretic issues arise in this setting.

We start in one dimension with an elementary identity (Lemma 4.3) involving roots of unity. The proof given reveals the power of geometric reasoning; one can also prove this identity by factoring $1 - t^m$ over the complex numbers.

DEFINITION 4.2. A complex number ω is called a *primitive m-th root of unity* if $\omega^m = 1$ and m is the smallest such positive integer.

The imaginary unit i is a primitive fourth root of unity. Given a primitive m-th root of unity ω, the powers ω^j for $j = 0, 1, \dots, m-1$ are equally spaced on the unit circle S^1. These m points define a cyclic subgroup of S^1 of order m. Note that the inverse of ω is ω^{m-1}, which also equals $\overline{\omega}$. Note also that $S^1 = \mathbf{U}(1)$.

LEMMA 4.3. *Let ω be a primitive m-th root of unity. Then*

$$1 - \prod_{j=0}^{m-1}(1 - \omega^j t) = t^m. \tag{14}$$

PROOF. The expression on the left-hand side is a polynomial in t of degree m. It is invariant under the map $t \mapsto \omega t$. The only invariant monomials of degree at most m are constants and constants times t^m. Hence this expression must be of the form $\alpha + \beta t^m$. Setting $t = 0$ shows that $\alpha = 0$ and setting $t = 1$ shows that $\beta = 1$.   □

This proof relies on the cyclic subgroup Γ_m of the unit circle, or of $\mathbf{U}(1)$, generated by ω. We will generalize this lemma and related ideas to higher dimensions, where things become more interesting.

We extend the notion of Hermitian symmetry (Definition 1.2) to higher dimensions in the natural way. A polynomial $R(z, \bar{\zeta})$ on $\mathbf{C}^n \times \mathbf{C}^n$ is *Hermitian symmetric* if $R(z, \bar{\zeta}) = \overline{R(\zeta, \bar{z})}$. The higher dimensional version of Lemma 1.3 holds; it is useful in the solution of Exercise 4.19.

Let Γ be a finite subgroup of $\mathbf{U}(n)$. The analogue of the left-hand side of (14) is the following Hermitian polynomial:

$$\Phi_\Gamma(z, \bar{\zeta}) = 1 - \prod_{\gamma \in \Gamma} (1 - \langle \gamma z, \zeta \rangle). \tag{15}$$

One can show (we do not use the result, and hence we omit the proof) that Φ_Γ is uniquely determined by the following properties:

(1) Φ_Γ is Hermitian symmetric.
(2) $\Phi_\Gamma(0, 0) = 0$.
(3) Φ_Γ is Γ-invariant.
(4) $\Phi_\Gamma(z, \bar{z})$ is of degree in z at most the order of Γ.
(5) $\Phi_\Gamma(z, \bar{z}) = 1$ for z on the unit sphere.

In the special case when Γ is the group generated by a primitive m-th root of unity times the identity operator, (14) generalizes to the identity (16):

$$1 - \prod_{j=0}^{m-1} \left(1 - \omega^j \sum_{k=1}^{n} t_k \right) = \left(\sum_{k=1}^{n} t_k \right)^m = \sum_{|\alpha|=m} \binom{m}{\alpha} t^\alpha. \tag{16}$$

In this case the multinomial coefficients $\binom{m}{\alpha}$ make an appearance:

$$\binom{m}{\alpha} = \frac{m!}{\alpha_1! \ldots \alpha_n!}.$$

See Sections 4 and 8 for more information about multi-index notation and the multinomial theorem, which is the far right equality in (16).

Interesting generalizations of (16) result from more complicated *representations* of cyclic groups. The product in (17) gives one collection of non-trivial examples:

$$1 - \prod_{j=0}^{m-1} \left(1 - \sum_{k=1}^{n} \omega^{kj} t_k \right). \tag{17}$$

The coefficients of the expansion are integers with many interesting properties.

EXERCISE 4.17. Prove Lemma 4.3 by factoring $1 - t^m$.

EXERCISE 4.18. Prove that $\Phi_\Gamma(z, \bar{w})$ is Hermitian symmetric.

EXERCISE 4.19. Let $R(z, \bar{z}) = \sum_{\alpha,\beta} c_{\alpha,\beta} z^\alpha \bar{z}^\beta$ be a Hermitian symmetric polynomial. Prove that there are linearly independent polynomials $A_j(z)$ and $B_k(z)$ such that

$$R(z, \bar{z}) = \sum_j |A_j(z)|^2 - \sum_k |B_k(z)|^2 = ||A(z)||^2 - ||B(z)||^2.$$

EXERCISE 4.20. Write $\Phi_\Gamma = ||A||^2 - ||B||^2$ as in the previous exercise. Show that we may choose A and B to be Γ-invariant.

In the rest of this section we consider several cyclic subgroups of $\mathbf{U}(2)$. Write (z, w) for a point in $\mathbf{C}^2$. Let η be a primitive p-th root of unity. We next study the mapping Φ_Γ when $\Gamma = \Gamma(p, q)$ is the cyclic group of $\mathbf{U}(2)$ of order p generated by the matrix

$$\begin{pmatrix} \eta & 0 \\ 0 & \eta^q \end{pmatrix}.$$

REMARK 4.4. The quotient space $S^3/\Gamma(p, q)$ is called a lens space. These spaces are important in topology.

The definition of $\Phi_{\Gamma(p,q)}$ yields

$$\Phi_{\Gamma(p,q)} = 1 - \prod_{j=0}^{p-1} (1 - \eta^j |z|^2 - \eta^{qj} |w|^2).$$

This expression depends only upon the expressions $|z|^2$ and $|w|^2$; we simplify notation by defining the polynomial $f_{p,q}(x, y)$ by

$$f_{p,q}(x, y) = 1 - \prod_{j=0}^{p-1} (1 - \eta^j x - \eta^{qj} y). \tag{18}$$

By taking $j = 0$ in the product, it follows that $f_{p,q}(x, y) = 1$ on the line $x + y = 1$.

LEMMA 4.4. $f_{p,1}(x, y) = (x + y)^p$.

PROOF. The result follows by replacing t by $x + y$ in Lemma 4.3. $\qquad \square$

The (binomial) coefficients of $f_{p,1}$ are integers which satisfy an astonishing number of identities and properties. More is true. For each q, the coefficients of $f_{p,q}$ are also integers, and they satisfy many interesting combinatorial and number-theoretic properties as well. We mention one of the properties now. Most people know the so-called *freshman's dream*, that $(x + y)^p \equiv x^p + y^p$ modulo p if and only if p is prime. The same result holds for each $f_{p,q}$, although we omit the proof here.

The polynomials $f_{p,2}$ are more complicated than $f_{p,1} = (x + y)^p$. When p is odd, all the coefficients of $f_{p,2}$ are non-negative. Here are the first few $f_{p,2}$:

$$f_{1,2}(x, y) = x + y$$
$$f_{2,2}(x, y) = x^2 + 2y - y^2$$
$$f_{3,2}(x, y) = x^3 + 3xy + y^3$$
$$f_{4,2}(x, y) = x^4 + 4x^2 y + 2y^2 - y^4$$
$$f_{5,2}(x, y) = x^5 + 5x^3 y + 5xy^2 + y^5. \tag{19}$$

We can find all these polynomials by solving a single difference equation. We offer two proofs of the following explicit formula for $f_{p,2}$. The key idea in the first proof is to interchange the order in a double product. See [D5] and its references for general results about group-invariant polynomials, proved by similar methods.

PROPOSITION 4.3. *For all non-negative integers p we have*

$$f_{p,2}(x,y) = (\frac{x + \sqrt{x^2 + 4y}}{2})^p + (\frac{x - \sqrt{x^2 + 4y}}{2})^p - (-y)^p. \qquad (20)$$

PROOF. Set $q = 2$ in (18). Each factor in the product is a quadratic in η^j, which we also factor. We obtain:

$$1 - f(x,y) = \prod_{j=0}^{p-1}(1 - \eta^j x - \eta^{2j} y) = \prod_{j=0}^{p-1}(1 - c_1(x,y)\eta^j)(1 - c_2(x,y)\eta^j)$$

$$= \prod_{j=0}^{p-1}(1 - c_1(x,y)\eta^j) \prod_{j=0}^{p-1}(1 - c_2(x,y)\eta^j).$$

Here c_1 and c_2 are the reciprocals of the roots of the quadratic $1 - x\eta - y\eta^2$. Each of the two products is familiar from Lemma 4.3. Using that result we obtain

$$1 - f(x,y) = (1 - c_1(x,y)^p)(1 - c_2(x,y)^p).$$

It follows that f has the following expression in terms of the c_j:

$$f(x,y) = c_1(x,y)^p + c_2(x,y)^p - (c_1(x,y)c_2(x,y))^p.$$

The product $c_1(x,y)c_2(x,y)$ equals $-y$. The sum $c_1(x,y)+c_2(x,y)$ equals x. Solving this system for c_1 and c_2 using the quadratic formula determines the expressions arising in (20). $\qquad \square$

We sketch a second proof based on recurrence relations.

PROOF. (Sketch). It follows by setting $x = 0$ in formula (18) that the term $-(-y)^p$ appears in $f_{p,2}$. Let $g_p(x,y)$ denote the other terms. The recurrence relation $g_{p+2}(x,y) = xg_{p+1}(x,y) + yg_p(x,y)$ also follows from (18). To solve this recurrence, we use the standard method. The characteristic equation is $\lambda^2 - x\lambda - y = 0$. Its roots are $\frac{x \pm \sqrt{x^2+4y}}{2}$. Using the initial conditions that $g_1(x,y) = x$ and $g_2(x,y) = x^2 + 2y$ we determine $g_p(x,y)$. Adding in the term $-(-y)^p$ yields (20). $\square$

These polynomials are related to some classical mathematics.

DEFINITION 4.3. The n-th Chebyshev polynomial T_n is defined by

$$T_n(x) = \cos(n \cos^{-1}(x)).$$

Although it is not instantly obvious, the n-th Chebyshev polynomial is a polynomial of degree n. Hence these polynomials are linearly independent.

EXAMPLE 4.3. The first few Chebyshev polynomials:
- $T_0(x) = 1$
- $T_1(x) = x$
- $T_2(x) = 2x^2 - 1$
- $T_3(x) = 4x^3 - 3x$
- $T_4(x) = 8x^4 - 8x^2 + 1$
- $T_5(x) = 16x^5 - 20x^3 + 5x.$

EXERCISE 4.21. Verify that $T_n(x)$ is a polynomial. (See Exercise 4.23 for one approach.) Verify the formulas for $T_j(x)$ for $j = 1, 2, 3, 4, 5$.

REMARK 4.5. The polynomials $T_n(x)$ are eigenfunctions of a Sturm-Liouville problem. The differential equation, (SL) from Chapter 2, is $(1-x^2)y'' - xy' + \lambda y = 0$. The T_n are orthogonal on the interval $[-1, 1]$ with respect to the weight function $w(x) = \frac{1}{\sqrt{1-x^2}}$. By Theorem 2.13, they form a complete orthogonal system for $L^2([-1,1], w)$.

EXERCISE 4.22. Verify that T_n is an eigenfunction as described in the remark; what is the corresponding eigenvalue λ?

PROPOSITION 4.4. The $f_{p,2}$ have the following relationship to the Chebyshev polynomials $T_p(x)$:

$$f_{p,2}\left(x, \frac{-1}{4}\right) + \left(\frac{1}{4}\right)^p = 2^{1-p}\left(\cos(p\,\cos^{-1}(x))\right) = 2^{1-p}T_p(x).$$

PROOF. See Exercise 4.23. □

REMARK 4.6. Evaluating the $f_{p,2}$ at other points also leads to interesting things. For example, let ϕ denote the golden ratio. Then

$$f_{p,2}(1, 1) = \left(\frac{1+\sqrt{5}}{2}\right)^p + \left(\frac{1-\sqrt{5}}{2}\right)^p + (-1)^{p+1} = \phi^p + (1-\phi)^p + (-1)^{p+1}.$$

The first two terms give the p-th Lucas number, and hence $f_{p,2}(1,1)$ differs from the p-th Lucas number by ± 1. The p-th Fibonacci number F_p has a similar formula:

$$F_p = \frac{1}{\sqrt{5}}\left((\frac{1+\sqrt{5}}{2})^p - (\frac{1-\sqrt{5}}{2})^p\right) = \frac{1}{\sqrt{5}}\left((\phi)^p - (1-\phi)^p\right).$$

It is remarkable that our considerations of group-invariant mappings connect so closely with classical mathematics. The polynomials $f_{p,2}$ arise for additional reasons in several complex variables. When p is odd, all the coefficients of $f_{p,2}$ are non-negative. Put $x = |z|^2$ and $y = |w|^2$ and write $p = 2r + 1$. Then

$$f_{2r+1,2}(|z|^2, |w|^2) = \sum_b c_b |z|^{2(2r+1-2b)}|w|^{2b} = ||g(z,w)||^2.$$

Since $f_{2r+1,2}(x, y) = 1$ on $x + y = 1$ we see that $||g(z,w)||^2 = 1$ on the unit sphere. Hence $g(z, w)$ maps the unit sphere S^3 to the unit sphere S^{2N-1}, where $N = r + 2$. Thus g provides a far from obvious example of a group-invariant mapping between spheres.

The functions $f_{p,2}$ satisfy an extremal property. If a polynomial f of degree d in x, y has N terms, all non-negative coefficients, and $f(x, y) = 1$ on $x + y = 1$, then the inequality $d \leq 2N - 3$ holds and is sharp. We omit the proof of this difficult result. Equality holds for the $f_{2r+1,2}$.

EXERCISE 4.23. Prove Proposition 4.4. Suggestion. First find a formula for $\cos^{-1}(s)$ using $\cos(t) = \frac{e^{it}+e^{-it}}{2} = s$ and solving a quadratic equation for e^{it}.

EXERCISE 4.24. Show that $T_{nm}(x) = T_n(T_m(x))$.

EXERCISE 4.25. Find a formula for the generating function $\sum_{n=0}^{\infty} T_n(x)t^n$. Do the same for $\sum_{n=0}^{\infty} f_{n,2}(x,y)t^n$.

The next exercise is intentionally a bit vague. See [D3] and the references there for considerably more information.

EXERCISE 4.26. Use Mathematica or something similar to find $f_{p,3}$ and $f_{p,4}$ for $1 \leq p \leq 11$. See what you can discover about these polynomials.

4. Proper mappings

Consider the group-invariant polynomial (15) above when $\zeta = z$. The factor $1 - \langle \gamma z, z \rangle$ vanishes on the sphere when γ is the identity of the group. Hence $\Phi_\Gamma(z, \overline{z}) = 1$ when z is on the sphere. By Exercises 4.19 and 4.20, we may write

$$\Phi_\Gamma(z, \overline{z}) = \sum_j |A_j(z)|^2 - \sum_k |B_k(z)|^2 = ||A(z)||^2 - ||B(z)||^2$$

where the polynomials A_j and B_k are invariant. If $B = 0$, (thus Φ_Γ is a squared norm), then Φ_Γ will be an invariant polynomial mapping between spheres. If $B \neq 0$, then the target is a *hyperquadric*.

The group-invariant situation, where the target is a sphere, is completely understood and beautiful. It is too restrictive for our current aims. In this section we therefore consider polynomial mappings between spheres, without the assumption of group-invariance.

In one dimension, the functions $z \mapsto z^m$ have played an important part in our story. On the circle, of course, $z^m = e^{im\theta}$. The function $z \mapsto z^m$ is complex analytic and maps the unit circle S^1 to itself. One of many generalizations of these functions to higher dimensions results from considering complex analytic functions sending the unit sphere S^{2n-1} into some unit sphere, perhaps in a different dimension. We discuss these ideas here and relate them to the combinatorial considerations from the previous section.

DEFINITION 4.4. Let Ω and Ω' be open, connected subsets of complex Euclidean spaces. Suppose $f : \Omega \to \Omega'$ is continuous. Then f is called *proper* if, whenever $K \subseteq \Omega'$ is compact, then $f^{-1}(K)$ is compact in Ω.

LEMMA 4.5. *A continuous map $f : \Omega \to \Omega'$ between bounded domains is proper if and only if the following holds: whenever $\{z_\nu\}$ is a sequence tending to the boundary $b\Omega$, then $\{f(z_\nu)\}$ tends to $b\Omega'$.*

PROOF. We prove both statements by proving their contrapositives. First let $\{z_\nu\}$ tend to $b\Omega$. If $\{f(z_\nu)\}$ does not tend to $b\Omega'$, then it has a subsequence which stays in a compact subset K of Ω'. But then $f^{-1}(K)$ is not compact, and f is not proper. Thus properness implies the sequence condition. Now suppose f is not proper. Find a compact set K such that $f^{-1}(K)$ is not compact in Ω. Then there is a sequence $\{z_\nu\}$ in $f^{-1}(K)$ tending to $b\Omega$, but the image sequence stays within a compact subset K. □

Lemma 4.5 states informally that f is proper if, whenever z is close to $b\Omega$, then $f(z)$ is close to $b\Omega'$. Hence it has an $\epsilon - \delta$ version which we state and use only when Ω and Ω' are open unit balls.

COROLLARY 4.3. *A continuous map $f : \mathbb{B}_n \to \mathbb{B}_N$ is proper if and only if, for all $\epsilon > 0$ there is a $\delta > 0$ such that $1 - \delta < ||z|| < 1$ implies $1 - \epsilon < ||f(z)|| < 1$.*

Our main interest is complex analytic mappings, especially such polynomial mappings, sending the unit sphere in $\mathbf{C}^n$ to the unit sphere in some $\mathbf{C}^N$. Consider mappings that are complex analytic on the open ball and continuous on the closed ball. The maximum principle implies that if such a mapping sends the unit sphere in the domain to some unit sphere, then it must actually be a proper mapping from ball to ball. On the other hand, a (complex analytic) polynomial mapping between balls is also defined on the boundary sphere, and Lemma 4.5 implies that such mappings send the boundary to the boundary. It would thus be possible never to mention the term *proper map* and we could still do everything we are going to do. We continue to work with proper mappings because of the intuition they provide.

REMARK 4.7. Proper complex analytic mappings must be finite-to-one, although not all points in the image must have the same number of inverse images. By definition of proper, the inverse image of a point must be a compact set. Because of complex analyticity, the inverse image of a point must also be a complex variety. Together these facts show that no point in the target can have more than a finite number of inverse images.

EXERCISE 4.27. Which of the following maps are proper from $\mathbf{R}^2 \to \mathbf{R}$?
 (1) $f(x,y) = x^2 + y^2$
 (2) $g(x,y) = x^2 - y^2$
 (3) $h(x,y) = x$.

EXERCISE 4.28. Under what circumstances is a linear map $L : \mathbf{C}^n \to \mathbf{C}^N$ proper?

Our primary concern will be complex analytic proper mappings between balls. We start with the unit disk B_1 contained in $\mathbf{C}$. Let us recall a simple version of the *maximum principle*. Suppose f is complex analytic in the open unit disk B_1, and $|f(z)| \leq M$ on the boundary of a closed subset K. Then the same estimate holds in the interior of K.

PROPOSITION 4.5. *Suppose $f : B_1 \to B_1$ is complex analytic and proper. Then f is a finite Blaschke product: there are points $a_1, ..., a_d$ in the unit disk, possibly repeated, and a point $e^{i\theta}$ on the circle, such that*

$$f(z) = e^{i\theta} \prod_{j=1}^{d} \frac{a_j - z}{1 - \overline{a}_j z}.$$

If also either $f^{-1}(0) = 0$ or f is a polynomial, then $f(z) = e^{i\theta} z^m$ for some positive integer m.

PROOF. Because f is proper, the set $f^{-1}(0)$ is compact. We first show that it is not empty. If it were empty, then both f and $\frac{1}{f}$ would be complex analytic on the unit disk, and the values of $\frac{1}{|f(z)|}$ would tend to 1 as z tends to the circle. The maximum principle would then force $|\frac{1}{f(z)}| \leq 1$ on the disk, which contradicts $|f(z)| < 1$ there.

Thus the compact set $f^{-1}(0)$ is not empty. Because f is complex analytic, this set must be discrete. Therefore it is finite, say $a_1, ..., a_d$ (with multiplicity allowed). Let $B(z)$ denote the product $\prod \frac{a_j - z}{1 - \overline{a}_j z}$. We show that $z \mapsto \frac{f(z)}{B(z)}$ is a constant map of modulus one. Then $f = e^{i\theta} B$.

By Corollary 4.3, applied to both f and B, for each $\epsilon > 0$ we can find a $\delta > 0$ such that $1 - \epsilon < |f(z)| \leq 1$ and $1 - \epsilon < |B(z)| \leq 1$ for $|z| > 1 - \delta$. It follows by the maximum principle that these estimates hold for all z with $|z| \leq 1 - \delta$ as well. The function $g = \frac{f}{B}$ is complex analytic in the disk, as the zeros of B and of f correspond and thus cancel in g. By the maximum principle applied to g, we have for all z that $1 - \epsilon < |g(z)| < \frac{1}{1-\epsilon}$. Since ϵ is arbitrary, we may let ϵ tend to 0 and conclude that $|g(z)| = 1$. It follows (by either Theorem 4.3 or the maximum principle) that g is a constant $e^{i\theta}$ of modulus one. Thus $f(z) = e^{i\theta} B(z)$. $\square$

EXERCISE 4.29. Suppose $f : B_1 \to B_1$ is complex analytic and proper. Find another proof that there is a z with $f(z) = 0$. One possible proof composes f with an automorphism of the disk, preserving properness while creating a zero.

Consider next the proper complex analytic self-mappings of the unit ball $\mathbb{B}_n$ in $\mathbf{C}^n$ for $n \geq 2$. We do not prove the following well-known result in several complex variables: the only proper complex analytic maps from the unit ball $\mathbb{B}_n$ to itself (when $n \geq 2$) are *automorphisms*. These mappings are analogues of the individual factors in Proposition 4.5. They have the form

$$f(z) = U \frac{z - L_a(z)}{1 - \langle z, a \rangle}.$$

Here U is unitary, and L_a is a linear transformation depending on a, for a an arbitrary point in $\mathbb{B}_n$. These rational maps were mentioned in Section 3; see the discussion near Exercises 4.15 and 4.16. They will play a major part in the next chapter. The only *polynomial* proper self-mappings of a ball are the unitary mappings $f(z) = Uz$. In order to obtain analogues of $z \mapsto z^d$, we must increase the target dimension.

The analogue of $z \mapsto z^d$ in one dimension will be the *tensor product* $z \mapsto z^{\otimes d}$. We will make things concrete, but completely rigorous, by first identifying $\mathbf{C}^M \otimes \mathbf{C}^N$ with $\mathbf{C}^{NM}$. The reader may simply regard the symbol $\otimes$ as notation.

DEFINITION 4.5. Let $f = (f_1, ..., f_M)$ and $g = (g_1, ..., g_N)$ be mappings taking values in $\mathbf{C}^M$ and $\mathbf{C}^N$. Their tensor product $f \otimes g$ is the mapping taking values in $\mathbf{C}^{MN}$ defined by $(f_1 g_1, ..., f_j g_k, ..., f_M g_N)$.

In Definition 4.5 we did not precisely indicate the order in which the terms $f_j g_k$ are listed. The reason is that we do not care; nearly everything we do in this section does not distinguish between h and Lh when $||Lh|| = ||h||$. The following formula suggests why the tensor product is relevant to proper mappings between balls:

$$||f \otimes g||^2 = ||f||^2 ||g||^2. \tag{21}$$

To verify (21), simply note that

$$||f||^2 \, ||g||^2 = \sum_j |f_j|^2 \sum_k |g_k|^2 = \sum_{j,k} |f_j g_k|^2.$$

Let m be a positive integer. We write $z^{\otimes m}$ for the tensor product of the identity map with itself m times. We show momentarily that $||z^{\otimes m}||^2 = ||z||^{2m}$; in particular the polynomial map $z \mapsto z^{\otimes m}$ takes the unit sphere in its domain to the unit sphere in its target. It exhibits many of the properties satisfied by the mapping $z \mapsto z^m$ in one dimension. The main difference is that the target dimension is much larger than the domain dimension when $n \geq 2$ and $m \neq 1$.

In much of what we do, the mapping $z \mapsto f(z)$ is less important than the real-valued function $z \mapsto ||f(z)||^2$. It is therefore sometimes worthwhile to introduce the concept of *norm equivalence*. Consider two maps f, g with the same domain, but with possibly different dimensional complex Euclidean spaces as targets. We say that f and g are *norm-equivalent* if the functions $||f||^2$ and $||g||^2$ are identical.

We are particularly interested in the norm equivalence class of the mapping $z \mapsto z^{\otimes m}$. One member of this equivalence class is the monomial mapping described in (22), and henceforth we *define* $z^{\otimes m}$ by the formula in (22). The target dimension is $\binom{n+m-1}{m}$, and the components are the monomials of degree m in n variables. Thus we put

$$H_m(z) = z^{\otimes m} = (..., c_\alpha z^\alpha, ...). \tag{22}$$

In (22), z^α is multi-index notation for $\prod_{j=1}^{n}(z_j)^{\alpha_j}$; each $\alpha = (\alpha_1, \ldots, \alpha_n)$ is an n-tuple of non-negative integers which sum to m, and all such α appear. There are $\binom{n+m-1}{m}$ such multi-indices; see Exercise 4.30. For each α, c_α is the positive square root of the multinomial coefficient $\binom{m}{\alpha}$. We write $|z|^{2\alpha}$ as an abbreviation for the product

$$\prod_j |z_j|^{2\alpha_j}.$$

See Section 10 for more information about multi-index notation and for additional properties of this mapping.

By the multinomial expansion we see that

$$||z^{\otimes m}||^2 = \sum_\alpha |c_\alpha|^2 |z|^{2\alpha} = \sum_\alpha \binom{m}{\alpha} |z|^{2\alpha} = (\sum_j |z_j|^2)^m = ||z||^{2m}.$$

The crucial formula $||z^{\otimes m}||^2 = ||z||^{2m}$ explains why c_α was defined as above. Furthermore, by Theorem 4.4 below, $\binom{n+m-1}{m}$ is the smallest possible dimension k for which there is a polynomial mapping $f : \mathbf{C}^n \to \mathbf{C}^k$ such that $||f(z)||^2 = ||z||^{2m}$. In other words, if f is norm-equivalent to $z^{\otimes m}$, then the target dimension must be at least $\binom{n+m-1}{m}$.

EXAMPLE 4.4. Put $n = 2$ and $m = 3$. We identify the map $z^{\otimes m}$ with the map H_3 defined by

$$(z_1, z_2) \to H_3(z_1, z_2) = (z_1^3, \sqrt{3}z_1^2 z_2, \sqrt{3}z_1 z_2^2, z_2^3).$$

Note that $||H_3(z_1, z_2)||^2 = (|z_1|^2 + |z_2|^2)^3$.

DEFINITION 4.6. Let $p : \mathbf{C}^n \to \mathbf{C}^N$ be a polynomial mapping. Then p is called *homogeneous of degree m* if, for all $t \in \mathbf{C}$, $p(tz) = t^m p(z)$.

Homogeneity is useful for many reasons. For example, a homogeneous polynomial is determined by its values on the unit sphere. Unless the degree of homogeneity is zero, in which case p is a constant, we have $p(0) = 0$. For $z \neq 0$ we have

$$p(z) = p\left(||z|| \frac{z}{||z||}\right) = ||z||^m p\left(\frac{z}{||z||}\right).$$

This simple fact leads to the next lemma, which we use in proving Theorem 4.6.

LEMMA 4.6. *Let p_j and p_k denote homogeneous polynomial mappings, of the indicated degrees, from $\mathbf{C}^n$ to $\mathbf{C}^N$. Assume that $\langle p_j(z), p_k(z) \rangle = 0$ for all z on the unit sphere. Then this inner product vanishes for all $z \in \mathbf{C}^n$.*

PROOF. The statement is trivial if $j = k = 0$, as p_0 is a constant. Otherwise the inner product vanishes at $z = 0$. For $z \neq 0$, put $w = \frac{z}{||z||}$. Homogeneity yields

$$\langle p_j(z), p_k(z) \rangle = ||z||^{j+k} \langle p_j(w), p_k(w) \rangle,$$

which vanishes by our assumption, because w is on the sphere.   □

EXERCISE 4.30. Show that the dimension of the vector space of homogeneous (complex-valued) polynomials of degree m in n variables equals $\binom{n+m-1}{m}$.

EXERCISE 4.31. Give an example of a polynomial $r(z, \overline{z})$ that vanishes on the sphere, also vanishes at 0, but does not vanish everywhere.

Recall formula (22) defining the mapping $z^{\otimes m}$. Thus $z^{\otimes m} : \mathbf{C}^n \to \mathbf{C}^N$, where N is the binomial coefficient $N = \binom{n+m-1}{m}$, the number of linearly independent monomials of degree m in n variables. This integer is the minimum possible dimension for any map f for which $||f(z)||^2 = ||z||^{2m}$.

THEOREM 4.4. Let $h_m : \mathbf{C}^n \to \mathbf{C}^N$ be a homogeneous polynomial mapping of degree m which maps S^{2n-1} to S^{2N-1}. Then $z^{\otimes m}$ and h_m are norm-equivalent. Assume in addition that the components of h_m are linearly independent. Then $N = \binom{n+m-1}{m}$ and there is a unitary transformation U such that

$$h_m(z) = U z^{\otimes m}.$$

PROOF. By linear independence of the components of h_m, the target dimension N of h_m is at most $\binom{n+m-1}{m}$. We claim that $N = \binom{n+m-1}{m}$. We are given that $||h_m(z)|| = ||z|| = 1$ on the sphere. Hence $||h_m(z)||^2 = ||z||^{2m} = ||z^{\otimes m}||^2$ on the sphere as well. By homogeneity, this equality holds everywhere, and the maps are norm-equivalent. Theorem 4.3 then implies the existence of an isometry V such that $z^{\otimes m} = V h_m(z)$. Since $z^{\otimes m}$ includes all the monomials of degree m, so does h_m. Hence the dimensions are equal, and V is unitary. Put $U = V^{-1}$.   □

A variant of the tensor product operation allows us to construct more examples of polynomial mappings between spheres. By also allowing an inverse operation we will find *all* polynomial mappings between spheres.

Let A be a subspace of $\mathbf{C}^N$, and let π_A be orthogonal projection onto A. Then we have $||f||^2 = ||\pi_A f||^2 + ||(1 - \pi_A)f||^2$ by the Pythagorean theorem. Combining this fact with (21) leads to the following:

PROPOSITION 4.6. Suppose $f : \mathbf{C}^n \to \mathbf{C}^M$ and $g : \mathbf{C}^n \to \mathbf{C}^N$ satisfy $||f||^2 = ||g||^2 = 1$ on some set S. Then, for any subspace A of $\mathbf{C}^M$, the map $E_{A,g}f = (1 - \pi_A)f \oplus (\pi_A f \otimes g)$ satisfies $||E_{A,g}f||^2 = 1$ on S.

PROOF. By definition of orthogonal sum and (21) we have

$$||E_{A,g}f||^2 = ||(1 - \pi_A)f \oplus (\pi_A f \otimes g)||^2 = ||(1 - \pi_A)f||^2 + ||\pi_A f||^2 ||g||^2. \quad (23)$$

If $||g||^2 = 1$ on S, then formula (23) becomes $||(1 - \pi_A)f||^2 + ||\pi_A f||^2 = ||f||^2 = 1$ on S.   □

When $g(z) = z$, we can write the computation in (23) as follows:

$$||E_A(f)||^2 = ||f||^2 + (||z||^2 - 1)||\pi_A(f)||^2.$$

This tensor operation evokes our discussion of spherical harmonics, where we multiplied polynomials by the squared norm in $\mathbf{R}^n$. The operation E_A is more subtle for several reasons; first of all, our map f is vector-valued. Second of all, we perform the multiplication (now a tensor product) on a proper subspace A of the target.

We will begin studying non-constant (complex-analytic) polynomial mappings taking S^{2n-1} to S^{2N-1}. By Proposition 4.5, when $n = N = 1$ the only possibilities are $z \mapsto e^{i\theta} z^m$. When $n = N \geq 2$, the only non-constant examples are unitary maps. When $N < n$, the only polynomial maps are constants. The proofs of these facts use several standard ideas in the theory of analytic functions of several complex variables, but we omit them here to maintain our focus and because we do not use them to prove any of our results. We therefore summarize these facts without proof. We also include a simple consequence of Proposition 4.5 in this collection of statements about polynomial mappings between spheres.

THEOREM 4.5. *Assume that $p : \mathbf{C}^n \to \mathbf{C}^N$ is a polynomial mapping with $p(S^{2n-1}) \subseteq S^{2N-1}$. If $N = n = 1$, then $p(z) = e^{i\theta} z^m$ for some m. If $N < n$, then p is a constant. If $n \leq N \leq 2n - 2$, then p is either a constant or an isometry.*

When N is much larger than n, there are many maps. We can understand them via a process of orthogonal homogenization.

Let $p : \mathbf{C}^n \to \mathbf{C}^N$ be a polynomial mapping. Let $\| \ \|$ denote the Euclidean norm in either the domain or target. We expand p in terms of *homogeneous parts*. Thus $p = \sum_{k=0}^{d} p_k$, where each $p_k : \mathbf{C}^n \to \mathbf{C}^N$ and p_k is *homogeneous of degree k*. That is, $p_k(tz) = t^k p_k(z)$ for all $t \in \mathbf{C}$. Suppose in addition that $p : S^{2n-1} \to S^{2N-1}$. Then, if $\|z\|^2 = 1$, we have

$$\|p(z)\|^2 = \left\| \sum p_k(z) \right\|^2 = \sum_{k,j} \langle p_k(z), p_j(z) \rangle = 1.$$

Replacing z by $e^{i\theta} z$ and using the homogeneity yields

$$1 = \sum_{k,j} e^{i\theta(k-j)} \langle p_k(z), p_j(z) \rangle. \tag{24}$$

But the right-hand side of (24) is a trig polynomial; hence all its coefficients vanish except for the constant term. We conclude that p must satisfy certain identities when $\|z\| = 1$:

$$\sum \|p_k\|^2 = 1, \tag{25.1}$$

$$\sum_k \langle p_k, p_{k+l} \rangle = 0. \quad (l \neq 0) \tag{25.2}$$

Let d be the degree of p. When $l = d$ in (25.2) the only term in the sum is when $k = 0$, and we conclude that p_0 and p_d are orthogonal. Let π_A denote the projection of $\mathbf{C}^N$ onto the span A of p_0. We can write

$$p = (1 - \pi_A)p \oplus \pi_A p. \tag{26}$$

Consider a new map g, defined by

$$g = E_A(p) = (1 - \pi_A)p \oplus (\pi_A p \otimes z).$$

By Proposition 4.6, $E_A(p)$ also takes the sphere to the sphere in a larger target dimension. The map $g = E_A(p)$ has no constant term and is of degree d. Thus $g_0 = 0$. Now we apply (25.2) to g, obtaining the following conclusion. Either g is homogeneous of degree 1, or its first order part g_1 is orthogonal to its highest order part g_d. We apply the same reasoning to g, letting π_B denote the orthogonal projection onto the span of the homogeneous part g_1. We obtain a map $E_B(E_A(p))$, still of degree d, whose homogeneous expansion now has no terms of order 0 or 1.

Proceeding in this fashion, we increase the order of vanishing without increasing the degree, stopping when the result is homogeneous. Thus we obtain a sequence of subspaces $A_0, \ldots, A_{d-1}$ such that composing these tensor product operations yields something homogeneous of degree d. As the last step, we compose with a linear map to guarantee that the components are linearly independent. Applying Theorem 4.3 we obtain the following result about orthogonal homogenization.

THEOREM 4.6. *Let p be a polynomial mapping such that $p(S^{2n-1}) \subseteq S^{2N-1}$ and p is of degree d. Then there is a linear L and a finite sequence of subspaces and tensor products such that*

$$z^{\otimes d} = L(E_{A_{d-1}}(\ldots(E_{A_0}(p))\ldots)). \tag{27}$$

Here $L = qU$ where U is unitary and q is a projection.

PROOF. We repeat the previous discussion in more concise language. If p is homogeneous, then the conclusion follows from Theorem 4.4. Otherwise, let ν denote the order of vanishing of p. Thus $\nu < d$ and $p = \sum_{j=\nu}^{d} p_j$, where p_j is homogeneous of degree j. By (25.2), p_ν is orthogonal to p_d on the sphere. By Lemma 4.6, they are orthogonal everywhere. Let A denote the span of the coefficient vectors in p_ν. By Proposition 4.2, the polynomial mapping $E_A(p)$ sends the unit sphere in its domain $\mathbf{C}^n$ to the unit sphere in its target. This mapping is also of degree d, but its order of vanishing exceeds ν. After finitely many steps of this sort we reach a homogeneous mapping of degree d. We then apply Theorem 4.4. □

In the next section we will use this result to prove a geometric inequality concerning the maximum volume (with multiplicity counted) of the image of the ball under a proper polynomial map, given its degree.

Next we illustrate Theorem 4.6 by way of a polynomial mapping S^3 to S^7.

EXAMPLE 4.5. Put $z = (w, \zeta)$ and $p(w, \zeta) = (w^3, w^2\zeta, w\zeta, \zeta)$. Then $A_0 = 0$. Also A_1 is the span of $(0, 0, 0, 1)$, and $E_{A_1}(p) = (w^3, w^2\zeta, w\zeta, w\zeta, \zeta^2)$. Now A_2 is the span of the three standard basis vectors e_3, e_4 and e_5 in $\mathbf{C}^5$. Tensoring on the subspace A_2 yields

$$f = E_2(E_1(p)) = (w^3, w^2\zeta, w^2\zeta, w\zeta^2, w^2\zeta, w\zeta^2, w\zeta^2, \zeta^3).$$

The image of f is contained in a 4-dimensional subspace of $\mathbf{C}^8$. We can apply a unitary map U to f to get

$$Uf = (w^3, \sqrt{3}w^2\zeta, \sqrt{3}w\zeta^2, \zeta^3, 0, 0, 0, 0).$$

Finally we project onto $\mathbf{C}^4$ and identify the result with the map $z^{\otimes 3}$. In the notation (27), $L = qU$ is the composition of the unitary map U and the projection q.

5. The derivative as a linear map

Our second proof of Corollary 4.2 used the differential 1-forms dz and $d\bar{z}$ in one complex dimension. In order to extend the result to higher dimensions, we must discuss complex vector fields and complex differential forms. We begin by reviewing the real case. See [Wa] for a definitive treatment of differential forms. See [Dar] for an alternative discussion of the basics and also for interesting applications.

In order to prepare for subsequent developments, we recall some multi-variable calculus. In doing so, we will clarify one of the most subtle points in elementary calculus. What do we mean by the differential dx in the first place? High school teachers often say that dx means an infinitesimal change in the x direction, but these words are too vague to have any meaning.

Let Ω be an open set in $\mathbf{R}^n$ and suppose that $f : \Omega \to \mathbf{R}^k$ is a function. Then f is differentiable at a point $p \in \Omega$ if it is *approximately linear* there. The easiest way to make this notion precise is to write

$$f(p + h) = f(p) + L(h) + e(p, h),$$

where L is a linear map from $\mathbf{R}^n$ to $\mathbf{R}^k$ and $e(p, h)$ is an error term, defined by the equation. Then f is differentiable at p if there is a linear map L for which

$$\lim_{h \to 0} \frac{||f(p + h) - f(p) - L(h)||}{||h||} = \lim_{h \to 0} \frac{||e(p, h)||}{||h||} = 0. \tag{28}$$

If such an L exists, then it is unique, and it is written $df(p)$ or $Df(p)$.

In the rest of this book we work almost exclusively with infinitely differentiable functions. As usual we call a function *smooth* on an open set if its partial derivatives of all orders exist and are continuous there.

EXERCISE 4.32. Put $n = k = 1$ in (28). Show that (28) agrees with the usual notion of the derivative from calculus.

EXERCISE 4.33. Put $n = k = 1$ and $f(x) = x^m$ for m a positive integer. Use (28) to show that $df(x) = mx^{m-1}$.

EXERCISE 4.34. Suppose that $\alpha > 1$ and that $||f(x)|| \le C||x||^\alpha$ for some constant C. Prove that f is differentiable at 0.

EXERCISE 4.35. Prove the chain rule: If f is differentiable at p and g is differentiable at $f(p)$, then $g \circ f$ is differentiable at p and $d(g \circ f)(p) = dg(f(p)) \circ df(p)$. Suggestion: estimate the error term for the composition.

This discussion becomes more interesting when we allow the point p to vary. Doing so leads to vector fields and differential forms. We begin by associating a copy of $\mathbf{R}^n$, written $T_p(\mathbf{R}^n)$, to each point $p \in \Omega$. Similarly we associate a copy of $\mathbf{R}^k$, written $T_{f(p)}(\mathbf{R}^k)$, to each point $f(p)$ in the image of f. We think of the derivative as *pushing forward* the vector $h \in T_p(\mathbf{R}^n)$ to the vector $df(p)h \in T_{f(p)}(\mathbf{R}^k)$. Some authors write $df(p)$ as f_* and call f_* the **pushforward** of f. The dual notion, called the **pullback**, will be crucial when we compute volumes of images of sets.

When the target dimension k equals 1, the derivative $df(p)$ assigns the scalar $df(p)h$ to the vector h. Since $df(p)$ is a linear map, it is an element of the dual space to $T_p(\mathbf{R}^n)$. As is customary, we write $T_p^*(\mathbf{R}^n)$ for the dual space. Then the map $p \mapsto df(p)$ is itself a map from Ω to $T_p^*(\mathbf{R}^n)$. Such an object is called a differential 1-form. We can think of df as a machine that inputs first a point p, then a tangent vector v based at p, and outputs the directional derivative:

$$df(p)(v) = \frac{\partial f}{\partial v}(p) = \lim_{t \to 0} \frac{f(p + tv) - f(p)}{t}. \tag{29}$$

We can also regard v as an operator; it is a machine which first inputs a point p, then a function f, and outputs $df(p)v$. Sometimes we write $v_p[f] = df(v)p$ when we wish to emphasize this perspective.

A *vector field* on $\mathbf{R}^n$ is simply a function $V : \mathbf{R}^n \to \mathbf{R}^n$. A vector field is called *smooth* if each component v_j is a smooth function. We think geometrically of placing the vector $V(x)$ at the point x. We make a conceptual leap by regarding the two copies of $\mathbf{R}^n$ as different spaces. (Doing so is analogous to regarding the x and y axes as different copies of the real line.) For $j = 1, \ldots, n$, we let e_j denote the j-th standard basis element of the first copy of $\mathbf{R}^n$. We write $\frac{\partial}{\partial x_j}$ for the indicated partial differential operator; $\frac{\partial}{\partial x_j}$ will be the j-th standard basis vector of the second copy of $\mathbf{R}^n$. The subsequent development will help clarify the sense in which a tangent vector behaves like a differential operator.

At each point $x = (x_1, \ldots, x_n)$ of $\mathbf{R}^n$, we call the real vector space $T_x(\mathbf{R}^n)$ the *tangent space* at x. The vector space $T_x(\mathbf{R}^n)$ is also n-dimensional; the operators $\frac{\partial}{\partial x_j}$ for $1 \le j \le n$ will form a basis for the tangent space. As is common, we don't include the base point x in the notation. Here is the precise definition of $\frac{\partial}{\partial x_j}$:

$$\frac{\partial}{\partial x_j}(f)(x) = \frac{\partial f}{\partial x_j}(x) = \lim_{t \to 0} \frac{f(x + te_j) - f(x)}{t}. \tag{30}$$

The $\frac{\partial}{\partial x_j}$, for $j = 1, \ldots, n$, form a basis for $T_x(\mathbf{R}^n)$. Thus an element of $T_x(\mathbf{R}^n)$ is a vector of the form $\sum_{j=1}^n c_j \frac{\partial}{\partial x_j}$.

Partial derivatives are special cases of directional derivatives. We could therefore avoid (30) and instead start with (31), the definition of the directional derivative of f in the direction $v = (v_1, \ldots, v_n)$:

$$\frac{\partial f}{\partial v}(x) = \lim_{t \to 0} \frac{f(x + tv) - f(x)}{t} = \sum_{j=1}^n v_j \frac{\partial f}{\partial x_j}(x) = V[f](x). \tag{31}$$

In this definition (31) of directional derivative, we do not assume that v is a unit vector. For us a vector field V is a function whose domain is a subset of $\mathbf{R}^n$ but whose value at p is an element of $T_p(\mathbf{R}^n)$. We therefore write $V = \sum v_j \frac{\partial}{\partial x_j}$. Then V can be applied to a differentiable function f and $V[f]$ means the directional derivative of f in the direction v, as suggested by the notation. Thus $T_x(\mathbf{R}^n)$ is the set of directions in which we can take a directional derivative at x.

REMARK 4.8. The viewpoint expressed by the previous sentence is useful when we replace $\mathbf{R}^n$ by a smooth submanifold M. The tangent space $T_x(M)$ is then precisely the set of such directions. See [Dar] and [Wa].

REMARK 4.9. The expression $\frac{\partial}{\partial x_j}$ is defined such that $\frac{\partial}{\partial x_j}(f)$ equals the directional derivative of f in the j-th coordinate direction. Warning! The expression $\frac{\partial}{\partial x_j}$ depends on the full choice of basis. We cannot define $\frac{\partial}{\partial x_1}$ until we have chosen all n coordinate directions. See Exercise 4.37.

We pause to restate the definition of vector field in modern language. Let $T(\mathbf{R}^n)$, called the *tangent bundle*, denote the disjoint union over x of all the spaces $T_x(\mathbf{R}^n)$. (To be precise, the definition of $T(\mathbf{R}^n)$ includes additional information, but we can safely ignore this point here.) A point in $T(\mathbf{R}^n)$ is a pair (x, v_x), where x is the base point and v_x is a (tangent) vector at x. A vector field is a map $V : \mathbf{R}^n \to T(\mathbf{R}^n)$ such that $V(x) \in T_x(\mathbf{R}^n)$ for all x. In other words, $V(x) = (x, v_x)$. In modern language, a vector field is a *section* of the *tangent bundle*

$T(\mathbf{R}^n)$. At each x, we regard $V(x)$ as a direction in which we can differentiate functions defined near x.

What is a differential 1-form? We begin by defining df for a smooth function f. Here *smooth* means infinitely differentiable.

Let $f : \mathbf{R}^n \to \mathbf{R}$ be a smooth function. Let V be a vector field; $v = V(x)$ is a vector based at x; thus $V(x) \in T_x(\mathbf{R}^n)$. We define df as follows:

$$df(x)[v] = (df(x), v) = \frac{\partial f}{\partial v}(x) = \lim_{t \to 0} \frac{f(x + tv) - f(x)}{t}. \tag{32}$$

The formula on the far right-hand side of (32) is the definition. The other expressions are different notations for the same quantity. In the first formula, $df(x)$ is a function, seeking a vector v as the input, and producing a real number as the output. In the second formula, $df(x)$ and v appear on equal footing. The third formula means the rate of change of f in the direction v at x. In coordinates, we have $V(x) = \sum v_j \frac{\partial}{\partial x_j}$, where $v = (v_1, \ldots, v_n)$, and

$$df(x)[v] = \sum_{j=1}^{n} v_j(x) \frac{\partial f}{\partial x_j}(x). \tag{33}$$

Formula (32) gives a precise, invariant definition of df for any smooth function f. In particular we can finally say what dx_k means. Let $f = x_k$ be the function that assigns to a point x in $\mathbf{R}^n$ its k-th coordinate, and consider df. The equation $dx_k = df$ gives a precise meaning to dx_k. (Confusion can arise because x_k denotes both the k-th coordinate and the function whose value is the k-th coordinate.)

The expression df is called the *exterior derivative* or *total differential* of f. We discuss the exterior derivative in detail in the next section. We can regard df as a function. Its domain consists of pairs (x, v), where $x \in \mathbf{R}^n$ and $v \in T_x(\mathbf{R}^n)$. By (32), $df(x)[v]$ is the directional derivative of f in the direction v at x. Since taking directional derivatives depends linearly on the direction, the object $df(x)$ is a linear functional on $T_x(\mathbf{R}^n)$. It is natural to call the space $T_x^*(\mathbf{R}^n)$ of linear functionals on $T_x(\mathbf{R}^n)$ the *cotangent space* at x. The cotangent space also has dimension n, but it is distinct both from the domain $\mathbf{R}^n$ and from the tangent space. The disjoint union of all the cotangent spaces is called the *cotangent bundle* and written $T^*(\mathbf{R}^n)$. A point in $T^*(\mathbf{R}^n)$ is a pair (x, ξ_x), where x is the base point and ξ_x is a *co-vector* at x. A differential 1-form is a section of the cotangent bundle. Not all 1-forms can be written in the form df for some function f. See the discussion after Stokes' theorem.

REMARK 4.10. Suppose $f : \mathbf{R}^n \to \mathbf{R}^k$ is differentiable at p. We then regard $df(p)$ as a linear map from the tangent space at p to the tangent space at $f(p)$. When f is differentiable on an open set Ω, we can think of df as a map whose value at p is a linear transformation from $T_p(\mathbf{R}^n)$ to $T_{f(p)}(\mathbf{R}^k)$.

We summarize the discussion, expressing things in an efficient order. For each $x \in \mathbf{R}^n$ we presume the existence of a vector space $T_x(\mathbf{R}^n)$, also of dimension n. The union $T(\mathbf{R}^n)$ over x of the spaces $T_x(\mathbf{R}^n)$ is called the tangent bundle. A vector field is a section of the tangent bundle. For each smooth real-valued function f, defined near x, we define df by (32). In particular, when f is the coordinate function x_j, we obtain a definition of dx_j. For each smooth f and each x, $df(x)$ is an element of the dual space $T_x^*(\mathbf{R}^n)$. The union of these spaces is the cotangent bundle. A 1-form is a section of the cotangent bundle.

We *define* the operators $\frac{\partial}{\partial x_j}$ by duality. Thus the differentials dx_j precede the operators $\frac{\partial}{\partial x_j}$ in the logical development. A 1-form is a combination $\sum b_j(x) dx_j$ and a vector field is a combination $\sum a_j(x) \frac{\partial}{\partial x_j}$.

6. Complex differential forms and vector fields

Our work requires complex vector fields and complex differential forms. In terms of real coordinates, a complex vector field on $\mathbf{R}^m$ can be written $\sum_{j=1}^m g_j(x) \frac{\partial}{\partial x_j}$ where the functions g_j are smooth and complex-valued. Similarly, a complex 1-form on $\mathbf{R}^m$ can be written $\sum_{j=1}^m h_j(x) dx_j$ where the functions h_j are smooth and complex-valued. Terminology such as *vector field with complex coefficients* or *complex-valued differential form* is sometimes used for clarity or emphasis.

We can identify complex Euclidean space $\mathbf{C}^n$ with $\mathbf{R}^{2n}$. Write $z = (z_1, \ldots, z_n)$, and put $z_j = x_j + iy_j$ (where i is the imaginary unit). We can express vector fields in terms of the $\frac{\partial}{\partial x_j}$ and $\frac{\partial}{\partial y_j}$, and differential forms in terms of the dx_j and dy_j. Complex geometry is magic; things simplify by working with complex (note the double entendre) objects. Everything follows easily from one obvious definition.

DEFINITION 4.7. Suppose Ω is an open set in $\mathbf{C}^n$, and $f : \Omega \to \mathbf{C}$ is smooth. Write $f = u + iv$ where u and v are real-valued. We define df by $df = du + idv$.

COROLLARY 4.4. *Let $z_j = x_j + iy_j$ denote the j-th coordinate function on $\mathbf{C}^n$. Then $dz_j = dx_j + idy_j$ and $d\bar{z}_j = dx_j - idy_j$.*

We define complex differentiation by duality as follows in Definition 4.8. We could also use the formulas in Corollary 4.5 as definitions.

DEFINITION 4.8. For $j = 1, \ldots n$, let $\{\frac{\partial}{\partial z_j}, \frac{\partial}{\partial \bar{z}_j}\}$ denote the dual basis to the basis $\{dz_j, d\bar{z}_j\}$. Thus $\frac{\partial}{\partial z_j}$ is defined by $dz_k[\frac{\partial}{\partial z_j}] = 0$ if $j \neq k$ and by $dz_k[\frac{\partial}{\partial z_k}] = 1$. Also, $\frac{\partial}{\partial \bar{z}_j}$ is defined by $dz_k[\frac{\partial}{\partial \bar{z}_j}] = 0$ for all j, k and $d\bar{z}_k[\frac{\partial}{\partial z_j}] = 0$ for $j \neq k$, but $d\bar{z}_k[\frac{\partial}{\partial \bar{z}_k}] = 1$.

Differentiable functions $g_1, \ldots, g_m$ form a *coordinate system* on an open set Ω in $\mathbf{R}^m$ if their differentials are linearly independent on Ω and the mapping $g = (g_1, \ldots, g_m)$ is injective there. This concept makes sense when these functions are either real or complex-valued. For example, the functions z and $\bar{z}$ define a coordinate system on $\mathbf{R}^2$, because $dx + idy$ and $dx - idy$ are linearly independent and the map $(x, y) \mapsto (x + iy, x - iy)$, embedding $\mathbf{R}^2$ into $\mathbf{C}^2$, is injective.

We can regard the $2n$ functions $z_1, \ldots, z_n, \bar{z}_1, \ldots, \bar{z}_n$ as complex-valued coordinates on $\mathbf{R}^{2n}$. The *exterior derivative df* is invariantly defined, independent of coordinate system, by (32) and Definition 4.7. Hence the following equality holds:

$$df = \sum_{j=1}^n \frac{\partial f}{\partial x_j} dx_j + \sum_{j=1}^n \frac{\partial f}{\partial y_j} dy_j = \sum_{j=1}^n \frac{\partial f}{\partial z_j} dz_j + \sum_{j=1}^n \frac{\partial f}{\partial \bar{z}_j} d\bar{z}_j. \tag{34}$$

The following formulas then follow by equating coefficients. See Exercise 4.36.

COROLLARY 4.5.

$$\frac{\partial}{\partial z_j} = \frac{1}{2} \left(\frac{\partial}{\partial x_j} - i \frac{\partial}{\partial y_j} \right) \tag{35.1}$$

$$\frac{\partial}{\partial \overline{z}_j} = \frac{1}{2} \left(\frac{\partial}{\partial x_j} + i \frac{\partial}{\partial y_j} \right). \tag{35.2}$$

Suppose f is differentiable on an open set in $\mathbf{C}^n$. By (34), we can decompose its exterior derivative df into two parts:

$$df = \partial f + \overline{\partial} f = \sum_{j=1}^{n} \frac{\partial f}{\partial z_j} dz_j + \sum_{j=1}^{n} \frac{\partial f}{\partial \overline{z}_j} d\overline{z}_j. \tag{36}$$

Formula (36) defines the splitting of the 1-form df into the sum of a $(1,0)$-form and a $(0,1)$-form. We next give the definition of complex analyticity in this language. The terms *complex analytic* and *holomorphic* are synonymous.

DEFINITION 4.9. Let Ω be an open subset of $\mathbf{C}^n$. Assume that $f : \Omega \to \mathbf{C}$ and f is continuously differentiable. Then f is *complex analytic* if and only if $\overline{\partial} f = 0$. Equivalently, if and only if $\frac{\partial f}{\partial \overline{z}_j} = 0$ for all j.

The differential equations in Definition 4.9 are called the Cauchy-Riemann equations. Thus complex analytic functions are the solutions to a first-order system of partial differential equations. As in one variable, complex analytic functions are given locally by convergent power series. In Theorem 4.3 we used the power series expansion of a complex analytic mapping in a ball. For most of what we do, the crucial point is that the Cauchy-Riemann equations have the simple expression $\overline{\partial} f = 0$. By (36), $\overline{\partial} f = 0$ means that f is independent of each $\overline{z}_j$. Part of the magic of complex analysis stems from regarding z and its conjugate $\overline{z}$ as independent variables. In the rest of this chapter most of the complex analytic functions we will encounter are polynomials in the variables $z_1, ..., z_n$. They will be independent of the complex conjugate variables. We emphasize the intuitive statement: f is complex analytic if and only if f is independent of the conjugate variable $\overline{z} = (\overline{z}_1, ..., \overline{z}_n)$.

COROLLARY 4.6. *A continuously differentiable function, defined on an open set in $\mathbf{C}^n$, is complex analytic if and only if $df = \partial f$.*

EXERCISE 4.36. Use (34) to verify (35.1) and (35.2).

EXERCISE 4.37. This exercise asks you to explain Remark 4.9. Consider the functions x and y as coordinates on $\mathbf{R}^2$. Then by definition, $\frac{\partial y}{\partial x} = 0$. Suppose instead we choose $u = x$ and $v = x + y$ as coordinates. Then we would have $\frac{\partial v}{\partial u} = 0$. But $\frac{\partial (x+y)}{\partial x} = 1$. Explain!

7. Differential forms of higher degree

The reader is surely familiar with the determinant. We can think of an n-by-n matrix as an n-tuple of row vectors $r_1, ..., r_n$. The determinant is a multi-linear function of these rows. In other words, for each j, the function $r_j \to \det(r_1, ..., r_n)$ is a linear functional. Here the other rows are held fixed. In addition, this function is *alternating*; its value gets multiplied by -1 when we interchange two rows. Differential forms have similar properties and are closely related.

Our work in higher dimensions relies on differential forms of higher degree. This discussion presumes that the reader has had some exposure to the wedge product of differential forms, and therefore knows intuitively what we mean by a k-form. We also use the modern Stokes' theorem, which in our setting expresses an integral of a

$2n$-form over the unit ball as an integral of a $(2n-1)$-form over the unit sphere. We develop enough of this material to enable us to do various volume computations.

DEFINITION 4.10. Let V be a (real or) complex vector space of finite dimension. A function $F : V \times \cdots \times V \to \mathbf{C}$ (with k factors) is called a *multi-linear form* if F is linear in each variable when the other variables are held fixed. We often say F is k-linear. It is called *alternating* if $F(v_1, \ldots, v_k) = 0$ whenever $v_i = v_j$ for some i, j with $i \neq j$.

EXAMPLE 4.6. Consider a k-by-k matrix M of (real or) complex numbers. Think of the rows (or columns) of M as elements of $\mathbf{C}^k$. The determinant function is an alternating k-linear form on $\mathbf{C}^k \times \cdots \times \mathbf{C}^k$.

EXAMPLE 4.7. Given vectors $a = (a_1, a_2, a_3)$ and $b = (b_1, b_2, b_3)$ in $\mathbf{R}^3$, define $F(a, b) = a_1 b_3 - a_3 b_1$. Then F is an alternating 2-linear form.

LEMMA 4.7. *A multi-linear form F (over $\mathbf{R}^n$ or $\mathbf{C}^n$) is alternating if and only if the following holds. For each pair i, j of distinct indices, the value of F is multiplied by -1 if we interchange the i-th and j-th slots:*

$$F(v_1, \ldots, v_i, \ldots, v_j, \ldots v_k) = -F(v_1, \ldots, v_j, \ldots v_i, \ldots, v_k). \qquad (37)$$

PROOF. It suffices to ignore all but two of the slots and then verify the result when F is 2-linear. By multi-linearity we have

$$F(v + w, v + w) = F(v, v) + F(v, w) + F(w, v) + F(w, w). \qquad (38)$$

If F is alternating, then all terms in (38) vanish except $F(v, w) + F(w, v)$. Hence this term must vanish as well. Conversely, if this term always vanishes, then (38) gives $F(v + w, v + w) = F(v, v) + F(w, w)$. Put $w = -v$. We get

$$0 = F(0, 0) = F(v, v) + F(-v, -v) = F(v, v) + (-1)^2 F(v, v) = 2F(v, v).$$

Hence $F(v, v) = 0$ for all v. $\qquad \square$

REMARK 4.11. The reader might wonder why we chose the definition of alternating to be the vanishing condition rather than the change of sign condition. The reason is suggested by the proof. Over $\mathbf{R}$ or $\mathbf{C}$, the conditions are the same. If we were working over more general fields, however, we could not rule out the possibility that $1 + 1 = 0$. In this case the two conditions are not equivalent.

We note that 0 is the only alternating k-linear form on V if k exceeds the dimension of V. When k equals the dimension of V, the only alternating k-linear form is a multiple of the determinant.

EXERCISE 4.38. Verify the statements in the previous paragraph.

We can now introduce differential forms of higher degree.

DEFINITION 4.11. Let V be a (real or) complex vector space of finite dimension n with dual space V^*. The collection $\Lambda^k(V^*)$ of all k-linear alternating forms on V is itself a vector space of dimension $\binom{n}{k}$. It is called the k-th exterior power of V^*.

Note that $\Lambda^1(V^*)$ consists of all 1-linear forms on V; thus it is the dual space of V and $\Lambda^1(V^*) = V^*$. By convention, $\Lambda^0(V^*)$ equals the ground field $\mathbf{R}$ or $\mathbf{C}$.

DEFINITION 4.12. Let Ω be an open subset of $\mathbf{R}^n$. A differential form of degree k on Ω (or a differential k-form) is a (smooth) section of the k-th exterior power $\Lambda^k(T^*(\mathbf{R}^n))$ of the cotangent bundle $T^*(\mathbf{R}^n)$.

At each point $x \in \Omega$, we have the vector space $T_x(\mathbf{R}^n)$ and its dual space $T_x^*(\mathbf{R}^n)$. A differential k-form assigns to each x an element of $\Lambda^k(T_x^*(\mathbf{R}^n))$. The value of the k-form at x is an alternating k-linear form.

By convention, a 0-form is a function. A 1-form assigns to each x a linear functional on $T_x(\mathbf{R}^n)$, as we have seen already. The value of a 2-form at x is a machine which seeks two vectors at x as inputs, and returns a number. If we switch the order of the two inputs, we multiply the output by -1.

Forms of all degrees can be generated from 1-forms using the wedge product. Before giving the definition of the wedge product, we express the idea informally using bases. Suppose $e_1, \ldots, e_n$ form a basis for the 1-forms at a point x. For each k with $1 \le k \le n$, and each increasing sequence of indices $i_1 < i_2 < \cdots < i_k$, we define a formal expression e_I, written

$$e_I = e_{i_1} \wedge e_{i_2} \wedge \cdots \wedge e_{i_k}. \tag{39}$$

Note that there are exactly $\binom{n}{k}$ such expressions. We decree that the collection of these objects form a basis for the space of k-forms. Thus the space of k-forms on an n-dimensional space has dimension $\binom{n}{k}$.

We can regard e_I as an alternating k-linear form. As written, the index I satisfies $i_1 < \cdots < i_k$. We extend the notation by demanding the alternating property. For example, when $k = 2$ and l, m are either 1 or 2, we put

$$(e_l \wedge e_m)(v, w) = e_l(v)e_m(w) - e_l(w)e_m(v).$$

Then $e_2 \wedge e_1 = -e_1 \wedge e_2$. More generally we put

$$(e_1 \wedge \cdots \wedge e_k)(v_1, \ldots, v_k) = \det(e_i(v_j)). \tag{40}$$

EXAMPLE 4.8. Consider $\mathbf{R}^3$ with basis e_1, e_2, e_3. The zero forms are spanned by the constant 1. The 1-forms are spanned by e_1, e_2, e_3. The 2-forms are spanned by $e_1 \wedge e_2$, $e_1 \wedge e_3$, and $e_2 \wedge e_3$. The 3-forms are spanned by $e_1 \wedge e_2 \wedge e_3$.

EXERCISE 4.39. For $0 \le k \le 4$, list bases for the k-forms on a 4-dimensional space.

A relationship between wedge products and determinants is evident. It is therefore no surprise that we *define* the wedge product in a manner similar to the Laplace expansion of a determinant.

Let us first recall the algebraic definition of the determinant. The motivation is geometric; $\det(v_1, \ldots, v_n)$ measures the oriented volume of the n-dimensional box spanned by these vectors. We normalize by assuming that the volume of the unit n-cube is 1.

DEFINITION 4.13. Let V be either $\mathbf{R}^n$ or $\mathbf{C}^n$. The determinant, written *det*, is the unique alternating n-linear form whose value on $e_1, \ldots, e_n$ is 1.

The Laplace expansion of the determinant follows from the definition. Suppose $v_j = \sum c_{jk}e_k$. We compute $\det(v_1, \ldots, v_n)$ by the definition. Multi-linearity yields:

$$\det(v_1, \ldots, v_n) = \sum_{k_1=1}^{n} \sum_{k_2=1}^{n} \cdots \sum_{k_n=1}^{n} \prod_{j=1}^{n} c_{jk_j} \det(e_{k_1}, \ldots, e_{k_n}).$$

Next we apply the alternating property to rewrite the determinant of each $(e_{k_1}, \ldots e_{k_n})$. If indices are repeated we get 0. Otherwise we get ± 1, depending on the signum of the permutation of the indices. We obtain the standard Laplace expansion of the determinant

$$\det(c_{jk}) = \sum_{\tau} \mathrm{sgn}(\tau) \prod_{j=1}^{n} c_{j\,\tau(j)}. \tag{41}$$

A permutation τ on n objects is a bijection on the set of these objects. The expression $\mathrm{sgn}(\tau)$ is either 1 or -1; it equals 1 when τ is an *even* permutation and -1 when τ is an odd permutation. Thus $\mathrm{sgn}(\tau)$ is the parity of the number of interchanges (of pairs of indices) required to put the indices in the order $1, 2, \ldots, n$.

EXERCISE 4.40. Show that $\mathrm{sgn}(\tau) = \prod_{1 \leq i < j \leq n} \frac{\tau(i) - \tau(j)}{i-j}$.

EXERCISE 4.41. Show that $\mathrm{sgn}(\tau_1 \circ \tau_2) = \mathrm{sgn}(\tau_1)\mathrm{sgn}(\tau_2)$. Suggestion: Use the previous exercise.

The wedge product is defined in a similar fashion:

DEFINITION 4.14. The wedge product of a k-form α and an l-form β is the $(k + l)$-form $\alpha \wedge \beta$ defined by

$$(\alpha \wedge \beta)(v_1, \ldots, v_{k+l}) = \sum_{\tau} \mathrm{sgn}(\tau)\alpha(v_{\tau(1)}, \ldots, v_{\tau(k)})\beta(v_{\tau(k+1)}, \ldots, v_{\tau(k+l)}). \tag{42}$$

The sum in (42) is taken over all permutations τ on $k + l$ objects.

PROPOSITION 4.7 (Properties of the wedge product). *Let $\alpha, \beta, \beta_1, \beta_2$ be differential forms. Then:*

(1) $\alpha \wedge (\beta_1 + \beta_2) = (\alpha \wedge \beta_1) + (\alpha \wedge \beta_2)$.
(2) $\alpha \wedge (\beta_1 \wedge \beta_2) = (\alpha \wedge \beta_1) \wedge \beta_2$.
(3) $\alpha \wedge \beta = (-1)^{kl}\beta \wedge \alpha$ *if α is a k-form and β is an l-form.*

PROOF. Left to the reader as Exercise 4.47.   □

The *exterior derivative d* is one of the most important and elegant operations in mathematics. When η is a k-form, $d\eta$ is a $(k+1)$-form. When η is a function (a 0-form), $d\eta$ agrees with our definition from (32). We can extend d to forms of all degrees by proceeding inductively on the degree of the form. After stating Theorem 4.7, we mention a more elegant approach.

If f is a function, then df is defined as in (32) by $df[v] = \frac{\partial f}{\partial v}$. In coordinates, $df = \sum \frac{\partial f}{\partial x_j}dx_j$. When $g = \sum_j g_j dx_j$ is an arbitrary 1-form, we define dg by

$$dg = \sum_j dg_j \wedge dx_j = \sum_j \sum_k \frac{\partial g_j}{\partial x_k} dx_k \wedge dx_j = \sum_{k<j} \left(\frac{\partial g_j}{\partial x_k} - \frac{\partial g_k}{\partial x_j} \right) dx_k \wedge dx_j. \tag{43}$$

On the far right-hand side of (43), we have rewritten dg using $dx_k \wedge dx_j = -dx_j \wedge dx_k$ to make the indices increase. The terms $dx_j \wedge dx_j$ drop out. For example,

$$d(Pdx + Qdy) = \frac{\partial P}{\partial y}dy \wedge dx + \frac{\partial Q}{\partial x}dx \wedge dy = (\frac{\partial Q}{\partial x} - \frac{\partial P}{\partial y})dx \wedge dy. \qquad (44)$$

Suppose in (44) that $Pdx + Qdy = df$ for some smooth function f. Then the equality of mixed second partial derivatives and (44) show that $d(df) = 0$. This statement in the language of differential forms is equivalent to the classical statement "the curl of a gradient is 0." In fact $d^2 = 0$ in general; see Theorem 4.7 and Exercise 4.42.

Let η be a k-form. We wish to define $d\eta$ in coordinates. To simplify the notation, write

$$dx^J = dx_{j_1} \wedge dx_{j_2} \wedge \cdots \wedge dx_{j_k}.$$

Then we can write $\eta = \sum_J \eta_J dx^J$ where the η_J are functions and each J is a k-tuple of indices. We proceed as we did for 1-forms and put

$$d\eta = \sum_J d\eta_J \wedge dx^J = \sum_J \sum_k \frac{\partial \eta_J}{\partial x_k} dx_k \wedge dx^J.$$

Thus $d\eta = \sum g_L dx^L$, where now L is a $(k+1)$-tuple of indices.

The following standard result, which applies in the setting of smooth manifolds, characterizes d. We omit the simple proof, which can be summarized as follows. Choose coordinates, use the properties to check the result in that coordinate system, and then use the chain rule to see that d is defined invariantly.

THEOREM 4.7. *There is a unique operator d mapping smooth k-forms to smooth $(k+1)$-forms satisfying the following properties.*

(1) *If f is a function, then df is defined by (32).*
(2) $d(\alpha + \beta) = d\alpha + d\beta$.
(3) $d(\alpha \wedge \beta) = d\alpha \wedge \beta + (-1)^p \alpha \wedge d\beta$ *if α is a p-form.*
(4) $d^2 = 0$.

It is possible to define d without resorting to a coordinate system. The definition on 0-forms is as in (32). We give the definition only for 1-forms. Let η be a 1-form; the 2-form $d\eta$ requires two vector fields as inputs; it must be alternating and multi-linear. Thus we will define $d\eta(v, w)$ for vector fields v and w.

We regard v and w as differential operators by recalling that $v(f) = df(v)$ for smooth functions f. Earlier we wrote $df[v]$, but henceforth we will use the symbol $[,]$ in another manner. We therefore use parentheses for the application of a 1-form on a vector field and for the action of a vector field on a function. We wish to define the expression $d\eta(v, w)$.

DEFINITION 4.15. Let v and w be vector fields. Their *Lie bracket*, or *commutator*, is the vector field $[v, w]$ defined by $[v, w](f) = v(w(f)) - w(v(f))$. Here f is a smooth function and we regard a vector field as a differential operator. (Exercise 4.43 asks you to check that the commutator is a vector field.)

We can now define $d\eta$. Given vector fields v and w, we put

$$d\eta(v, w) = v(\eta(w)) - w(\eta(v)) - \eta([v, w]).$$

The notation $v(\eta(w))$ here means the derivative of the function $\eta(w)$ in the direction v. The full expression is alternating in v and w. The term involving commutators is required to make certain that $d\eta$ is linear over the functions. See Exercise 4.44. This formula (and its generalization to forms of all degrees) is known as the Cartan formula for the exterior derivative.

EXERCISE 4.42. Show that $d^2 = 0$. Recall, for smooth functions f, we have

$$\frac{\partial^2 f}{\partial x_j \partial x_k} = \frac{\partial^2 f}{\partial x_k \partial x_j}.$$

EXERCISE 4.43. Verify that the commutator of two vector fields is a vector field. Suggestion: Use coordinates.

EXERCISE 4.44. Suppose we tried to define a 2-form ζ by

$$\zeta(v, w) = v(\eta(w)) - w(\eta(v)).$$

Show that $\zeta(fv, w) \neq f\zeta(v, w)$ in general, and thus linearity fails. Then show that the commutator term in the definition of $d\eta$ enables linearity to hold.

Equation (44) fits nicely with Green's theorem. The line integral of the 1-form $\eta = Pdx + Qdy$ around a simple closed curve equals the double integral of $d\eta$ over the curve's interior. The generalization of this result to forms of all degrees is known as the modern Stokes' theorem. This theorem subsumes many results, including the fundamental theorem of calculus, Green's theorem, Gauss's divergence theorem, the classical Stokes' theorem, etc., and it illuminates results such as Maxwell's equations from the theory of electricity and magnetism. We state it only for domains in $\mathbf{R}^N$, but it holds much more generally. We will apply Stokes' theorem only when the surface in question is the unit sphere, which is oriented by the outward normal vector. In the next section we discuss pullbacks and indicate how line and surface integrals are defined. For now, we state one of the great achievements of modern mathematics in the setting we require. See [Wa], for example, for a complete treatment.

THEOREM 4.8 (Stokes' theorem). *Let* $S = b\Omega$ *be a piecewise-smooth oriented* $(N-1)$-*dimensional surface bounding an open subset* Ω *of* $\mathbf{R}^N$. *Let* η *be an* $(N-1)$-*form that is smooth on* Ω *and continuous on* $\Omega \cup b\Omega$. *Then*

$$\int_{b\Omega} \eta = \int_{\Omega} d\eta.$$

COROLLARY 4.7. *If* $d\eta = 0$, *then* $\int_{b\Omega} \eta = 0$.

Theorem 4.8 holds whether or not $b\Omega$ is connected, as long as one is careful with orientation. If Ω is the region between concentric spheres, for example, then the spheres must be oppositely oriented.

Each 1-form η on an open subset of $\mathbf{R}^N$ can be written $\eta = \sum_{j=1}^{N} g_j dx_j$, where the g_j are smooth functions. A 1-form η is called *exact* if there is a smooth function f such that $\eta = df$; thus $g_j = \frac{\partial f}{\partial x_j}$. Readers who are familiar with using line integrals to compute *work* will recognize that exact 1-forms correspond to conservative force fields. More generally, a k-form η is exact if there is a $(k-1)$-form α with $d\alpha = \eta$. A necessary condition for being exact arises from the equality of mixed partial derivatives. A form η is called *closed* if $d\eta = 0$. That *exact* implies *closed* follows directly from $d^2 = 0$.

If a form is *closed* on an open set, it need not be *exact* there. The standard examples are of course

$$\eta = \frac{-y\,dx + x\,dy}{x^2 + y^2} \tag{45}$$

$$\eta = \frac{x\,dy \wedge dz + y\,dz \wedge dx + z\,dx \wedge dy}{(x^2 + y^2 + z^2)^{\frac{3}{2}}}. \tag{46}$$

These are defined on the complement of the origin in $\mathbf{R}^2$ and $\mathbf{R}^3$, respectively. The form in (46) provides essentially the same information as the electrical or gravitational field due to a charge or mass at the origin.

Such forms lead to the subject of deRham cohomology. One relates the existence and number of holes in a space to whether closed forms are exact.

8. Pullbacks and integrals

Let g be a smooth map from an open set in $\mathbf{R}^m$ to $\mathbf{R}^n$ and let ω be a k-form on the image of g in $\mathbf{R}^n$. We wish to pull ω back to a form in $\mathbf{R}^m$; the motivation arises from the definitions of line and surface integrals.

DEFINITION 4.16. Let $g : \Omega \subseteq \mathbf{R}^m \to \mathbf{R}^n$ be smooth. Let ω be a k-form on an open set in $\mathbf{R}^n$ containing the image $g(\Omega)$. Consider a k-tuple of vectors in $T_p(\mathbf{R}^n)$. We define the *pullback* g^*w by

$$g^*\omega(v_1, ..., v_k) = \omega(g_*v_1, ..., g_*v_k).$$

Here g_* is the pushforward $dg(p)$ at each point $p \in \Omega$. When $k = 0$, that is, ω is a function f, we define the pullback g^*f to be $f \circ g$.

EXAMPLE 4.9. Define g in two dimensions by $g(u_1, u_2) = (u_1^3, u_2^3)$. Define ω by $\omega = x\,dy$. Then $g^*(\omega) = u_1^3 3u_2^2 du_2$.

Before doing a harder example we state without proof some simple formal properties of pullbacks.

PROPOSITION 4.8. *Let g be as in the definition of pullback; let u denote the coordinates in $\mathbf{R}^m$. Let ω and η differential forms in $\mathbf{R}^n$. Then*

- $g^*(\omega + \eta) = g^*(\omega) + g^*(\eta)$
- $g^*(\omega \wedge \eta) = g^*(\omega) \wedge g^*(\eta)$
- $g^*(dx_a) = d(g^*(x_a)) = dg_a = \sum_{j=1}^m \frac{\partial g_a}{\partial u_j} du_j.$
- $g^*(d\omega) = d(g^*(\omega)).$

Pulling back amounts to careful substitution. In computing pullbacks we must also remember that the 1-forms $\{dx_k\}$ form a basis dual to the vector fields $\{\frac{\partial}{\partial x_j}\}$.

EXAMPLE 4.10. Put $g(r, \theta) = (r\cos(\theta), r\sin(\theta))$ and put $\omega = -y\,dx + x\,dy$. Here $\{dr, d\theta\}$ is the dual basis to $\{\frac{\partial}{\partial r}, \frac{\partial}{\partial \theta}\}$. Hence, for example, $dr(\frac{\partial}{\partial r}) = 1$ and so on. We compute the pullback $g^*\omega$ in two ways to illustrate the idea.

First we use substitution. The derivative matrix of dg is given by

$$\begin{pmatrix} \cos(\theta) & r\sin(\theta) \\ \sin(\theta) & r\cos(\theta) \end{pmatrix}.$$

We compute and simplify, obtaining

$$g^*\omega = -y(\cos(\theta)dr - r\sin(\theta)d\theta) + x(\sin(\theta)dr + r\cos(\theta)d\theta) = r^2\,d\theta.$$

Thus

$$g^* \omega \left(\frac{\partial}{\partial r} \right) = 0$$

$$g^* \omega \left(\frac{\partial}{\partial \theta} \right) = r^2.$$

Next we use directly the definition involving g_*:

$$g^* \omega \left(\frac{\partial}{\partial r} \right) = \omega \left(g_* \left(\frac{\partial}{\partial r} \right) \right) = (x \, dy - y \, dx) \left(\cos(\theta) \frac{\partial}{\partial x} + \sin(\theta) \frac{\partial}{\partial y} \right) = 0 \quad (47.1)$$

$$g^* \omega \left(\frac{\partial}{\partial \theta} \right) = \omega \left(g_* \frac{\partial}{\partial r} \right) = (x \, dy - y \, dx) \left(-r \sin(\theta) \frac{\partial}{\partial x} + r \cos(\theta) \frac{\partial}{\partial y} \right) = r^2. \quad (47.2)$$

Combining (47.1) and (47.2) verifies that $g^* \omega = r^2 \, d\theta$.

The next example of pullbacks is surely the most often used. It illustrates the change of variables formula for multiple integrals. It also makes the method for changing variables in multiple integrals seem the same as in one variable.

EXAMPLE 4.11. Suppose $g : \mathbf{R}^n \to \mathbf{R}^n$ and $\omega = f dx_1 \wedge ... \wedge dx_n$. Then

$$g^*(\omega) = f \circ g \, \det(dg) \, du_1 \wedge ... \wedge du_n.$$

Example 4.11 allows us to state the change of variables formula for multiple integrals as

$$\int_{g(\Omega)} f(x) \, dV(x) = \int_\Omega g^*(f \, dV) = \int_\Omega f(g(u)) \, \det(dg)(u) \, dV(u). \quad (48)$$

REMARK 4.12. In (48) one often replaces $\det(dg)$ with its absolute value; orientation is a subtle point which we ignore for now.

Example 4.11 also motivates the definitions of line and surface integrals via pulling back. We give a simple but important complex variable example.

EXAMPLE 4.12. Consider the line integral of $\bar{z} \, dz$ over the unit circle, traversed counterclockwise. We consider the circle as the image of $[0, 2\pi]$ under the map $t \mapsto \gamma(t) = e^{it}$. Then

$$\int_\gamma \bar{z} \, dz = \int_0^{2\pi} \gamma^*(\bar{z} \, dz) = \int_0^{2\pi} e^{-it} i e^{it} dt = 2\pi i.$$

We conclude this section with some comments about the exterior derivative in the complex setting. We can write $d = \partial + \bar{\partial}$, where the operator ∂ differentiates only in the z_j directions and the operator $\bar{\partial}$ differentiates only in the $\bar{z}_j$ directions. We easily see that $\partial^2 = 0$ and $\bar{\partial}^2 = 0$. A function f defined on an open subset of $\mathbf{C}^n$ is complex analytic if and only if $\bar{\partial} f = 0$. Various simple identities follow from these considerations. For example, suppose that f is complex analytic. Then $df = \partial f$ and hence $0 = d^2 f = d(\partial f)$. Furthermore suppose that $f_1, ..., f_n$ are complex analytic. Then the differential form

$$\eta = \partial f_1 \wedge \bar{\partial} f_1 \wedge ... \wedge \partial f_n \wedge \bar{\partial} f_n$$

is exact. For example, by Exercise 4.52, it satisfies

$$d(f_1 \wedge \bar{\partial} f_1 \wedge ... \wedge \partial f_n \wedge \bar{\partial} f_n) = \eta. \quad (49)$$

EXERCISE 4.45. Verify the last step of the computations in (47.1) and (47.2).

EXERCISE 4.46. Prove that $d(g^*\omega) = g^*(d\omega)$. Suggestion. Use induction on the degree of the form. By linearity it suffices to prove it when $\omega = f dx^J$, where $dx^J = dx_{j_1} \wedge \ldots \wedge dx_{j_k}$ is a k form.

EXERCISE 4.47. Prove Proposition 4.7.

EXERCISE 4.48. For $0 < r < \infty$ and $0 \leq \theta < 2\pi$, put $(x,y)=(r\cos(\theta), r\sin(\theta))$. Show that $dx \wedge dy = r dr \wedge d\theta$.

EXERCISE 4.49. For $0 < \rho < \infty$, for $0 \leq \theta < 2\pi$, and for $0 \leq \phi < \pi$, put

$$(x, y, z) = (\rho\cos(\theta)\sin(\phi), \rho\sin(\theta)\sin(\phi), \rho\cos(\phi)).$$

Compute $dx \wedge dy \wedge dz$ in terms of $\rho, \theta, \phi, d\rho, d\theta, d\phi$.

EXERCISE 4.50. Express the complex 1-form $\frac{dz}{z}$ in terms of x, y, dx, dy. Express the form in (45) in terms of dz and $d\bar{z}$.

EXERCISE 4.51. Show that $dz \wedge d\bar{z} = -2i dx \wedge dy$. Put $z = re^{i\theta}$. Compute $dz \wedge d\bar{z}$.

EXERCISE 4.52. Prove formula (49).

EXERCISE 4.53. Put $\eta = dx_1 \wedge dx_2 + dx_3 \wedge dx_4$. Find $\eta \wedge \eta$. The answer is not 0. Explain.

EXERCISE 4.54. Verify that the forms in (45) and (46) are closed but not exact. (To show they are not exact, use Stokes' theorem on concentric circles or concentric spheres.) For $n \geq 3$, what is the analogue of (46) for the complement of the origin in $\mathbf{R}^n$?

EXERCISE 4.55. Use wedge products to give a test for deciding whether a collection of 1-forms is linearly independent.

EXERCISE 4.56. For $n \geq k \geq 2$, let $r_1, \ldots r_k$ be smooth real-valued functions on $\mathbf{C}^n$. Show that it is possible for $dr_1, \ldots, dr_k$ to be linearly independent while $\partial r_1, \ldots, \partial r_k$ are linearly dependent. Here $\partial r = \sum \frac{\partial r}{\partial z_j} dz_j$. This problem is even easier if we drop the assumption that the r_j are real-valued. Why?

9. Volumes of parametrized sets

Our next geometric inequality extends the ideas of Proposition 4.2 to higher dimensions. Things are more complicated for several reasons, but we obtain a sharp inequality on volumes of images of proper polynomial mappings between balls. We will also perform some computations from multi-variable calculus which are useful in many contexts.

We begin with a quick review of higher dimensional volume. Let Ω be an open subset of $\mathbf{R}^k$. Let $u_1, \ldots, u_k$ be coordinates on $\mathbf{R}^k$. The ordering of the u_j, or equivalently the du_j, defines the orientation on $\mathbf{R}^k$. We write

$$dV = dV_k = dV_k(u) = du_1 \wedge \cdots \wedge du_k$$

for the Euclidean volume form. When $u = F(x)$ is a change of variables, preserving the orientation, we obtain

$$dV(u) = \det(DF(x)) dV(x).$$

Suppose $F : \Omega \to \mathbf{R}^N$ is continuously differentiable and injective, except perhaps on a small set. Let us also assume that the derivative map $DF : \mathbf{R}^k \to \mathbf{R}^N$ is injective, again except perhaps on a small set. At each x, $DF(x)$ is a linear map from $T_x(\mathbf{R}^k) \to T_{F(x)}(\mathbf{R}^N)$. Let $(DF)(x)^*$ denote the transpose of $DF(x)$. Then $(DF)(x)^* : T_{F(x)}(\mathbf{R}^N) \to T_x(\mathbf{R}^k)$. The composition $(DF)^*(x)DF(x)$ is then a linear mapping from the space $T_x(\mathbf{R}^k)$ to itself, and hence its determinant is defined. The k-dimensional volume of the set $F(\Omega)$ is then given by an integral:

$$\mathrm{Vol}(F(\Omega)) = \int_\Omega \sqrt{\det((DF)^*DF)}dV_k. \tag{50}$$

EXAMPLE 4.13. Let Ω denote the unit disk in $\mathbf{R}^2$. Define $F_\alpha : \Omega \to \mathbf{R}^4$ by

$$F_\alpha(x, y) = (\cos(\alpha)x, \cos(\alpha)y, \sin(\alpha)(x^2 - y^2), \sin(\alpha)2xy).$$

Computation shows that

$$DF_\alpha = \begin{pmatrix} \cos(\alpha) & 0 \\ 0 & \cos(\alpha) \\ 2x\sin(\alpha) & -2y\sin(\alpha) \\ 2y\sin(\alpha) & 2x\sin(\alpha) \end{pmatrix}.$$

Matrix multiplication shows that $DF_\alpha^*(x, y)DF_\alpha(x, y)$ is the matrix in (51):

$$\begin{pmatrix} \cos^2(\alpha) + 4(x^2 + y^2)\sin^2(\alpha) & 0 \\ 0 & \cos^2(\alpha) + 4(x^2 + y^2)\sin^2(\alpha) \end{pmatrix}. \tag{51}$$

Hence $\sqrt{\det(DF_\alpha^*DF_\alpha)} = \cos^2(\alpha) + 4(x^2 + y^2)\sin^2(\alpha)$. Thus the area of the image of the unit disk B_1 under F_α is the integral:

$$\int_{B_1} (\cos^2(\alpha) + 4(x^2 + y^2)\sin^2(\alpha))dxdy = \pi(1 + \sin^2(\alpha)). \tag{52}$$

EXAMPLE 4.14. To anticipate a later development, we find the 3-dimensional volume of S^3. Let Ω denote the open subset of $\mathbf{R}^3$ defined by the inequalities: $0 < r < 1$, $0 < \theta < 2\pi$, $0 < \phi < 2\pi$. We parametrize (most of) S^3 by

$$(r, \theta, \phi) \mapsto F(r, \theta, \phi) = (r\cos(\theta), r\sin(\theta), s\cos(\phi), s\sin(\phi)).$$

Here $s = \sqrt{1 - r^2}$. Note that both θ and ϕ range from 0 to 2π; they are **not** the usual spherical coordinates on S^2. Computing DF and DF^* gives

$$(DF)^* = \begin{pmatrix} \cos(\theta) & \sin(\theta) & \frac{-r}{s^2}\cos(\phi) & \frac{-r}{s^2}\sin(\phi) \\ -r\sin(\theta) & r\cos(\theta) & 0 & 0 \\ 0 & 0 & -s\sin(\phi) & s\cos(\phi). \end{pmatrix}$$

Multiplying $(DF)^*$ by DF and computing determinants yields the 3-dimensional volume form $rdrd\theta d\phi$ on the sphere. Thus

$$\mathrm{Vol}(S^3) = \int_0^{2\pi} \int_0^{2\pi} \int_0^1 rdrd\theta d\phi = (2\pi)^2 \frac{1}{2} = 2\pi^2.$$

We are interested in images of sets in $\mathbf{C}^n$ under complex analytic mappings. When f is a complex-analytic and equi-dimensional mapping, we write f' for its derivative and Jf for its Jacobian determinant. Thus

$$Jf = \det\left(\frac{\partial f_j}{\partial z_k}\right).$$

Volume computations simplify in the complex-analytic case, even when f is not equi-dimensional. We could express Example 4.13 using the complex analytic map

f_α defined by $f_\alpha(z) = (\cos(\alpha)z, \sin(\alpha)z^2)$ and we easily obtain (51). The following result in the equi-dimensional case explains why:

LEMMA 4.8. *Suppose* $f : \Omega \subseteq \mathbf{C}^n \to \mathbf{C}^n$ *is complex analytic. Define* $F : \mathbf{R}^{2n} \to \mathbf{R}^{2n}$ *by* $F(x, y) = (\mathrm{Re}(f(x + iy)), \mathrm{Im}(f(x + iy)))$. *Then* $\det(DF) = |\det(f')|^2 = |Jf|^2$. *In particular,* F *preserves orientation.*

PROOF. When $\mathbf{u} = F(\mathbf{x})$ is a change of variables on $\mathbf{R}^k$, then $dV(\mathbf{u}) = \pm\det((DF)(\mathbf{x}))dV(\mathbf{x})$. The proof amounts to rewriting this equality using complex variables and their conjugates, and using the relationship between wedge products and determinants.

Put $w = f(z)$, where both z and w are in $\mathbf{C}^n$. Put $w = u + iv$ and $z = x + iy$. In real variables we have

$$dV_{2n}(u, v) = du_1 \wedge dv_1 \wedge \cdots \wedge du_n \wedge dv_n = \det(DF)dx_1 \wedge dy_1 \wedge \cdots \wedge dx_n \wedge dy_n. \quad (53)$$

We will write the volume forms in the $z, \overline{z}$ variables in the domain and the $w, \overline{w}$ variables in the target. Note that

$$dw_j = \sum \frac{\partial f_j}{\partial z_k} dz_k.$$

Hence $dw_1 \wedge \cdots \wedge dw_n = \det(\frac{\partial f_j}{\partial z_k}) \, dz_1 \wedge \cdots \wedge dz_n = (Jf) \, dz_1 \wedge \cdots \wedge dz_n$.

Recall from Exercise 4.51 that $dz_j \wedge d\overline{z}_j = (-2i)dx_j \wedge dy_j$ and similarly for the w variables. Putting everything together we get

$$dV_{2n}(u, v) = du_1 \wedge dv_1 \wedge \cdots \wedge du_n \wedge dv_n = (\frac{i}{2})^n dw_1 \wedge d\overline{w}_1 \wedge \cdots \wedge dw_n \wedge d\overline{w}_n$$

$$= |\det(f'(z))|^2 (\frac{i}{2})^n dz_1 \wedge d\overline{z}_1 \wedge \cdots \wedge dz_n \wedge d\overline{z}_n$$

$$= |\det(f'(z))|^2 dx_1 \wedge dy_1 \wedge \cdots \wedge dx_n \wedge dy_n = |\det(f'(z))|^2 dV_{2n}(x, y). \quad (54)$$

Comparing (53) and (54) finishes the proof.   □

EXERCISE 4.57. Prove (54) using the real form of the Cauchy-Riemann equations. The computation is somewhat punishing; do it only in two complex variables where you will deal with four-by-four matrices.

We continue discussing higher dimensional volumes of complex analytic images. Let Ψ denote the differential form on $\mathbf{C}^N$ defined by

$$\Psi = \frac{i}{2} \sum_{j=1}^{N} d\zeta_j \wedge d\overline{\zeta}_j.$$

The factor $\frac{i}{2}$ arises because $dz \wedge d\overline{z} = -2idx \wedge dy$ in one dimension. See Exercise 4.45. The form Ψ^k, where we wedge Ψ with itself k times, is used to define $2k$-dimensional volume. As before we take multiplicity into account.

DEFINITION 4.17. ($2k$-dimensional volume) Let Ω be an open subset in $\mathbf{C}^k$, and suppose that $f : \Omega \to \mathbf{C}^N$ is complex analytic. We define $V_{2k}(f, \Omega)$, the $(2k)$-dimensional volume with multiplicity counted, by (55):

$$V_{2k}(f, \Omega) = \int_\Omega \frac{(f^*\Psi)^k}{k!} = \frac{1}{k!}(\frac{i}{2})^k \int_\Omega (\sum_{j=1}^{N} \partial f_j \wedge \overline{\partial f_j})^k. \quad (55)$$

REMARK 4.13. Equation (55) is the natural definition based on our L^2 perspective. When f is not injective, the formula takes multiplicity into account. For $w \in \mathbf{C}^N$, let $\#(f, w)$ denote the number of points in $\Omega \cap f^{-1}(w)$. Then we could define $V_{2k}(f, \Omega)$ by

$$V_{2k}(f, \Omega) = \int_{\mathbf{C}^N} \#(f, w) d\mathbf{h}^{2k}(w).$$

Here $d\mathbf{h}^{2k}(w)$ is the $2k$-dimensional Hausdorff measure. The so-called *area formula* from Geometric Measure Theory shows under rather general hypotheses, met in our context, that this computation agrees with (55).

We are primarily interested in the case when Ω is the unit ball B_k; in this case we abbreviate $V_{2k}(f, \Omega)$ by V_f. In (55) the upper star notation denotes pullback, and the $k!$ arises because there are $k!$ ways to permute the indices from 1 to k. The form $\frac{(f^*\Psi)^k}{k!}$ is rdV, where $dV = dV_{2k}$ is the Euclidean volume form in k complex dimensions, for some function r depending on f. The next section provides techniques for evaluating the resulting integrals.

REMARK 4.14. Caution! In the complex 2-dimensional case, the volume form is $h \, dV_4$, where $h = EG - |F|^2$, and

$$E = \|\frac{\partial f}{\partial z}\|^2,$$

$$G = \|\frac{\partial f}{\partial w}\|^2,$$

$$F = \langle \frac{\partial f}{\partial z}, \frac{\partial f}{\partial w} \rangle.$$

No square root appears here. By contrast, in the real case, the classical formula for the surface area form is $\sqrt{EG - F^2}$, where E, G, F have analogous definitions.

EXAMPLE 4.15. We consider several maps from B_2 to $\mathbf{C}^3$. Using (55) and the methods of the next section we obtain the following values:

(1) Put $g(z, w) = (z, 0, w)$. Then $V_g = \frac{\pi^2}{2}$.
(2) For $0 \le \lambda \le \sqrt{2}$, put $f(z, w) = (z^2, \lambda zw, w^2)$. Then $V_f = \frac{2(\lambda^2+1)}{3}\pi^2$.

The first map is injective, and V_f gives the volume of the image. For $\lambda \neq 0$, the second map is generically two-to-one. If (a, b, c) is in the image of f, and (a, b, c) is not the origin, then $f^{-1}(a, b, c)$ has precisely two points. When $\lambda^2 = 2$, we obtain 4 times the volume of the unit ball. This volume is computed in Corollary 4.8. When $\lambda = 0$, the answer is $\frac{4}{3}$ times the volume of the unit ball.

EXAMPLE 4.16. Define $h : \mathbf{C}^2 \to \mathbf{C}^3$ by $h(z, w) = (z, zw, w^2)$. This map and its generalization to higher dimensions will play an important role in our work, because h maps the unit sphere in $\mathbf{C}^2$ into the unit sphere in $\mathbf{C}^3$. Here it illustrates the subtleties involved in computing multiplicities. Let $p = (a, b, c)$ be a point in $\mathbf{C}^3$. Suppose first that $a \neq 0$. Then $h^{-1}(p)$ is empty unless $b^2 = ca^2$, in which case $h^{-1}(p)$ is a single point. When $a = 0$, things change. If $b \neq 0$, then $h^{-1}(p)$ is empty. If $a = b = 0$, then $h^{-1}(p)$ consists of two points for $c \neq 0$ and one point with multiplicity two if $c = 0$.

We will use the expanded version of the far right-hand side of (56) to compute volumes. Let Ω be an open set in $\mathbf{C}^k$, and assume that $f : \Omega \to \mathbf{C}^N$ is complex analytic. Here we allow the target dimension to differ from the domain dimension. We define the pointwise squared Jacobian $||Jf||^2$ by

$$||Jf||^2 = \sum |J(f_{i_1}, ..., f_{i_k})|^2 = \sum |J(f_I)|^2. \tag{56}$$

The sum in (56) is taken over all increasing k-tuples. Equivalently, we form all possible Jacobian determinants of k of the component functions, and sum their squared moduli. Recall, in the equi-dimensional case, that

$$Jg = \det\left(\frac{\partial g_j}{\partial z_k}\right).$$

EXERCISE 4.58. Let $\alpha = \sum_{j=1}^{3} \partial f_j \wedge \overline{\partial f_j}$. Find $\alpha \wedge \alpha \wedge \alpha$ by expanding and compare with (56).

The next lemma provides another method for finding V_f. Let r be a twice differentiable function of several complex variables. The complex Hessian of r is the matrix $(r_{j\overline{k}}) = \left(\frac{\partial^2 r}{\partial z_j \partial \overline{z_k}}\right)$. Lemma 4.9 relates the determinant of the Hessian of $||f||^2$ to the Jacobian Jf, when f is a complex analytic mapping. This lemma allows us to compute one determinant, rather than many, even when $N > n$.

LEMMA 4.9. If $f : \mathbf{C}^k \to \mathbf{C}^N$ is complex analytic, then $||Jf||^2 = \det\left((||f||^2)_{j\overline{k}}\right)$.

PROOF. See Exercise 4.59. $\qquad\qquad\qquad\qquad\qquad\qquad\qquad\qquad\qquad\qquad\square$

To find the volume (with multiplicity accounted for) of the image of a complex analytic mapping $f : \Omega \subseteq \mathbf{C}^k \to \mathbf{C}^N$, we must either integrate the determinant of the Hessian of $||f||^2$, or sum the L^2 norms of each Jacobian $J(f_{j_1}, ..., f_{j_k})$ formed from the components of f:

$$V_{2k}(f, \Omega) = \int_\Omega ||Jf||^2 dV_{2k} = \int_\Omega \det\left((||f||^2)_{j\overline{k}}\right) dV_{2k}. \tag{57}$$

EXERCISE 4.59. Put $r(z, \overline{z}) = \sum_{j=1}^{N} |f_j(z)|^2 = ||f(z)||^2$. Use differential forms to prove Lemma 4.9.

10. Volume computations

Our next goal is to compute the $2n$-dimensional volume of the image of the unit ball in $\mathbf{C}^n$ under the mapping $z \mapsto z^{\otimes m}$. As a warm-up, suppose $n = 1$. Then the map $z \mapsto z^m$ covers the ball m times, and hence the area of the image with multiplicity counted is πm. We get the same answer using integrals:

$$A = \int_{B_1} |mz^{m-1}|^2 dV = m^2 \int_0^{2\pi} \int_0^1 r^{2m-1} dr d\theta = m^2 \frac{2\pi}{2m} = \pi m. \tag{58}$$

In order to help us do computations and to simplify the notation, we recall and extend our discussion of multi-index notation from Section 4. A multi-index α is an n-tuple $\alpha = (\alpha_1, \dots, \alpha_n)$ of non-negative numbers, not necessarily integers. When the α_j are integers, we write $|\alpha| = \sum_{j=1}^{n} \alpha_j$ and $\alpha! = \prod_{j=1}^{n} \alpha_j!$. In case $d = |\alpha|$, we write multinomial coefficients using multi-indices:

$$\binom{d}{\alpha} = \frac{d!}{\alpha!} = \frac{d!}{\alpha_1! \dots \alpha_n!}.$$

Multi-indices are especially useful for writing polynomials and power series. If $z \in \mathbf{C}^n$, we write

$$z^\alpha = \prod_{j=1}^n (z_j)^{\alpha_j}$$

$$|z|^{2\alpha} = \prod_{j=1}^n |z_j|^{2\alpha_j}.$$

The multinomial theorem gives the following result from Section 4:

$$||z||^{2d} = (\sum_{j=1}^n |z_j|^2)^d = \sum_{|\alpha|=d} \binom{d}{\alpha} |z|^{2\alpha}.$$

In order to help us find volumes in higher dimensions we introduce the Γ-function. For $x > 0$, we let $\Gamma(x)$ denote the Gamma function:

$$\Gamma(x) = \int_0^\infty e^{-t} t^{x-1} dt.$$

The integral is improper at $t = 0$ for $x < 1$, but it converges there for $x > 0$. When n is an integer and $n \geq 0$, then $\Gamma(n+1) = n!$. More generally, $\Gamma(x+1) = x\Gamma(x)$. This property enables one to extend the definition of the Γ-function. The integral defining $\Gamma(x)$ converges when x is complex and $\mathrm{Re}(x) > 0$. The formula $\Gamma(x+1) = x\Gamma(x)$ provides a definition when $-1 < \mathrm{Re}(x) < 0$, and by induction, a definition whenever $\mathrm{Re}(x)$ is not a negative integer or zero.

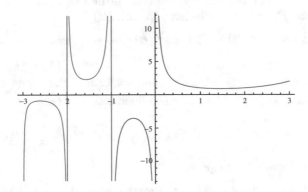

FIGURE 5. The Gamma function

It is useful to know that $\Gamma(\frac{1}{2}) = \sqrt{\pi}$. Exercise 4.61 asks for a proof; the result is equivalent to the evaluation of the Gaussian integral from Proposition 3.4. One squares the integral and changes variables appropriately.

Let K_+ denote the part of the unit ball in $\mathbf{R}^n$ lying in the first orthant; that is, $K_+ = \{x : \sum x_j^2 \leq 1 \text{ and } x_j \geq 0 \text{ for all } j\}$. Let α be an n-tuple of positive real numbers. We define an n-dimensional analogue of the Euler Beta function by

$$\mathcal{B}(\alpha) = \frac{\prod \Gamma(\alpha_j)}{\Gamma(|\alpha|)}. \tag{59}$$

The expression (59) is the value of a certain integral:

$$\mathcal{B}(\alpha) = 2^n |\alpha| \int_{K_+} \mathbf{r}^{2\alpha-1} dV(\mathbf{r}). \tag{60}$$

Note the use of multi-index notation in (60); $2\alpha - 1$ means the multi-index whose j-th entry is $2\alpha_j - 1$. Thus $\mathbf{r}^{2\alpha-1}$ means

$$\prod_{j=1}^{n} \mathbf{r}_j^{2\alpha_j - 1}.$$

The notation $\mathbf{r} = (\mathbf{r}_1, \ldots, \mathbf{r}_n)$ has a specific purpose. Certain integrals over balls in $\mathbf{C}^n$ (See Lemma 4.10) reduce to integrals such as (60) when we use polar coordinates in each variable separately; that is, $z_j = \mathbf{r}_j e^{i\theta_j}$.

COROLLARY 4.8. *The volume of the unit ball in* $\mathbf{R}^n$ *is* $\frac{\Gamma(\frac{1}{2})^n}{\Gamma(\frac{n}{2}+1)}$.

PROOF. Put $\alpha = (\frac{1}{2}, \frac{1}{2}, \ldots, \frac{1}{2})$ in (60) and use (59). □

EXERCISE 4.60. Verify that $\Gamma(x+1) = x\Gamma(x)$ and $\Gamma(n+1) = n!$.

EXERCISE 4.61. Show that $\Gamma(\frac{1}{2}) = \sqrt{\pi}$.

EXERCISE 4.62. Express the formula for the volume of the unit ball in $\mathbf{R}^n$ in the form $c_n \pi^n$. (Use the previous two exercises.)

EXERCISE 4.63. Put $\beta(a,b) = \int_0^1 t^{a-1}(1-t)^{b-1} dt$ for $a, b > 0$. This integral is the classical Euler Beta function. By first computing $\Gamma(a)\Gamma(b)$, evaluate it in terms of the Γ-function. Explain the relationship with (60).

EXERCISE 4.64. Prove that (59) and (60) are equivalent.

REMARK 4.15. Integrals of the form $\int_0^{2\pi} \cos^k(\theta)\sin^l(\theta) d\theta$ (for integer exponents) are easily evaluated by using the complex form of the exponential. Integrals of the form $\int_0^{\frac{\pi}{2}} \cos^k(\theta)\sin^l(\theta) d\theta$ are harder. Such integrals reduce to Beta functions:

$$\beta(a,b) = \int_0^1 t^{a-1}(1-t)^{b-1} dt = 2 \int_0^{\frac{\pi}{2}} \sin^{2a-1}(\theta)\cos^{2b-1}(\theta) d\theta,$$

even when a and b are not integers.

EXERCISE 4.65. Use the Euler Beta function to verify the following *duplication formula* for the Γ function.

$$\frac{\Gamma(x)}{\Gamma(2x)} = 2^{1-2x} \frac{\Gamma(\frac{1}{2})}{\Gamma(x+\frac{1}{2})}. \tag{61}$$

Suggestion. First multiply both sides by $\Gamma(x)$. The left-hand side of the result is then $\beta(x,x)$. Write it as a single integral over $[0,1]$ as in Exercise 4.63. Rewrite by symmetry as twice the integral over $[0, \frac{1}{2}]$. Then change variables by $2t = 1 - \sqrt{s}$. You will obtain $2^{1-2x}\beta(x, \frac{1}{2})$ and (61) follows.

EXERCISE 4.66. Put $\phi(x,y) = \frac{\Gamma(x)\Gamma(x+y)}{\Gamma(2x)\Gamma(y)}$. Find $\phi(x, \frac{1}{2})$ and $\phi(x, \frac{3}{2})$. Show that

$$\phi(x, \frac{5}{2}) = 2^{1-2x} \frac{(1+2x)(3+2x)}{3}.$$

EXERCISE 4.67. (Difficult) Verify the following formula for $\Gamma(z)\Gamma(1-z)$:

$$\Gamma(z)\Gamma(1-z) = \frac{\pi}{\sin(\pi z)}.$$

Suggestion: First write the left-hand side as a Beta function integral over $[0,1]$. Convert it to an integral over $[0,\infty)$ by setting $t = \frac{s}{s+1}$. Then use contour integration. The computation is valid for all complex numbers except the integers. See also Exercise 3.45.

The Γ function also arises naturally in the following exercise.

EXERCISE 4.68. (For those who know probability). Let X be a Gaussian random variable with mean 0 and variance σ^2. Use the fundamental theorem of calculus to find the density of the random variable X^2. The answer is called the Γ-density with parameters $\frac{1}{2}$ and $\frac{1}{2\sigma^2}$. Use this method to show that $\Gamma(\frac{1}{2}) = \sqrt{\pi}$.

We will evaluate several integrals using the n-dimensional Beta function. Recall the notation $|z|^{2\alpha} = \prod |z_j|^{2\alpha_j}$ used in (60).

LEMMA 4.10. Let d be a non-negative integer, and let α be a multi-index of non-negative real numbers. Let $\mathbb{B}_n$ denote the unit ball in $\mathbf{C}^n$. Then

$$\int_{\mathbb{B}_n} ||z||^{2d} dV = \frac{\pi^n}{(n-1)!(n+d)}. \tag{62.1}$$

$$\int_{\mathbb{B}_n} |z|^{2\alpha} dV = \frac{\pi^n}{(n+|\alpha|)} \mathcal{B}(\alpha+1). \tag{62.2}$$

PROOF. We use polar coordinates in each variable separately; to evaluate (62.1) we have

$$I = \int_{\mathbb{B}_n} ||z||^{2d} dV_{2n} = (2\pi)^n \int_{K_+} ||\mathbf{r}||^{2d} \prod \mathbf{r}_j dV_n.$$

We then expand $||\mathbf{r}||^{2d}$ using the multinomial theorem to obtain (63):

$$I = \pi^n 2^n \sum_{|\gamma|=d} \binom{d}{\gamma} \int_{K_+} \mathbf{r}^{2\gamma+1} dV_n. \tag{63}$$

Using formulas (59) and (60) for the Beta function in (63) we obtain

$$I = \pi^n \sum_{|\gamma|=d} \binom{d}{\gamma} \frac{\mathcal{B}(\gamma+1)}{|\gamma+1|} = \pi^n \sum_{|\gamma|=d} \frac{d!}{\prod \gamma_j} \frac{\prod \gamma_j}{(d+n)\Gamma(d+n)} = \pi^n \frac{d!}{(d+n)!} \sum_{|\gamma|=d} 1.$$

By Exercise 4.30, the number of independent homogeneous monomials of degree d in n variables is $\binom{n+d-1}{d}$. We replace the sum in the last term with this number to obtain the desired result:

$$I = \pi^n \frac{d!}{(d+n)!} \frac{(n+d-1)!}{(n-1)!d!} = \frac{\pi^n}{(n-1)!(n+d)}. \tag{64}$$

The calculation of (62.2) is similar but easier as there is no summation to compute:

$$\int_{\mathbb{B}_n} |z|^{2\alpha} dV_{2n} = (2\pi)^n \int_{K_+} \mathbf{r}^{2\alpha+1} dV_n = \pi^n \frac{\mathcal{B}(\alpha+1)}{|\alpha|+n}. \qquad \square$$

For convenience we write (62.2) when $n = 2$ and a, b are integers:

$$\int_{\mathbb{B}_2} |z|^{2a} |w|^{2b} dV_4 = \frac{\pi^2 a! b!}{(a+b+2)!}.$$

We return to the homogeneous mapping $H_m(z)$. We consider $H_m : \mathbb{B}_k \to \mathbf{C}^N$, where $N = \binom{k+m-1}{k-1}$, the dimension of the space of homogeneous polynomials of degree m in k variables. We use the following lemma to find (Theorem 4.9) an explicit formula for the $2k$-dimensional volume (with multiplicity counted) of the image of the unit ball under H_m.

LEMMA 4.11. *The pullback k-th power $(H_m^*(\Psi))^k$ satisfies the following:*

$$(H_m^*(\Psi))^k = m^{k+1} k! ||z||^{2k(m-1)} dV_{2k}. \tag{65}$$

PROOF. Note first that $(H_m^*(\Psi))^k$ is a smooth $(2k)$-form, and hence a multiple τ of dV_{2k}. Note next that H_m is invariant under unitary transformations, and therefore τ must be a function of $||z||^2$. Since H_m is homogeneous of degree m, each first derivative is homogeneous of degree $m - 1$. The $(1,1)$ form $H_m^*(\Psi)$ must then have coefficients that are bihomogeneous of degree $(m-1, m-1)$. The coefficient τ of its k-th power must be homogeneous of degree $2k(m-1)$. Combining the homogeneity with the dependence on $||z||^2$ gives the desired expression, except for evaluating the constant $m^{k+1} k!$.

For simplicity we write $|dz_j|^2$ for $dz_j \wedge d\bar{z}_j$. To evaluate the constant it suffices to compute the coefficient of $|z_1|^{2k(m-1)}$. To do so, we compute dH_m, and then work modulo $z_2, ..., z_n$. Thus, in the formula for $(H_m^*(\Psi))^k$ we set all variables equal to zero except the first. Doing so yields

$$H_m^*(\Psi) = m^2 |z_1|^{2m-2} |dz_1|^2 + m |z_1|^{2m-2} \sum_{j=2}^{k} |dz_j|^2. \tag{66}$$

From (66) it suffices to compute

$$(m^2 |dz_1|^2 + m \sum_{j=2}^{k} |dz_j|^2)^k. \tag{67}$$

Expanding (67) yields

$$k! m^{k+1} dz_1 \wedge d\bar{z}_1 \wedge ... \wedge dz_k \wedge d\bar{z}_k,$$

and (65) follows by putting the factor $|z_1|^{(2m-2)k}$ from (66) back in. $\qquad\square$

THEOREM 4.9. *Let $f : \mathbb{B}_n \to \mathbb{B}_K$ be a proper complex analytic homogeneous polynomial mapping of degree m. The $2n$-dimensional volume V_f (with multiplicity counted) is given by*

$$V_f = m^n \pi^n \frac{1}{n!}. \tag{68}$$

PROOF. Consider the function $||f||^2$. Since

$$||f(z)||^2 = 1 = ||z||^{2m} = ||H_m(z)||^2$$

on the unit sphere, and both f and H_m are homogeneous, this equality holds everywhere. Hence $||f||^2 = ||H_m||^2$ and these two functions have the same complex Hessian determinant. By Lemma 4.9 they determine the same volume form:

$$\sum_I |J(f_I)|^2 = \sum_I |J((H_m)_I))|^2,$$

and hence by Lemma 4.11

$$V_f = \int_{\mathbb{B}_n} \frac{(H_m^*(\Psi))^n}{n!} = \int_{\mathbb{B}_n} m^{n+1}||z||^{2n(m-1)} dV_{2n}.$$

Lemma 4.10 yields

$$V_f = m^{n+1} \frac{\pi^n}{(n(m-1)+n)} \frac{1}{(n-1)!} = \frac{m^n \pi^n}{n!}.$$

As a check we observe, when $m = 1$, that $V_f = \frac{\pi^n}{n!}$, which is the volume of the unit ball. When $n = 1$, we obtain $V_f = \pi m$, also the correct result, as noted in (58). $\square$

The factor of m^n in (68) arises because the image of the unit sphere in $\mathbf{C}^n$ covers m times a subset of the unit sphere in the target. Compare with item (2) of Example 4.11.

11. Inequalities

We are now ready to state a sharp inequality in Theorem 4.10. The proof of this volume comparison result combines Theorems 4.6, 4.9, and Theorem 4.11 (proved below). Theorem 4.11 generalizes Proposition 4.2 to higher dimensions. Our proof here uses differential forms; the result can also be proved by elaborate computation. See [D4] for the computational proof.

THEOREM 4.10. *Let* $p : \mathbf{C}^n \to \mathbf{C}^N$ *be a polynomial mapping of degree* m. *Assume that* $p(S^{2n-1}) \subseteq S^{2N-1}$. *Then* $V_p \le \frac{m^n \pi^n}{n!}$. *Equality happens if and only if* p *is homogeneous of degree* m.

PROOF. If p is a constant mapping, then $m = 0$ and the conclusion holds. When p is homogeneous of degree m, the result is Theorem 4.9. When p is not homogeneous, we apply the process from Theorem 4.6 until we obtain a homogeneous mapping. The key point is that the operation of tensoring with z on a subspace A increases the volume of the image, in analogy with Proposition 4.2. Since tensoring on a k-dimensional subspace gives the same result as tensoring k times on one-dimensional subspaces, we need only show that the volume of the image increases if we tensor on a one-dimensional space.

We must therefore establish the following statement, which we state and prove as Theorem 4.11 below. Put $f = (f_1, \ldots, f_N)$. Put

$$g = (z_1 f_1, \ldots, z_n f_1, f_2, \ldots, f_N). \tag{69}$$

Then $V_f \le V_g$, with equality only if $f_1 = 0$.

Each tensor operation from Theorem 4.6 then increases the volume. We stop when we reach a homogeneous map. Theorem 4.9 then gives the volume $\frac{m^n \pi^n}{n!}$, the stated upper bound. $\square$

With g as in (69), we need to verify that $V_f \le V_g$. We proved this result (Corollary 4.2) when $n = N = 1$, in two ways. As noted above, one can prove the general result in both fashions. We give the proof involving a boundary integral.

Let us first recall what we mean by the volume form on the unit sphere in $\mathbf{R}^N$. It is convenient to introduce the notion of interior multiplication. Assume η is a k-form, and write

$$\eta = dx_j \wedge \tau + \mu,$$

where μ does not contain dx_j. The *contraction* in the j-th direction, or *interior product* with $\frac{\partial}{\partial x_j}$, is the $(k-1)$-form $I_j(\eta)$, defined by $I_j(\eta) = \tau$. Informally speaking, we are eliminating dx_j from η. More precisely, we define $I_j(\eta)$ by its action on vectors $v_2, \ldots, v_k$:

$$I_j(\eta)(v_2, \ldots, v_k) = \eta(\frac{\partial}{\partial x_j}, v_2, \ldots, v_k).$$

We use this notation to write a standard expression from calculus. The Euclidean $(N-1)$-dimensional volume form on the sphere is given by:

$$\sigma_{N-1} = \sum_{j=1}^{N} x_j(-1)^{j+1} I_j(dx_1 \wedge \cdots \wedge dx_N).$$

For example, when $N = 2$ (and x, y are the variables), we have $\sigma_1 = x\,dy - y\,dx$. When $N = 3$ (and x, y, z are the variables), we have

$$\sigma_2 = x\,dy \wedge dz - y\,dx \wedge dz + z\,dx \wedge dy.$$

Note that $d\sigma_{N-1} = N\,dV_N$, where dV_N is the volume form on Euclidean space. It follows immediately from Stokes' theorem that the $(N-1)$-dimensional volume of the unit sphere is N times the N-dimensional volume of the unit ball.

REMARK 4.16. In the previous paragraph, σ_{N-1} is a differential form, and $d\sigma_{N-1}$ is its exterior derivative. Calculus books often write $d\sigma$ or dS for the surface area form (and ds for the arc-length form), even though these objects are not differential forms. The symbol d is simply irresistible.

EXERCISE 4.69. Verify the following formulas for the $(N-1)$-dimensional volume W_N of the unit sphere in $\mathbf{R}^N$:

- $W_1 = 2$
- $W_2 = 2\pi$
- $W_3 = 4\pi$
- $W_4 = 2\pi^2$
- $W_5 = \frac{8}{3}\pi^2$.

Put $\rho(z) = ||z||^2$. The unit sphere S^{2n-1} is the set of points where $\rho = 1$. The differential form $d\rho$ is orthogonal to the sphere at each point, and the cotangent space to the sphere is the orthogonal complement to $d\rho$. The decomposition $d\rho = \partial\rho + \bar{\partial}\rho$ will be crucial to our proof. Since $d\rho$ is orthogonal to the sphere, we may use the relation $\partial\rho = -\bar{\partial}\rho$ when doing integrals over the sphere.

We can express the form σ_{2n-1} in terms of complex variables. Let $W_{j\bar{j}}$ denote the $(2n-2)$-form defined by eliminating $dz_j \wedge d\bar{z}_j$ from $dz_1 \wedge d\bar{z}_1 \wedge \cdots \wedge dz_n \wedge d\bar{z}_n$. For $1 \le j \le n$, put $z_j = x_j + iy_j$. Write $x_j = \frac{z_j + \bar{z}_j}{2}$ and $y_j = \frac{z_j - \bar{z}_j}{2i}$. Substituting in the form σ_{2n-1} and collecting terms, we obtain

$$\sigma_{2n-1} = (\frac{i}{2})^n \sum_{j=1}^{n} (z_j\,d\bar{z}_j - \bar{z}_j\,dz_j) \wedge W_{j\bar{j}}. \tag{70}$$

As a check, we note when $n = 1$ that this expression equals $\frac{i}{2}(z d\bar{z} - \bar{z} dz)$. Putting $z = e^{i\theta}$ then yields $d\theta$, as expected. As a second check, we compute d of the right-hand side of (70), using $dz_j \wedge d\bar{z}_j = -2i \, dx_j \wedge dy_j$, obtaining

$$(\frac{i}{2})^n (2n)(-2i)^n dV_{2n} = 2n \, dV_{2n},$$

as expected (since we are in $2n$ real dimensions).

With these preparations we can finally show that the tensor product operation increases volumes; in other words, $V_{Ef} > V_f$ (unless $f_1 = 0$).

THEOREM 4.11. *Assume that $f = (f_1, ..., f_N)$ is complex analytic on the unit ball $\mathbb{B}_n$ in $\mathbf{C}^n$. Define the partial tensor product Ef by*

$$Ef = (z_1 f_1, z_2 f_1, \ldots, z_n f_1, f_2, \ldots, f_N).$$

Then $V_{Ef} > V_f$ unless $f_1 = 0$.

PROOF. We prove the result assuming f has a continuously differentiable extension to the boundary sphere. [D4] has a proof without this assumption.

Recall that $V_f = \int_{\mathbb{B}_n} ||Jf||^2 dV$. Here, as in (53), Jf denotes all possible Jacobians formed by selecting n of the components of f. In case f is an equidimensional mapping, we also have

$$V_f = c_n \int_{\mathbb{B}_n} \partial f_1 \wedge \overline{\partial f_1} \wedge \partial f_2 \wedge \overline{\partial f_2} \wedge \cdots \wedge \partial f_n \wedge \overline{\partial f_n}. \tag{71}$$

In general V_f is a sum of integrals, as in (71), over all choices of n components. The constant c_n equals $(\frac{i}{2})^n$; see the discussion near Definition 4.16.

We want to compute $V_{Ef} = \int ||J(Ef)||^2$. Many terms arise. We partition these terms into three types. Type I terms are those for which the n functions selected among the components of Ef include none of the functions $z_j f_1$ for $1 \leq j \leq n$. These terms also arise when computing V_f. Hence terms of type I drop out when computing the difference $V_{Ef} - V_f$, and we may ignore them. Type II terms are those for which we select at least two of the functions $z_j f_1$. These terms arise in the computation of V_{Ef}, but not in the computation of V_f. All of these terms thus contribute non-negatively. The type III terms remain. They are of the form $(z_j f_1, f_{i_2}, \ldots, f_{i_n})$. We will show, for each choice $(f_{i_2}, \ldots, f_{i_n})$ of $n - 1$ of the functions $f_2, ..., f_N$, that the sum on j of the volumes of the images of $(z_j f_1, f_{i_2}, \ldots, f_{i_n})$ is at least as large as the volume of the image of the map $(f_1, f_{i_2}, \ldots, f_{i_n})$. Combining these conclusions shows that $V_{Ef} \geq V_f$.

For simplicity of notation, let us write the $(n-1)$-tuple as $(f_2, \ldots, f_n)$. By the above paragraph, it suffices to prove the result when $f = (f_1, \ldots, f_n)$ is an equi-dimensional mapping. In the rest of the proof we let f denote this n-tuple.

Since f_1 is complex analytic, $df_1 = \partial f_1$. By the closing paragraph of Section 6, we can write the form in (71) as an exact form. We then apply Stokes' theorem to get

$$V_f = c_n \int_{\mathbb{B}_n} d(f_1 \wedge \overline{\partial f_1} \wedge \partial f_2 \wedge \overline{\partial f_2} \wedge \cdots \wedge \partial f_n \wedge \overline{\partial f_n})$$

$$= c_n \int_{S^{2n-1}} f_1 \wedge \overline{\partial f_1} \wedge \partial f_2 \wedge \overline{\partial f_2} \wedge \cdots \wedge \partial f_n \wedge \overline{\partial f_n}. \tag{72}$$

For $1 \leq j \leq n$ we replace f_1 in (72) with $z_j f_1$ and sum, obtaining

$$V_{Ef} \geq c_n \sum_{j=1}^{n} \int_{S^{2n-1}} z_j f_1 \wedge \overline{\partial(z_j f_1)} \wedge \partial f_2 \wedge \overline{\partial f_2} \wedge \cdots \wedge \partial f_n \wedge \overline{\partial f_n}. \qquad (73)$$

Note that $\partial(z_j f_1) = f_1 dz_j + z_j df_1$ by the product rule. Using this formula in (73) and then subtracting (72) from (73) shows that the excess is at least

$$V_{Ef} - V_f \geq c_n \int_{S^{2n-1}} (\sum_{j=1}^{n} |z_j|^2 - 1) f_1 \overline{\partial f_1} \wedge \partial f_2 \wedge \overline{\partial f_2} \wedge \cdots \wedge \partial f_n \wedge \overline{\partial f_n}$$

$$+ c_n \int_{S^{2n-1}} |f_1|^2 (\sum_{j=1}^{n} z_j d\overline{z}_j) \wedge \partial f_2 \wedge \overline{\partial f_2} \wedge \cdots \wedge \partial f_n \wedge \overline{\partial f_n}. \qquad (74)$$

Since $\sum |z_j|^2 = 1$ on the sphere, the expression in the top line of (74) vanishes. We claim that the other term is non-negative. We will show that the form

$$c_n |f_1|^2 (\sum_{j=1}^{n} z_j d\overline{z}_j) \wedge \partial f_2 \wedge \overline{\partial f_2} \wedge \cdots \wedge \partial f_n \wedge \overline{\partial f_n}$$

arising in (74) is a non-negative multiple of the real $(2n - 1)$-dimensional volume form on the sphere, and hence its integral is non-negative.

It suffices to prove that the form

$$\eta = c_n \overline{\partial} \rho \wedge \partial f_2 \wedge \overline{\partial f_2} \wedge \cdots \wedge \partial f_n \wedge \overline{\partial f_n} \qquad (75)$$

is a non-negative multiple of the volume form on the sphere.

Note that $\partial f_j = df_j$, because f_j is complex analytic. We wish to write df_j in terms a particular basis of 1-forms. We would like to find independent differential 1-forms $\omega_1, ..., \omega_{n-1}$, with the following properties. Each of these forms involves only the dz_j (not the $d\overline{z}_j$). Each ω_j is in the cotangent space to the sphere. Finally, these forms, their conjugates, and the additional forms $\partial \rho$ and $\overline{\partial} \rho$ are linearly independent at each point. Doing so is not generally possible, but we can always find $\omega_1, \ldots, \omega_{n-1}$ such that linear independence holds except on a small set. After the proof (Remark 4.17), we explain how to do so.

Given these forms, we work on the set U where linear independence holds. We compute the exterior derivatives of the f_j for $2 \leq j \leq n$ in terms of this basis:

$$df_j = \partial f_j = \sum_{k=1}^{n-1} B_{jk} \omega_k + B_j \partial \rho.$$

On the intersection of U and the sphere, we obtain

$$df_j = \partial f_j = \sum_{k=1}^{n-1} B_{jk} \omega_k + B_j \partial \rho = \sum_{k=1}^{n-1} B_{jk} \omega_k - B_j \overline{\partial} \rho.$$

$$\overline{\partial f_j} = \sum_{k=1}^{n-1} \overline{B_{jk}} \overline{\omega}_k + \overline{B}_j \overline{\partial} \rho.$$

In these formulas, B_{jk} denotes the coefficient function; B_{jk} can be written $L_k(f_j)$ for complex vector fields L_k dual to the ω_k.

These formulas allow us compute the wedge product in (75) very easily. We can ignore all the functions B_j, because the wedge product of $\overline{\partial}\rho$ with itself is 0. We obtain

$$\eta = c_n |\det(B_{jk})|^2 \overline{\partial}\rho \wedge \omega_1 \wedge \overline{\omega}_1 \wedge \cdots \wedge \omega_{n-1} \wedge \overline{\omega}_{n-1}. \tag{76}$$

In (76), the index k runs from 1 to $n-1$, and the index j runs from 2 to n. Hence it makes sense to take the determinant of the square matrix B_{jk} of functions. Since the ω_k and their conjugates are orthogonal to the normal direction $d\rho$, the form in (76) is a nonnegative multiple of σ_{2n-1}.

We have verified that $V_{Ef} - V_f \geq 0$.   □

REMARK 4.17. Let f and Ef be as in Theorem 4.11. Assume f_1 is not identically 0. For all z in the ball, $||(Ef)(z)||^2 \leq ||f(z)||^2$, with strict inequality except where $f_1(z) = 0$. There is no pointwise inequality relating $\det\left((||Ef||^2)_{j\overline{k}}\right)$ and $\det\left((||f||^2)_{j\overline{k}}\right)$. But, Theorem 4.11 and Lemma 4.9 yield

$$\int \det\left((||Ef||^2)_{j\overline{k}}\right) dV > \int \det\left((||f||^2)_{j\overline{k}}\right) dV.$$

Thus $||Ef||^2$ is (pointwise) smaller than $||f||^2$, there is no pointwise inequality between their Hessian determinants, but the average value (integral) of the Hessian determinant of $||Ef||^2$ is *larger* than the average value of the Hessian determinant of $||f||^2$.

REMARK 4.18. We show how to construct the 1-forms used in the proof. First consider $S^3 \subseteq \mathbf{C}^2$. We can put $\omega_1 = z\,dw - w\,dz$. Then, except at the origin, the four 1-forms $\omega_1, \overline{\omega}_1, \partial\rho, \overline{\partial}\rho$ do the job. The three 1-forms $\omega_1, \overline{\omega}_1, \partial\rho - \overline{\partial}\rho$ form a basis for the cotangent space at each point of the unit sphere.

In the higher dimensional case, we work on the set U where $z_n \neq 0$. The complement of U in the sphere is a lower dimensional sphere, and hence a small set as far as integration is concerned. For $1 \leq j \leq n-1$, we define ω_j by

$$\omega_j = \frac{z_n\,dz_j - z_j\,dz_n}{|z_j|^2 + |z_n|^2}$$

The forms ω_j are linearly independent on U, and each is orthogonal to $d\rho$. See the next Chapter and Exercise 5.1 for their role in CR geometry.

We now discuss in more detail why η is a non-negative multiple of the $(2n-1)$-dimensional volume form on the sphere. One way to verify this fact is to introduce polar coordinates in each variable separately and compute. Thus $z_j = \mathbf{r}_j e^{i\theta_j}$, where each $\mathbf{r}_j$ is non-negative. On the unit sphere we have the relation $\sum \mathbf{r}_j^2 = 1$; it follows that $\sum \mathbf{r}_j d\mathbf{r}_j = 0$ on the sphere. We therefore use all the θ_j as coordinates, but we use only $\mathbf{r}_1, \ldots, \mathbf{r}_{n-1}$. The $(2n-1)$-dimensional volume form on the sphere turns out to be (where the product is a wedge product)

$$\left(\prod_{j=1}^{n-1} \mathbf{r}_j d\mathbf{r}_j \wedge d\theta_j\right) \wedge d\theta_n.$$

We continue this geometric approach by noting the following simple Lemma, expressing the Cauchy-Riemann equations in polar coordinates.

LEMMA 4.12. *Assume h is complex analytic in one variable. Use polar coordinates* $z = re^{i\theta}$. *Then* $\frac{\partial h}{\partial \theta} = ri\frac{\partial h}{\partial r}$.

PROOF. We will use subscripts to denote partial derivatives in this proof. Since h is complex analytic, $h_{\bar{z}} = \frac{\partial h}{\partial \bar{z}} = 0$. It follows that

$$h_r = \frac{\partial h}{\partial r} = \frac{\partial h}{\partial z}\frac{\partial z}{\partial r} = h_z e^{i\theta}.$$

Similarly,

$$h_\theta = \frac{\partial h}{\partial \theta} = \frac{\partial h}{\partial z}\frac{\partial z}{\partial \theta} = h_z rie^{i\theta} = rih_r.$$

$\square$

REMARK 4.19. One can also prove Lemma 4.12 by observing that it suffices to check it for $h(z) = z^k$, for each k.

EXERCISE 4.70. Prove Lemma 4.12 as suggested in the Remark.

A continuously differentiable function of several complex variables is complex analytic if and only if it is complex analytic in each variable separately. (The same conclusion holds without the hypothesis of continuous differentiability, but this result, which we do not need, is much harder to prove.) The geometry of the sphere suggests, and the easier implication justifies, working in polar coordinates in each variable separately.

Put $z_j = \mathbf{r}_j e^{i\theta_j}$ for $1 \leq j \leq n$. Computation yields

$$dz_j = e^{i\theta_j}d\mathbf{r}_j + i\mathbf{r}_j e^{i\theta_j}d\theta_j.$$

Note that $\sum_1^n \mathbf{r}_j d\mathbf{r}_j = 0$ on the sphere. We compute $\bar{\partial}\rho = \sum_1^n z_j d\bar{z}_j$ as follows:

$$\bar{\partial}\rho = \sum_{j=1}^n z_j d\bar{z}_j = \sum_{j=1}^n \mathbf{r}_j d\mathbf{r}_j - i\sum_{j=1}^n \mathbf{r}_j^2 d\theta_j = -i(\sum_{j=1}^n \mathbf{r}_j^2 d\theta_j).$$

We can express the form η from (75) in terms of these new variables. We provide the details only when $n = 2$. For ease of notation, we write $z = re^{i\theta}$ and $w = se^{i\phi}$. We obtain

$$z d\bar{z} + w d\bar{w} = -i(r^2 d\theta + s^2 d\phi). \tag{77}$$

We compute $\partial g \wedge \bar{\partial}g$, where $g = f_2$ in (75). Now that we do not have subscripts on the functions, we can use subscripts to denote partial derivatives. Since g is complex analytic, we have

$$\partial g = dg = g_r dr + g_\theta d\theta + g_s ds + g_\phi d\phi.$$

The Cauchy-Riemann equations in polar coordinates give $g_\theta = rig_r$ and $g_\phi = sig_s$. From these equations we find

$$\partial g = g_r(dr + ird\theta) + g_s(ds + isd\phi). \tag{78}$$

We need to compute $\partial g \wedge \bar{\partial}g$. We obtain

$$\partial g \wedge \bar{\partial}g = |g_r|^2(-2irdr \wedge d\theta) + |g_s|^2(-2isds \wedge d\phi)$$

$$+ g_r\bar{g}_s(-isdr \wedge d\phi + ird\theta ds + rsd\theta \wedge d\phi)$$

$$+ g_s\bar{g}_r(-isdr \wedge d\phi + ird\theta ds - rsd\theta \wedge d\phi). \tag{79}$$

We wedge (77) with (79) and collect terms in the order $dr d\theta d\phi$. The result is

$$(z d\overline{z} + w d\overline{w}) \wedge \partial g \wedge \overline{\partial} g = -2r|sg_r - rg_s|^2 dr d\theta d\phi. \tag{80}$$

The form η in question is $(\frac{i}{2})^2$ times the expression in (80). Hence we see that

$$\eta = |sg_r - rg_s|^2 \frac{r}{2} dr d\theta d\phi, \tag{81}$$

which is a non-negative multiple of the volume form $r dr d\theta d\phi$ for the sphere.

We gain considerable insight by expressing $sg_r - rg_s$ in terms of g_z and g_w. Using the chain rule and some manipulation we get

$$|sg_r - rg_s|^2 = |sg_z z_r - rg_w w_s|^2 = |se^{i\theta} g_z - re^{i\phi} g_w|^2 = |\overline{w} g_z - \overline{z} g_w|^2. \tag{82}$$

We can interpret (82) geometrically. Define a complex vector field L by

$$L = \overline{w} \frac{\partial}{\partial z} - \overline{z} \frac{\partial}{\partial w}. \tag{83}$$

Then L is tangent to the unit sphere, and (81) and (82) yield $\eta = \frac{1}{2}|L(g)|^2 \, \sigma_3$. In the next section we will interpret L in the context of CR geometry.

EXERCISE 4.71. Use polar coordinates to compute the form η from (75) in 3 complex dimensions.

EXERCISE 4.72. Show that $\{z^\alpha\}$, as α ranges over all non-negative integer multi-indices, is a complete orthogonal system for $\mathcal{A}^2$. Here $\mathcal{A}^2$ denotes the complex analytic functions in $L^2(\mathbb{B}_n)$.

EXERCISE 4.73. Let $c_\alpha = ||z^\alpha||_{L^2}^2$ for the unit ball $\mathbb{B}_n$. Find a simple formula for the Bergman kernel $B(z, \overline{z})$ for the ball, defined by

$$B(z, \overline{z}) = \sum_\alpha \frac{|z|^{2\alpha}}{c_\alpha}.$$

EXERCISE 4.74. Compute V_f if $f(z, w) = (z^a, w^b)$. Also compute V_g if $g(z) = (z^a, zw^b, w^{b+1})$.

EXERCISE 4.75. Express the $(2n-1)$-dimensional volume of the unit sphere S^{2n-1} in terms of the $2n$-dimensional volume of $\mathbb{B}_n$. Suggestion: Use (71) and (72) when $f(z) = z$.

EXERCISE 4.76. Consider the Hilbert space $\mathcal{H}$ consisting of complex analytic functions on $\mathbf{C}^n$ that are square integrable with respect to the Gaussian weight function $\exp(-||z||^2)$. Show that the monomials form a complete orthogonal system for $\mathcal{H}$. Compute $c_\alpha = ||z^\alpha||_{L^2}^2$. Finally, analogously to Exercise 4.73, compute

$$\sum_\alpha \frac{|z|^{2\alpha}}{c_\alpha}.$$

12. Unifying remarks

Section 2 made considerable use of the Hilbert space of complex analytic functions in the unit disk. The paragraph after Lemma 4.2 indicated to a small extent why Hilbert spaces whose elements are complex analytic (holomorphic) functions have such nice properties. In this section we expand the basic idea hoping both to unify many topics already discussed and to anticipate Theorems 5.1 and 5.2 from the next chapter. These ideas also play a significant role in mathematical physics, but we say very little in this direction.

We begin by considering two different Hilbert spaces.

DEFINITION 4.18. Let $\mathcal{A}^2(\mathbb{B}_n)$ denote the collection of holomorphic functions $f : \mathbb{B}_n \to \mathbf{C}$ for which

$$\int_{\mathbb{B}_n} |f(z)|^2 dV(z) < \infty. \tag{84.1}$$

Let $\mathcal{A}^2(\mathbf{C}^n, G)$ denote the space of holomorphic maps $f : \mathbf{C}^n \to \mathbf{C}$ for which

$$\int_{\mathbf{C}^n} |f(z)|^2 e^{-|z|^2} dV(z) < \infty. \tag{84.2}$$

In (84.2) the expression $e^{-|z|^2}$ denotes (up to a constant) a Gaussian:

$$e^{-|z|^2} = e^{-(\sum_{j=1}^n |z_j|^2)}.$$

DEFINITION 4.19. The inner product on $\mathcal{A}^2(\mathbb{B}_n)$ is given by

$$\langle f, g \rangle = \int_{\mathbb{B}_n} f(z)\, \overline{g(z)} dV(z). \tag{85.1}$$

The inner product on $\mathcal{A}^2(\mathbf{C}^n, G)$ is given by

$$\langle f, g \rangle = \int_{\mathbf{C}^n} f(z)\, \overline{g(z)} e^{-|z|^2} dV(z). \tag{85.2}$$

We write $|z|^2$ rather than $||z||^2$ in (84.2) and (85.2) for the Euclidean squared norm in order to avoid confusion with the squared norms on these Hilbert spaces. In Chapter 5 we return to using $||z||^2$ for the Euclidean squared norm. We also note that the letter G is used to suggest the Gaussian factor.

In the literature from mathematical physics, the space $\mathcal{A}^2(\mathbf{C}^n, G)$ is also known as the Segal-Bargmann space, the Bargmann space, and the Bargmann Fock space. We refer to [F3] for more information. The Wikipedia page on the Segal-Bargmann space is also reliable.

We leave it to the reader to complete the sketch of the following crucial fact.

THEOREM 4.12. *Both* $\mathcal{A}^2(\mathbb{B}_n)$ *and* $\mathcal{A}^2(\mathbf{C}^n, G)$ *are Hilbert spaces.*

PROOF. That these sets are vector spaces is obvious. That formulas (85.1) and (85.2) define inner products is also obvious. The main issue is completeness. The crucial point in each case is that these spaces are closed subspaces of the corresponding L^2 spaces.

First consider $\mathcal{A}^2(\mathbb{B}_n)$. Using the mean-value property for holomorphic functions and the Cauchy-Schwarz inequality, one can prove for each compact subset K of the ball that there is a constant C for which

$$\sup_{z \in K} |f(z)| \leq C ||f||_{L^2}. \tag{86}$$

It follows that a Cauchy sequence of holomorphic functions converges uniformly on compact subsets. It is a standard fact in complex analysis (in one or several variables) that the uniform limit on compact subsets of a sequence of holomorphic functions is itself holomorphic. Hence $\mathcal{A}^2(\mathbb{B}_n)$ is a closed subspace of $L^2(\mathbb{B}_n)$ and hence a Hilbert space.

The discussion for $\mathcal{A}^2(\mathbf{C}^n, G)$ is similar. □

REMARK 4.20. The proof that $\mathcal{A}^2(\mathbb{B}_n)$ is a Hilbert space works when the ball is replaced by an arbitrary bounded domain. The ball, however, is particularly nice because the monomials form an orthogonal system and hence the Taylor expansion of a function is also an orthonormal expansion.

We also note the following crucial properties:

LEMMA 4.13. *In each case, the evaluation map δ_p defined by $\delta_p(f) = f(p)$ is a continuous linear functional. Hence, by Theorem 2.4 (Riesz lemma), there is an element K_p of the space for which*

$$f(p) = \delta_p(f) = \langle f, K_p \rangle.$$

PROOF. First consider $\mathcal{A}^2(\mathbb{B}_n)$. Since a point is a compact subset, (86) yields the desired result for the ball. Again the proof for $\mathcal{A}^2(\mathbf{C}^n, G)$ is similar. $\quad\square$

LEMMA 4.14. *In both cases, the monomials form a complete orthogonal system for the Hilbert space.*

PROOF. Lemma 4.2 proves the result for $\mathcal{A}^2(\mathbb{B}_1)$. The proof for general n is virtually the same. Simply use polar coordinates in each variable separately to check orthogonality. To check completeness one needs to know that a function holomorphic in the unit ball has a power series expansion that converges uniformly on compact subsets of the ball.

The proof for $\mathcal{A}^2(\mathbf{C}^n, G)$, requested above in Exercise 4.76, is similar. $\quad\square$

Lemma 4.14 has several useful consequences. For example, in either setting a holomorphic function has a power series representation whose individual terms are orthogonal. By the lemma, the decomposition into homogeneous parts is therefore an orthogonal sum. The reader might revisit Theorems 2.15 and 2.16 from Chapter 2 to gain additional insight into those results. Note also that the function $z \mapsto K_p(z)$ from Lemma 4.13 can then be computed by summing an explicit orthonormal series. Such functions are known as *reproducing kernels*. See Exercise 4.12 from Section 2, Exercise 4.73, Exercise 4.76, and Exercise 4.86 below.

Next we make a connection to quantum mechanics. Consider $\mathcal{H} = \mathcal{A}^2(\mathbf{C}^n, G)$. For each j we may consider the unbounded linear operator $a_j = \frac{\partial}{\partial z_j}$ on $\mathcal{H}$. This operator a_j is called an *annihilation operator*. The operator M_k of multiplication by z_k is called a *creation operator*. The following facts are simple to verify:

LEMMA 4.15. *The adjoint a_j^* of $\frac{\partial}{\partial z_j}$ is M_j. The identities $[a_j, M_k] = \delta_{jk}$ hold. Here $[a, M]$ is the commutator $aM - Ma$.*

PROOF. The commutator identities are immediate. See Exercise 4.78 for the proof finding a_j^*. $\quad\square$

REMARK 4.21. In quantum mechanics, one defines **position** by $A_j = \frac{a_j + M_j}{2}$ and **momentum** by $B_j = \frac{a_j - M_j}{2i}$.

These operators are defined on $\mathcal{H} = \mathcal{A}^2(\mathbf{C}^n, G)$. We have seen something similar. Versions of D and M on $\mathcal{A}^2(\mathbb{B}_1)$ were used in Section 2 to compute volumes of holomorphic images. When f is injective, the squared L^2 norm $||f'||^2$ equals the volume of the image of f. Proposition 4.2 and Corollary 4.2 give a geometric interpretation of the operator $M^*D^*DM - D^*D$. In this case, where the domain is the ball, D and M are not adjoints.

EXERCISE 4.77. Fill in the details of the proofs of Theorem 4.12, Lemma 4.13, and Lemma 4.14.

EXERCISE 4.78. Suppose $f(z) = \sum c_\alpha z^\alpha$ is entire. Give the necessary and sufficient condition on the c_α for $f \in \mathcal{A}^2(\mathbf{C}^n, G)$. Do the same for a first derivative of f. Use this information to prove the first part of Lemma 4.15.

EXERCISE 4.79. Show that there are no finite-dimensional linear maps A, B with $[A, B] = I$. Harder: Show that there are no bounded operators on a Hilbert space satisfying $[A, B] = I$. A hint is given in Exercise 2.42 of Chapter 2.

EXERCISE 4.80. Generalize Exercises 4.12 and 4.73 to $\mathcal{A}^2(\mathbf{C}^n, G)$.

EXERCISE 4.81. Consider $\mathcal{A}^2(\mathbf{C}, G)$. Find f in this space for which f' is not in the space. Find g in this space for which $zg(z)$ is not in this space.

We conclude this section by sketching a proof of Theorem 5.1 from the next chapter. The starting point is a bihomogeneous polynomial $r(z, \overline{w})$. Here each of z, w lies in $\mathbf{C}^n$ and r is homogeneous of degree m in each of these variables. We suppose that $r(z, \overline{z})$ is real-valued; see section 4 for additional discussion about such Hermitian symmetric polynomials. We write

$$r(z, \overline{z}) = \sum_{|a|=|b|=m} c_{ab} z^a \overline{z}^b, \tag{87}$$

where the matrix (c_{ab}) of coefficients is Hermitian. We call this matrix the *under-lying matrix of coefficients*.

We wish to analyze several possible positivity conditions for r. If the matrix (c_{ab}) is positive semi-definite, then there is a holomorphic polynomial map g, taking values in $\mathbf{C}^N$ and homogeneous of degree m, such that

$$r(z, \overline{z}) = ||g(z)||^2 = \sum_{j=1}^N |g_j(z)|^2. \tag{88}$$

When (88) holds we say that r is a *Hermitian squared norm*. In this case, $r(z, \overline{z}) \geq 0$ for all z. We say r is non-negative as a function. Exercise 4.82 asks for the easy proof of (88).

If r is non-negative as a function, however, the underlying matrix of coefficients need not be positive semi-definite. The simplest example is given by

$$r(z, \overline{z}) = |z_1|^4 + \lambda |z_1 z_2|^2 + |z_2|^4.$$

It is routine to check that $r(z, \overline{z}) \geq 0$ for all z if and only if $\lambda \geq -2$. Hence, for $-2 \leq \lambda < 0$, we have a non-negative function whose matrix of coefficients has a negative eigenvalue. For $-2 < \lambda$, the function r is *strictly* positive away from the origin and yet there is still a negative eigenvalue.

Theorem 5.1 from the next chapter, stated here as Theorem 4.13, clarifies the relationship between the two notions of positivity. The theorem considers bihomogeneous polynomials that are strictly positive away from the origin. The conclusion need not hold for bihomogeneous polynomials that are non-negative away from the origin. See Example 5.4 from the next chapter. We also remark that one cannot choose the integer d in terms of the dimension n and the degree of homogeneity alone. This integer depends on the actual values of $r(z, \overline{z})$. See [D1] for considerable additional information about this matter.

THEOREM 4.13. *Let $r(z, \overline{z})$ be a bihomogeneous polynomial such that $r(z, \overline{z}) > 0$ for $z \neq 0$. Then there are a positive integer d and a holomorphic polynomial map h such that*

$$||z||^{2d} r(z, \overline{z}) = \sum E_{\mu\nu} z^\mu \overline{z}^\nu = ||h(z)||^2. \tag{89}$$

Furthermore, the matrix $(E_{\mu\nu})$ is strictly positive definite.

If (89) holds for some d_0, then it holds whenever $d \geq d_0$. The reason is that

$$||z||^{2k} \, ||h(z)||^2 = ||z^{\otimes k} \otimes h(z)||^2.$$

As we have done all chapter, we do not indicate the dimension of the space in the notation for the squared norm.

Theorem 4.13 was first proved by Quillen [Q] and later reproved by Catlin-D'Angelo as one step in the general set of ideas from Chapter 5. The proof from [Q] uses the space $\mathcal{A}^2(\mathbf{C}^n, G)$. The proof from [CD] uses $\mathcal{A}^2(\mathbb{B}_n)$. See [D1] for a detailed exposition of a slightly simplified version of the proof from [CD]. To help unify ideas in this book, we sketch the proof.

We first note, by linear algebra, that a bihomogeneous polynomial is a Hermitian squared norm if and only if the underlying matrix of coefficients is non-negative definite. This statement holds if and only if a certain integral operator is non-negative definite. Namely, given a bihomogeneous polynomial r, we define a linear operator $\mathcal{K}_r$ on $\mathcal{A}^2(\mathbb{B}_n)$ by

$$\mathcal{K}_r f(z) = \int_{\mathbb{B}_n} r(z, \overline{w}) \overline{f(w)} \, dV(w).$$

Since r is bihomogeneous and monomials of different degrees are orthogonal, this operator is 0 except on the subspace of homogeneous polynomials of degree m. On this space, however, we have $\langle \mathcal{K}_r f, f \rangle \geq c||f||^2$ for all f and $c > 0$ if and only if the matrix of coefficients of r is positive definite. We say that $\mathcal{K}_r$ *corresponds* to r.

We see that our goal is to prove that there is a d_0 with the following property. For each $d \geq d_0$, the integral operator corresponding to $||z||^{2d} r(z, \overline{z})$ is positive definite on the space of homogeneous polynomials of degree $m + d$. We consider all these operators at the same time by introducing the Bergman kernel function. The Bergman kernel function for the ball, computed in Exercise 4.73, is given by

$$B(z, \overline{w}) = \frac{n!}{\pi^n} \left(1 - \langle z, w \rangle\right)^{-n-1}. \tag{90}$$

Two things about B matter. First, B is the integral kernel of orthogonal projection $P : L^2(\mathbb{B}_n) \to \mathcal{A}^2(\mathbb{B}_n)$. Thus, for $f \in L^2(\mathbb{B}_n)$,

$$Pf(z) = \int_{\mathbb{B}_n} B(z, \overline{w}) f(w) \, dV(w).$$

Second, B serves as a generating function for powers of the inner product:

$$B(z, \overline{w}) = \sum_{j=0}^{\infty} c_j \, \langle z, w \rangle^j.$$

Here all the coefficients c_j are positive. We also recall from Lemma 4.14 that $\mathcal{A}^2(\mathbb{B}_n)$ is the orthogonal sum of the spaces of homogeneous polynomials.

Choose a smooth non-negative function χ that is 1 near the origin and has compact support in the ball. We write, using the powerful technique of adding and subtracting,

$$r(z,\overline{w})B(z,\overline{w}) = (r(z,\overline{w}) - r(z,\overline{z}))\,B(z,\overline{w}) + (r(z,\overline{z}) + \chi(z))\,B(z,\overline{w}) - \chi(z)B(z,\overline{w}).$$

We think of this expression on the operator level as

$$Pr = [P,r] + (r + \chi)P - \chi P = T_1 + T_2 + T_3. \tag{91}$$

We are now close to where we wish to be. The operator T_3 is compact on $L^2(\mathbb{B}_n)$, because χ is smooth and compactly supported. (See Proposition 2.12 from Chapter 2.) The operator T_2 is positive definite on $\mathcal{A}^2(\mathbb{B}_n)$ by Exercise 4.84. If we knew that T_1 were a compact operator, then Pr would be a compact operator plus an operator that is positive definite on $\mathcal{A}^2(\mathbb{B}_n)$. Because of compactness, there would be a finite-dimensional subspace away from which the operator Pr is positive definite. See Exercise 4.85. In other words, for d_0 sufficiently large, the operator corresponding to

$$\sum_{j \geq d_0} r(z,\overline{w})c_j\langle z,w\rangle^j$$

is positive definite. Recall that polynomials of different degrees are orthogonal. Since $c_j > 0$ for all j, we conclude (when T_1 is compact) that the operator corresponding to $r(z,\overline{w})\langle z,w\rangle^d$ is positive definite on the space of homogeneous polynomials of degree $m + d$, when $m + d \geq d_0$.

Thus the remaining issue is the compactness of the commutator $[P,r]$ of the Bergman projection and multiplication by r. See [D1] for the details of a long but elementary proof. The statement also follows from general facts in operator theory. See for example [SZ] and [Ha]. See also [CeS] for recent related work. The operator $[P,r]$ is called a Hankel operator.

COROLLARY 4.9. *Let r be a bihomogeneous polynomial of degree $2m$. The following are equivalent:*

- *There is a positive constant c such that $r(z,\overline{z}) \geq c\|z\|^{2m}$.*
- *There is an integer d and a holomorphic homogeneous polynomial mapping h such that the set of common zeroes of the components of h is $\{0\}$ and*

$$\|z\|^{2d}\,r(z,\overline{z}) = \|h(z)\|^2.$$

- *For a possibly larger d, the operator $\mathcal{K}$ corresponding to $\|z\|^{2d}r(z,\overline{z})$ is positive definite on the space of holomorphic polynomials of degree $m + d$.*

EXAMPLE 4.17. Consider the polynomial $r(z,\overline{z}) = |z_1|^6 + |z_2|^6$. It is already a Hermitian squared norm, but its matrix of coefficients is only positive semi-definite. If we multiply by $\|z\|^4$, then we obtain a Hermitian squared norm whose underlying matrix is positive-definite.

EXAMPLE 4.18. Put $r(z,\overline{z}) = |z_1|^4 + |z_1z_2|^2$. Then r is a Hermitian squared norm, and hence non-negative, but r has a large zero-set. The strict positivity condition of Corollary 4.9 does not hold. For every d, the underlying matrix for the function $\|z\|^{2d}r(z,\overline{z})$, although positive semi-definite, is not positive definite.

See also Examples 5.4 and 5.6 from the next chapter for more insight into how the zero set of r matters.

EXERCISE 4.82. Use elementary linear algebra to prove (88).

EXERCISE 4.83. In formula (91), it is stated that $[P, r]$ is the operator T_1 corresponding to $(r(z, \overline{w}) - r(z, \overline{z})) B(z, \overline{w})$. Prove this claim. Note the distinction between the $\overline{w}$ and the $\overline{z}$. The integration takes place with respect to $dV(w)$.

EXERCISE 4.84. Prove that the operator $(r + \chi)P$ is positive definite on $\mathcal{A}^2(\mathbb{B}_n)$. (Note that we multiply by $r + \chi$ after we project.)

EXERCISE 4.85. Let T be a compact operator on a Hilbert space. Let P be a positive definite operator. Prove that there is a finite-dimensional subspace away from which $P + T$ is positive definite.

EXERCISE 4.86. Suppose that ϕ_j is a complete orthonormal system for a Hilbert space $\mathcal{H}$ of holomorphic functions. For $f \in \mathcal{H}$, find the value of

$$\langle f(w), \sum_j \phi_j(w) \overline{\phi_j(z)} \rangle.$$

Here the inner product is taken with respect to the w variable. Compare with Lemma 4.13 and the remarks after Lemma 4.14.

EXERCISE 4.87. Put $r(z, \overline{z}) = |z_1|^4 + \lambda |z_1 z_2|^2 + |z_2|^4$. For each $\lambda > -2$, Theorem 4.13 guarantees that there is a minimum $d = d_\lambda$ such that $(|z_1|^2 + |z_2|^2)^d r(z, \overline{z})$ is a Hermitian squared norm. Prove that d_λ tends to infinity as λ tends to -2.

CHAPTER 5

The unit sphere and CR geometry

CR geometry considers the interplay between real and complex spaces. The name itself has an interesting history, which we do not discuss here, other than to say that CR stands both for Cauchy-Riemann and for Complex-Real. See [DT] for a survey of CR Geometry and its connections with other branches of mathematics. See [J] for a good exposition of the theory of CR structures, and see [BER] for a definitive treatment of CR mappings.

In this chapter we consider simple aspects of the CR geometry of the unit sphere in $\mathbf{C}^n$ and relate them to the holomorphic automorphism group of the unit ball. The unit sphere S^{2n-1} in $\mathbf{R}^{2n}$ becomes an object in CR Geometry *after* we identify $\mathbf{R}^{2n}$ with $\mathbf{C}^n$. Given this identification, we discover that the tangent directions to the sphere do not all behave the same way from the point of view of complex analysis. This issue lies at the foundation of CR Geometry and we will develop it in Section 1.

Much of the action from Chapter 4 involved the unit sphere. Chapter 5 goes further. We include a generalization of the Riesz-Fejer theorem on non-negative trigonometric polynomials to a result on Hermitian polynomials that are positive on the unit sphere. We apply this result to the study of proper mappings between balls. We then study groups associated with holomorphic and CR mappings. This chapter thus provides many ways to extend results from the unit circle to higher dimensions, all informed by orthogonality and Hermitian analysis.

The theorems, examples, and geometric considerations in this chapter illustrate the following theme. When passing from analysis on the unit circle to analysis in higher dimensions, the mathematics becomes both more complicated and more beautiful. Ideas revolving around Hermitian symmetry appear throughout. This perspective leads naturally to CR geometry. We refer again to [DT] for an introduction to CR geometry, and to its references for viewing the many directions in which Hermitian analysis is developing.

1. Geometry of the unit sphere

Let S^{2n-1} denote the unit sphere in $\mathbf{R}^{2n}$. Consider a point p in S^{2n-1}. If we regard p as a unit vector v (from 0 to p) in $\mathbf{R}^{2n}$, then v is orthogonal to the sphere at p. Hence any vector w orthogonal to v is tangent to the sphere. Put $r(x) = \sum_{j=1}^{2n} x_j^2 - 1$. Then the unit sphere is the zero-set of r, and furthermore $dr(x) \neq 0$ for x on the sphere. We call such a function a *defining function* for the sphere. The 1-form dr annihilates the tangent space $T_p(S^{2n-1})$ at each point. It defines the co-normal direction to the sphere.

In this Chapter we write $\langle \eta, L \rangle$ for the contraction of a 1-form η with a vector field L. Previously we have been writing $\eta(L)$. A vector field $L = \sum_{j=1}^{2n} a_j \frac{\partial}{\partial x_j}$ on

© Springer Nature Switzerland AG 2019
J. P. D'Angelo, *Hermitian Analysis*, Cornerstones,
https://doi.org/10.1007/978-3-030-16514-7_5

$\mathbf{R}^{2n}$ is tangent to S^{2n-1} if and only if

$$0 = \langle dr, L \rangle = dr(L) = L(r) = \sum_{j=1}^{2n} a_j \frac{\partial r}{\partial x_j}$$

on the sphere.

Given the focus of this book, we regard $\mathbf{R}^{2n}$ as $\mathbf{C}^n$ and express these geometric ideas using complex vector fields. A new phenomenon arises. Not all directions in the tangent space behave the same way, from the complex variable point of view.

Let X be a complex vector field on $\mathbf{C}^n$. We can write

$$X = \sum_{j=1}^{n} a_j \frac{\partial}{\partial z_j} + \sum_{j=1}^{n} b_j \frac{\partial}{\partial \overline{z}_j}$$

where the coefficient functions a_j, b_j are smooth and complex-valued. Each complex vector field is the sum of two vector fields, one of which involves differentiations in only the unbarred directions, the other involves differentiations in only the barred directions. Let $T^{1,0}(\mathbf{C}^n)$ denote the bundle whose sections are vector fields of the first kind and $T^{0,1}(\mathbf{C}^n)$ the bundle whose sections are of the second kind. The only vector field of both kinds is the 0 vector field. We therefore write

$$T(\mathbf{C}^n) \otimes \mathbf{C} = T^{1,0}(\mathbf{C}^n) \oplus T^{0,1}(\mathbf{C}^n). \tag{1}$$

The tensor product on the left-hand side of (1) arises because we are considering complex (rather than real) vector fields. The left-hand side of (1) means the bundle whose sections are the complex vector fields on $\mathbf{C}^n$. We next study how the decomposition in (1) applies to vector fields tangent to S^{2n-1}.

Let $T^{1,0}(S^{2n-1})$ denote the bundle whose sections are complex vector fields of type $(1,0)$ and tangent to S^{2n-1}. Then $T^{0,1}(S^{2n-1})$ denotes the complex conjugate bundle. For p on the sphere, each of the vector spaces $T_p^{1,0}(S^{2n-1})$ and $T_p^{0,1}(S^{2n-1})$ has complex dimension $n-1$. But $T_p(S^{2n-1}) \otimes \mathbf{C}$ has dimension $2n-1$. Hence there is a *missing direction*. How can we describe and interpret this missing direction?

Observe first that the commutator $[L, K]$ of vector fields L, K, each of type $(1,0)$ and tangent to S^{2n-1}, also satisfies these properties. That $[L, K]$ is of type $(1,0)$ follows easily from the formula $[L, K] = LK - KL$. That $[L, K]$ is tangent follows by applying this formula to a defining function r:

$$[L, K](r) = L(K(r)) - K(L(r)) = 0 - 0 = 0.$$

Since K is tangent, $K(r) = 0$ on the sphere. Since L is tangent, $L(K(r)) = 0$ there. By symmetry, $K(L(r)) = 0$ as well. Note Remark 5.1. By symmetry considerations, the commutator of two $(0,1)$ tangent vector fields is also of type $(0,1)$ and tangent. On the sphere, however, the commutator of each non-zero $(1,0)$ vector field L with its conjugate $\overline{L}$ has a non-vanishing component in the missing direction.

REMARK 5.1. Warning! Is the derivative of a constant zero? The function $R(x,y) = x^2 + y^2 - 1$ equals 0 everywhere on the unit circle, but $\frac{\partial R}{\partial x} = 2x$ and hence is NOT zero at most points. The problem is that the differentiation with respect to x is not tangent to the unit circle.

We can abstract the geometry of the sphere as follows:

DEFINITION 5.1. The *CR structure* on S^{2n-1} is given by $V = T^{1,0}(S^{2n-1})$, a subbundle of $T(S^{2n-1}) \otimes \mathbf{C}$ with the following properties:

(1) $V \cap \overline{V} = \{0\}$.

(2) The set of smooth sections of V is closed under the Lie bracket.

(3) $V \oplus \overline{V}$ has codimension one in $T(S^{2n-1}) \otimes \mathbf{C}$.

A bundle whose set of sections is closed under the Lie bracket is called **integrable** or **involutive**. The famous Frobenius theorem, which we do not use in this book, states that a (real) sub-bundle of the tangent bundle of a real manifold is the tangent bundle of a submanifold if and only if it is integrable. See for example [Wa]. There is also a complex Frobenius theorem. See for example [HT].

DEFINITION 5.2. A *CR manifold of hypersurface type* is a real manifold M for which there is a subbundle $V \subseteq T(M) \otimes \mathbf{C}$ satisfying the three properties from Definition 5.1. Thus V is integrable, $V \cap \overline{V} = 0$, and $V \oplus \overline{V}$ has codimension one in $T(M) \otimes \mathbf{C}$.

Any real hypersurface M in $\mathbf{C}^n$ is a CR manifold of hypersurface type. Since $V \oplus \overline{V}$ has codimension one in $T(M) \otimes \mathbf{C}$, there is a non-vanishing 1-form η, defined up to a multiple, annihilating $V \oplus \overline{V}$. By convention, we assume that this form is purely imaginary. Exercise 5.4 and Remark 5.2 partially explain this convention. Thus $\langle \eta, L \rangle = 0$ whenever L is a vector field of type $(1,0)$, and similarly for vector fields of type $(0,1)$.

DEFINITION 5.3. Let M be a CR manifold of hypersurface type. The *Levi form* λ is the Hermitian form on sections of $T^{1,0}(M)$ defined by

$$\lambda(L, \overline{K}) = \langle \eta, [L, \overline{K}] \rangle.$$

Let us return to the unit sphere. Near a point where $z_n \neq 0$, for $1 \leq j \leq n-1$, we define $n-1$ vector fields of type $(1,0)$ by

$$L_j = \overline{z}_n \frac{\partial}{\partial z_j} - \overline{z}_j \frac{\partial}{\partial z_n}. \tag{2}$$

A simple check shows that each L_j is tangent to the sphere. Similarly the complex conjugate vector fields $\overline{L}_j$ are tangent. These vector fields are linearly independent (as long as we are working where $z_n \neq 0$). There are $2n - 2$ of them. The missing direction requires both unbarred and barred derivatives. We can fill out the complex tangent space by setting

$$\mathbf{T} = z_n \frac{\partial}{\partial z_n} - \overline{z}_n \frac{\partial}{\partial \overline{z}_n}. \tag{3}$$

Then $L_1, \ldots, L_{n-1}, \overline{L}_1, \ldots, \overline{L}_{n-1}, \mathbf{T}$ span the complex tangent space to S^{2n-1} at each point where $z_n \neq 0$.

EXERCISE 5.1. Verify that the L_j from (2) and $\mathbf{T}$ from (3) are tangent to the sphere. Let ω_j be as in Remark 4.18 from Chapter 4. Verify that $\langle L_j, \omega_j \rangle = 1$.

EXERCISE 5.2. Find a purely imaginary 1-form annihilating $T^{1,0} \oplus T^{0,1}$ on the sphere.

EXERCISE 5.3. Compute the commutator $[L_j, \overline{L}_k]$.

EXERCISE 5.4. Use the previous two exercises to show that the Levi form on the sphere is positive definite.

EXERCISE 5.5. Show that translating the sphere leads to the defining function

$$r(\zeta, \overline{\zeta}) = \sum_{j=1}^{n-1} |\zeta_j|^2 + |\zeta_n|^2 + 2\mathrm{Re}(\zeta_n). \tag{4.1}$$

Show that a more elaborate holomorphic change of variables and multiplication by a non-vanishing factor lead to the defining function:

$$r(w, \overline{w}) = \sum_{j=1}^{n-1} |w_j|^2 + 2\mathrm{Re}(w_n). \tag{4.2}$$

Suggestion: When $n = 1$, (4.1) defines a disk and (4.2) defines a half-plane. First do this case.

EXERCISE 5.6. Show that $\lambda(L, \overline{K}) = \overline{\lambda(K, \overline{L})}$.

EXERCISE 5.7. Let r be a smooth real-valued function on $\mathbf{C}^n$. Assume that dr does not vanish on M, the zero-set of r. Then M is a real hypersurface, and hence a CR manifold. Compute the Levi form λ on M in terms of derivatives of r. The answer, in terms of the basis $\{L_j\}$ given below in (5.2) for sections of $T^{1,0}(M)$, is the following formula:

$$\lambda_{jk} = r_{j\overline{k}}|r_n|^2 - r_{j\overline{n}}r_n r_{\overline{k}} - r_{n\overline{k}}r_j r_{\overline{n}} + r_{n\overline{n}}r_j r_{\overline{k}}. \tag{5.1}$$

Suggestion: Work near a point where $r_{z_n} \neq 0$. For $1 \leq j \leq n - 1$, define L_j by

$$L_j = \frac{\partial}{\partial z_j} - \frac{r_{z_j}}{r_{z_n}} \frac{\partial}{\partial z_n} \tag{5.2}$$

and define $\overline{L}_k$ in a similar manner. Find the 1-form η, and compute $[L_j, \overline{L}_k]$.

REMARK 5.2. The Levi form plays a crucial role in complex analysis and CR geometry. A smooth real hypersurface is called *pseudoconvex* at a point p when all the eigenvalues of the Levi form there have the same sign (allowing 0). It is *strongly pseudoconvex* when it is pseudoconvex and there are no zero eigenvalues. A strongly pseudoconvex hypersurface is locally biholomorphically equivalent to a strongly convex hypersurface. Readers familiar with the differential geometry of convex surfaces in real Euclidean spaces should compare formula (5.1) with its real variable analogue. This formula exhibits the Levi form as the restriction of the complex Hessian of r to the space $T^{1,0}(M)$.

EXERCISE 5.8. Find the Levi form on the hyperplane defined by $\mathrm{Re}(z_n) = 0$.

EXERCISE 5.9. Consider the hypersurface defined by

$$r(z) = 2\,\mathrm{Re}(z_n) + \sum_{j=1}^{K} |f_j(z')|^2.$$

Here $z' = (z_1, ..., z_{n-1})$ and the functions f_j are holomorphic. Find the determinant of the Levi form on the hypersurface defined by $r = 0$. The same computation appears, in a different context, in Chapter 4.

The zero-set of the function r from formula (4.2) in Exercise 5.5, a biholomorphic image of the sphere, is an unbounded object H, commonly known as the Heisenberg group. Put $n = 2$ and define A by

$$A = \frac{\partial}{\partial w_1} - \overline{w}_1 \frac{\partial}{\partial w_2}.$$

Then A, $\overline{A}$, and $[A, \overline{A}]$ form a basis for the sections of $T(H) \otimes \mathbf{C}$ at each point. See [DT] and its references for considerable information about the role of the Heisenberg group in complex analysis, geometry, and PDE.

We next use the CR geometry of the unit sphere to briefly study harmonic polynomials. For simplicity we work on S^3, where a single vector field L defines the CR structure. Recall that (z, w) denotes the variable in $\mathbf{C}^2$. Put $L = \overline{w}\frac{\partial}{\partial z} - \overline{z}\frac{\partial}{\partial w}$. We also recall from Section 11 of Chapter 1 that a smooth function is harmonic if its Laplacian is 0. We can express the Laplace operator in terms of complex partial derivatives; a (possibly complex-valued) smooth function u is harmonic on $\mathbf{C}^2$ if and only if

$$u_{z\overline{z}} + u_{w\overline{w}} = 0.$$

As in Section 13 from Chapter 2, it is natural to consider harmonic homogeneous polynomials. Here we allow our harmonic functions to be complex-valued. The complex vector space V_d, consisting of homogeneous polynomials of degree d (with complex coefficients) in the underlying $2n$ real variables, decomposes into a sum of spaces $V_{p,q}$. Here $p + q = d$ and the elements of $V_{p,q}$ are homogeneous of degree p in z and of degree q in $\overline{z}$. We obtain a decomposition $\mathbf{H_d} = \sum \mathbf{H}_{p,q}$ of the space of *harmonic* homogeneous polynomials.

EXAMPLE 5.1. Put $n = 2$ and $d = 2$. By our work in Chapter 2, the space $\mathbf{H_2}$ is 9-dimensional. We have the following:

- $\mathbf{H}_{2,0}$ is spanned by z^2, zw, w^2.
- $\mathbf{H}_{1,1}$ is spanned by $z\overline{w}, \overline{z}w, |z|^2 - |w|^2$.
- $\mathbf{H}_{0,2}$ is spanned by $\overline{z}^2, \overline{zw}, \overline{w}^2$.

As in Chapter 2, the sum of these three spaces is the orthogonal complement of the (span of the) function $|z|^2 + |w|^2$ in the space of polynomials of degree 2.

Let us briefly consider eigenvalues and the CR vector fields. For each pair a, b of non-negative integers, observe that the monomials $z^a\overline{w}^b$ and $\overline{z}^a w^b$ are harmonic. Elementary calculus yields:

$$L(z^a\overline{w}^b) = az^{a-1}\overline{w}^{b+1}$$
$$\overline{L}(z^a\overline{w}^b) = -bz^{a+1}\overline{w}^{b-1}.$$

Combining these results shows that

$$L\overline{L}(z^a\overline{w}^b) = -b(a+1)z^a\overline{w}^b.$$
$$\overline{L}L(z^a\overline{w}^b) = -a(b+1)z^a\overline{w}^b$$

Thus the harmonic monomials $z^a\overline{w}^b$ are eigenfunctions of the differential operators $L\overline{L}$ and $\overline{L}L$, with eigenvalues $-b(a+1)$ and $-a(b+1)$. Hence they are also eigenfunctions of the commutator $\mathbf{T} = [L, \overline{L}]$, with eigenvalue $a - b$.

2. Positivity conditions for Hermitian polynomials

This section applies Theorem 4.13 from Chapter 4 (an analogue of the Riesz-Fejer theorem) to the study of proper mappings between balls. We restate Theorem 4.13 as Theorem 5.1 here, and then extend it in Theorem 5.2 to polynomials that are not necessarily bihomogeneous.

The Riesz-Fejer theorem (Theorem 1.1) characterizes non-negative trig polynomials; each such polynomial agrees on the circle with the squared absolute value of a single polynomial in the complex variable z. We naturally seek to extend this result from the unit circle to the unit sphere in $\mathbf{C}^n$. Things become more complicated but also more interesting.

We start with a Hermitian symmetric polynomial $r(z, \overline{z}) = \sum_{\alpha, \beta} c_{\alpha\beta} z^\alpha \overline{z}^\beta$ of degree m in $z \in \mathbf{C}^n$. Because of Hermitian symmetry, the matrix $(c_{\alpha\beta})$ of coefficients is Hermitian symmetric. In particular, r is also of degree m in $\overline{z}$. The total degree of r can range from m to $2m$. We can always *bihomogenize* r by adding a variable as follows. We put $r_H(0, 0) = 0$. For $z \neq 0$ we put

$$r_H(z, t, \overline{z}, \overline{t}) = |t|^{2m} r\left(\frac{z}{t}, \frac{\overline{z}}{\overline{t}}\right).$$

Then r_H is homogeneous of degree m in the variables z, t and also homogeneous of degree m in their conjugates. The polynomial r_H is thus determined by its values on the unit sphere in $\mathbf{C}^{n+1}$. Conversely we can dehomogenize a bihomogeneous polynomial in two or more variables by setting one of its variables (and its conjugate!) equal to the number 1.

EXAMPLE 5.2. Put $n = 1$ and put $r(z, \overline{z}) = z^2 + \overline{z}^2$. We compute r_H:

$$r_H(z, t, \overline{z}, \overline{t}) = |t|^4 \left(\left(\frac{z}{t}\right)^2 + \left(\frac{\overline{z}}{\overline{t}}\right)^2\right) = \overline{t}^2 z^2 + \overline{z}^2 t^2.$$

EXAMPLE 5.3. Put $r = (|zw|^2 - 1)^2 + |z|^2$. Then r is positive everywhere, but r_H, while nonnegative, has many zeroes. Here

$$r_H(z, w, t, \overline{z}, \overline{w}, \overline{t}) = (|zw|^2 - |t|^4)^2 + |zt^3|^2.$$

Because of the bihomogenization, there is no loss in generality in our discussion if we restrict our attention to the bihomogeneous case. Let R be a bihomogeneous polynomial in n variables and their conjugates. Assume $R(z, \overline{z}) \geq 0$ on the unit sphere. As a generalization of the Riesz-Fejer theorem, we naturally ask if there exist homogeneous polynomials $f_1(z), ..., f_K(z)$ such that

$$R(z, \overline{z}) = ||f(z)||^2 = \sum_{j=1}^{K} |f_j(z)|^2.$$

We call such an R a *Hermitian sum of squares* or *Hermitian squared norm*. Of course we cannot expect K to be any smaller than the dimension. For example, the polynomial $\sum_{j=1}^{n} |z_j|^4$ is positive on the sphere, but cannot be written as a Hermitian squared norm with fewer terms. Furthermore, not every non-negative R is a Hermitian squared norm. Even restricted to the unit sphere, such a result fails in general, and hence the analogue of the Riesz-Fejer theorem is more subtle.

EXAMPLE 5.4. Put $R(z, \overline{z}) = (|z_1|^2 - |z_2|^2)^2$. Then R is bihomogeneous and non-negative. Its underlying matrix $C_{\alpha\beta}$ of coefficients is diagonal with eigenvalues $1, -2, 1$. Suppose for some f that $R(z, \overline{z}) = ||f(z)||^2$. Then f would vanish on the subset of the unit sphere defined by $|z_1|^2 = |z_2|^2 = \frac{1}{2}$ (a torus), because R does. A complex analytic function vanishing there would also vanish for $|z_1|^2 \leq \frac{1}{2}$ and $|z_2|^2 \leq \frac{1}{2}$ by the maximum principle. Hence f would have to be identically zero. Thus R does not agree with a squared norm of any complex analytic mapping. The zero-set of R does not satisfy appropriate necessary conditions here.

The following elaboration of Example 5.4 clarifies the matter. Consider the family of polynomials R_ϵ defined by

$$R_\epsilon(z,\overline{z}) = (|z_1|^2 - |z_2|^2)^2 + \epsilon|z_1|^2|z_2|^2.$$

For each $\epsilon > 0$, we have $R_\epsilon(z,\overline{z}) > 0$ on the sphere. By Theorem 5.2 below there is a polynomial mapping f_ϵ such that $R_\epsilon = ||f_\epsilon||^2$ on the sphere. Both the degree and the number of components of f_ϵ must tend to infinity as ϵ tends to 0. See [D1] for a lengthy discussion of this sort of issue.

From Example 5.4 we discover that non-negativity is too weak of a condition to imply that R agrees with a Hermitian squared norm. See also Example 5.6. On the other hand, when $R(z,\overline{z}) > 0$ on the sphere, the conclusion does hold. See [D1] for detailed proofs of Theorem 5.1 and Theorem 5.2 below. The proof of Theorem 5.1 there, sketched at the end of Chapter 4 in this book, uses the theory of compact operators. The proof in [Q] does not mention compactness, but it shares some of the same ideas.

THEOREM 5.1. *Let r be a Hermitian symmetric bihomogeneous polynomial in n variables and their conjugates. Suppose $r(z,\overline{z}) > 0$ on S^{2n-1}. Then there are positive integers d and K, and a polynomial mapping $g : \mathbf{C}^n \to \mathbf{C}^K$, such that*

$$||z||^{2d}r(z,\overline{z}) = ||g(z)||^2.$$

We can remove the assumption of bihomogeneity if we want equality to hold only on the unit sphere.

THEOREM 5.2. *Let r be a Hermitian symmetric polynomial in n variables and their conjugates. Assume that $r(z,\overline{z}) > 0$ on S^{2n-1}. Then there are an integer N and a polynomial mapping $h : \mathbf{C}^n \to \mathbf{C}^N$ such that, for $z \in S^{2n-1}$,*

$$r(z,\overline{z}) = ||h(z)||^2.$$

PROOF. We sketch the derivation of Theorem 5.2 from Theorem 5.1. First we bihomogenize r to get $r_H(z,t,\overline{z},\overline{t})$, bihomogeneous of degree m in the z,t variables. We may assume m is even. The polynomial r_H could have negative values on the sphere $||z||^2 + |t|^2 = 1$. To correct for this possibility, we define a bihomogeneous polynomial F_C by

$$F_C(z,\overline{z},t,\overline{t}) = r_H(z,t,\overline{z},\overline{t}) + C(||z||^2 - |t|^2)^m.$$

It is easy to show that we can choose C large enough to make F_C strictly positive away from the origin. By Theorem 5.1, we can find an integer d such that

$$(||z||^2 + |t|^2)^d F_C(z,\overline{z},t,\overline{t}) = ||g(z,t)||^2.$$

Setting $t = 1$ and then $||z||^2 = 1$ shows, for $z \in S^{2n-1}$, that

$$2^d r(z,\overline{z}) = ||g(z,1)||^2.$$

$\square$

REMARK 5.3. Suppose $n \geq 2$ and $f : \mathbb{B}_n \to \mathbb{B}_N$ is a proper holomorphic mapping. If f has a continuously differentiable extension to the unit sphere, then the restriction to the unit sphere defines a *CR mapping* between spheres. Thus the CR geometry of the unit sphere is closely related to properties of proper mappings between balls.

The following Corollary of Theorem 5.2 connects this Theorem with proper complex analytic mappings between balls.

COROLLARY 5.1. *Let $f = \frac{p}{q} : \mathbf{C}^n \to \mathbf{C}^N$ be a rational mapping. Assume that the image of the closed unit ball under f lies in the open unit ball in $\mathbf{C}^N$. Then there are an integer K and a polynomial mapping $g : \mathbf{C}^n \to \mathbf{C}^K$ such that $\frac{p \oplus g}{q}$ maps the unit sphere S^{2n-1} to the unit sphere $S^{2(N+K)-1}$.*

PROOF. The hypothesis implies that $|q|^2 - ||p||^2$ is strictly positive on the sphere. By Theorem 5.2 there is a polynomial map g such that $|q|^2 - ||p||^2 = ||g||^2$ on the sphere. Then $\frac{p \oplus g}{q}$ does the job.   □

This corollary implies that there are many rational mappings taking the unit sphere in the domain into the unit sphere in some target. We choose the first several components to be anything we want, as long as the closed ball gets mapped to the open ball. Then we can find additional components, using the same denominator, such that the resulting map takes the sphere to the sphere. The following simple example already indicates the depth of these ideas.

EXAMPLE 5.5. Consider the maps $p_\lambda : \mathbf{C}^2 \to \mathbf{C}$ given by $p_\lambda(z, w) = \lambda z w$. Then p_λ maps the closed ball in $\mathbf{C}^2$ inside the unit disk if $|\lambda|^2 < 4$. If this condition is met, then we can include additional components to make p_λ into a component of a polynomial mapping sending S^3 to some unit sphere. In case $\lambda = \sqrt{3}$, we obtain the map $(\sqrt{3}zw, z^3, w^3)$, which is one of the group-invariant examples from Chapter 4. If $\sqrt{3} < \lambda < 2$, then we must map into a dimension higher than 3. As λ approaches 2, the minimum possible target dimension approaches infinity.

The following surprising example combines ideas from many parts of this book.

EXAMPLE 5.6 ([D1]). There exists a bihomogeneous polynomial $r(z, \overline{z})$, in three variables, with the following properties:

- $r(z, \overline{z}) \geq 0$ for all z.
- The zero set of r is a copy of $\mathbf{C}$ (a one-dimensional subspace of $\mathbf{C}^3$).
- 0 is the only polynomial s for which rs is a Hermitian squared norm.

We put $r(z, \overline{z}) = (|z_1 z_2|^2 - |z_3|^4)^2 + |z_1|^8$. The non-negativity is evident. The zero-set of r is the set of z of the form $(0, z_2, 0)$, and hence a copy of $\mathbf{C}$. Assume that rs is a Hermitian squared norm $||A||^2$. Consider the map from $\mathbf{C}$ to $\mathbf{C}^3$ given by $t \mapsto (t^2, 1 + t, t) = z(t)$. Pulling back yields the equation

$$z^*(rs) = r(z(t), \overline{z(t)}) \, s(z(t), \overline{z(t)}) = ||c_m t^m + \cdots||^2,$$

where $\cdots$ denotes higher order terms. Hence the product of the lowest order terms in the pullback of s with the lowest order terms in the pullback of r is $||c_m||^2 |t|^{2m}$. A simple computation shows that the lowest order terms in the pullback of r are

$$t^4 \overline{t}^6 + 2|t|^{10} + t^6 \overline{t}^4 = 2|t|^{10}(1 + \cos(2\theta)). \tag{6}$$

There is no trig polynomial p other than 0 for which multiplying the right-hand side of (6) by an expression of the form $|t|^{2k} p(\theta)$ yields a result independent of θ.

No such example is possible in one dimension, because the only bihomogeneous polynomials are of the form $c|t|^{2m}$. It is easy to find a non-negative polynomial $g(t, \overline{t})$ that doesn't divide any Hermitian squared norm (other than 0); for example,

$$2|t|^2 + t^2 + \overline{t}^2 = 2|t|^2(1 + \cos(2\theta))$$

does the job. Example 5.6 is surprising because r is bihomogeneous.

3. Groups associated with holomorphic mappings

We return to the study of proper holomorphic mappings between balls. We pay particular attention to groups associated with such mappings, following [DX1] and [DX2]. Many of the ideas hold more generally, so we begin with a general framework.

Let Ω be a set and let $\mathrm{Aut}(\Omega)$ denote the group of its automorphisms. In this Chapter, Ω will be a domain in complex Euclidiean space $\mathbf{C}^n$ and we will consider only *holomorphic* automorphisms. A mapping $f : \Omega \to \Omega$ is a holomorphic automorphism if $f : \Omega \to \Omega$ is holomorphic, injective, surjective, and the inverse function is also holomorphic. We note that holomorphicity of the inverse is automatic. By contrast, however, things differ for smooth functions. The mapping $x \mapsto x^3$ on the real line is of class C^∞, injective, and surjective, but the inverse function is not smooth at 0. For holomorphic maps, this kind of example does not arise.

Consider a pair of domains Ω_1 and Ω_2, not necessarily in the same dimension. We start with the group $\mathrm{Aut}(\Omega_1) \times \mathrm{Aut}(\Omega_2)$. We will associate various groups with a holomorphic mapping $f : \Omega_1 \to \Omega_2$.

DEFINITION 5.4. (Five groups). Let $f : \Omega_1 \to \Omega_2$ be a holomorphic map.

- Let $\mathbf{A}_f = \{(\gamma, \psi) : f \circ \gamma = \psi \circ f\}$; here $\gamma \in \mathrm{Aut}(\Omega_1)$ and $\psi \in \mathrm{Aut}(\Omega_2)$. Thus $\mathbf{A}_f$ is a subgroup of $\mathrm{Aut}(\Omega_1) \times \mathrm{Aut}(\Omega_2)$.
- Let Γ_f denote the projection of $\mathbf{A}_f$ onto its first factor.
- Let T_f denote the projection of $\mathbf{A}_f$ onto its second factor.
- Let $G_f = \{\gamma \in \mathrm{Aut}(\Omega_1) : f \circ \gamma = f\}$.
- Let $H_f = \{\psi \in \mathrm{Aut}(\Omega_2) : \psi \circ f = f\}$.

Note that an automorphism γ is in Γ_f if and only if there exists an automorphism ψ of the target for which $f \circ \gamma = \psi \circ f$.

DEFINITION 5.5. Let f and g be maps from Ω_1 to Ω_2. We say that f and g are *equivalent* if there are automorphisms ψ and φ for which

$$\psi \circ f = g \circ \varphi.$$

When Ω_1 and Ω_2 are unit balls (in possibly different dimensions), and f and g are equivalent, we say that f and g are *spherically equivalent*.

LEMMA 5.1. *Assume that f and g are equivalent; that is, there are automorphisms for which*

$$\psi \circ f = g \circ \varphi.$$

Let $(\gamma, \zeta) \in \mathbf{A}_f$. Then

$$(\varphi \circ \gamma \circ \varphi^{-1}, \psi \circ \zeta \circ \psi^{-1}) \in \mathbf{A}_g.$$

In particular, Γ_f and Γ_g are conjugate by φ:

$$\Gamma_g = \varphi \circ \Gamma_f \circ \varphi^{-1}.$$

Also T_f and T_g are conjugate by ψ.

PROOF. The proof is a formal computation. We are given that $g = \psi \circ f \circ \varphi^{-1}$. Assuming $f \circ \gamma = \zeta \circ f$ we must show that

$$g \circ (\varphi \circ \gamma \circ \varphi^{-1}) = (\psi \circ \zeta \circ \psi^{-1}) \circ g. \tag{7}$$

Starting with $f \circ \gamma = \zeta \circ f$ we obtain

$$\psi \circ f \circ \gamma \circ \varphi^{-1} = \psi \circ \zeta \circ f \circ \varphi^{-1}.$$

Inserting $\varphi^{-1} \circ \varphi$ and $\psi^{-1} \circ \psi$ in the right places gives

$$(\psi \circ f \circ \varphi^{-1}) \circ (\varphi \circ \gamma \circ \varphi^{-1}) = (\psi \circ \zeta \circ \psi^{-1}) \circ (\psi \circ f \circ \varphi^{-1}). \qquad (8)$$

Since $g = \psi \circ f \circ \varphi^{-1}$, equation (8) implies equation (7).   □

We start with some easy general examples. Later we will focus on these groups for holomorphic mappings between balls.

EXAMPLE 5.7. Let $f : \Omega \to \Omega$ be the identity function. Then $\mathbf{A}_f$ is isomorphic to $\mathrm{Aut}(\Omega)$, expressed as the diagonal of $\mathrm{Aut}(\Omega) \times \mathrm{Aut}(\Omega)$.

EXAMPLE 5.8. Let $\Omega_1 = \Omega_2 = \mathbf{C}$. The automorphism group consists of affine maps $z \mapsto az + b$ with $a \neq 0$. We find Γ_f and T_f for three explicit functions f.

- Put $f(z) = z^k$, for $k \geq 2$. Then $\Gamma_f = \mathbf{C}^*$, the multiplicative group of non-zero complex numbers. To see this fact, given the automorphism $az + b$ we need to find c, d such that

$$(az + b)^k = cz^k + d.$$

  For $k > 1$, finding such c, d is possible only when $b = 0$. Since a is an arbitrary element of $\mathbf{C}^*$ we conclude that $\Gamma_f = \mathbf{C}^*$. Also, $b = 0$ implies $d = 0$, and hence $c = a^k$ is an arbitrary non-zero complex number. Thus $T_f = \mathbf{C}^*$ as well.
- Put $f(z) = e^z$, then Γ_f consists of translations and is isomorphic to $\mathbf{C}$. The reason is that the equation

$$e^{az+b} = ce^z + d$$

  forces $a = 1$ and $d = 0$, but allows b to be arbitrary. Here $c = e^b$, and hence $T_f = \mathbf{C}^*$.
- Put $f(z) = z + e^z$. To determine Γ_f, we consider the equation

$$(az + b) + e^{az+b} = c(z + e^z) + d. \qquad (9)$$

  Differentiating (9) twice shows that $a = 1$ and $e^b = c$. Putting $a = 1$ in (9) gives $c = 1$ and hence $b = 0$. We conclude that Γ_f is the trivial group. It follows that T_f is trivial as well.

REMARK 5.4. Things are much more complicated when $\Omega = \mathbf{C}^n$ for $n \geq 2$. The group $\mathrm{Aut}(\mathbf{C}^n)$ is very large. We illustrate when $n = 2$, using z and w for the variables. For any entire function p in one variable, the map $(z, w) \mapsto (z, w + p(z))$ is an automorphism. Such maps are called *shears*. We can of course interchange the roles of z and w. A finite composition of such maps is also an automorphism. Andersen and Lempert introduced a collection of automorphisms called *overshears*; such maps have the form

$$(z, w) \mapsto (z, we^{h(z)} + g(z))$$
$$(z, w) \mapsto (ze^{h(w)} + g(w), w).$$

There are analogous definitions in higher dimensions. The Andersen-Lempert [AL] theorems for $n \geq 2$ provide the following information. First, there are auto-morphisms that are not finite compositions of overshears. Second, the subgroup generated by overshears is dense in $\mathrm{Aut}(\mathbf{C}^n)$ in the topology of uniform limits on compact subsets. Finally, an analogue of this second result holds for the group of diffeomorphisms of real Euclidean space.

We next provide additional easy examples of the groups.

EXAMPLE 5.9. Suppose that Ω is the unit disk, and $f(z) = z^m$, where m is an integer at least two. Then G_f is the cyclic group generated by a primitive m-th root of unity.

EXAMPLE 5.10. As in Example 5.8, let $\Omega = \mathbf{C}$ and $f(z) = z^m$ for $m \geq 2$. Then $\Gamma_f = \mathbf{C}^*$. As noted there, T_f is also $\mathbf{C}^*$. There is a group homomorphism $\Phi : \Gamma_f \to T_f$ given by $\Phi(a) = a^m$. The kernel of Φ is the set of m-th roots of unity, a cyclic group of order m. The kernel is thus the group G_f. This relationship holds rather generally. See Example 5.11 and [DX2]. See also Example 5.17 and Exercise 5.15 for another case where G_f is interesting.

EXAMPLE 5.11. Let $f : \Omega_1 \to \Omega_2$ be holomorphic. The following result holds. The group H_f is trivial if and only if, for each $\gamma \in \Gamma_f$, there is a *unique* $\psi \in T_f$ for which $f \circ \gamma = \psi \circ f$. Here is the proof: Assume first that H_f is trivial. If

$$f \circ \gamma = \psi_1 \circ f = \psi_2 \circ f$$

then $\psi_2^{-1} \circ \psi_1 \circ f = f$ and hence $\psi_2^{-1} \circ \psi_1 \in H_f$. Therefore $\psi_2 = \psi_1$. Conversely, suppose that uniqueness holds. Let I_2 denote the identity in the target and I_1 the identity in the source. Assume $\psi \in H_f$. Then

$$\psi \circ f = f = f \circ I_1 = I_2 \circ f.$$

By uniqueness, $\psi = I_2$ and H_f is trivial. In this setting the map Φ that assigns ψ to γ is a group homomorphism. See Exercise 5.14. For holomorphic maps between balls, these conditions are equivalent to a geometric condition; namely, the target dimension of f is *minimal*. In other words, the image of the map f lies in no affine subspace of lower dimension. See [DX2].

EXAMPLE 5.12. Let Ω_1 be the unit disk and let Ω_2 be the unit ball in $\mathbf{C}^2$. Put $f(z) = \frac{1}{2}(z + z^2, z^2 - z^3)$. Then $f : \mathbb{B}_1 \to \mathbb{B}_2$ is a proper holomorphic mapping and Γ_f is the trivial group. We will prove this fact and more in Theorem 5.8.

EXAMPLE 5.13. For $m \geq 2$, put $f(z) = z^{\otimes m}$. Then $f : \mathbb{B}_n \to \mathbb{B}_N$ is a proper mapping between balls. The group Γ_f is the unitary group $\mathbf{U}(n)$.

EXERCISE 5.10. Show that $\mathbf{A}_f$ is a subgroup of $\mathrm{Aut}(\Omega_1) \times \mathrm{Aut}(\Omega_2)$.

EXERCISE 5.11. Put $f(z) = e^z$ as a map from $\mathbf{C}$ to itself. Compute T_f and G_f. Find the group homomorphism $\Phi : \Gamma_f \to T_f$ whose kernel is G_f. Compare with Example 5.10. Do the same problem when $f(z) = e^{z^2} + z^2$.

EXERCISE 5.12. Put $f(\zeta) = (\zeta_1^3, \zeta_1\zeta_2, \zeta_2^3)$. For ζ real, show that f is injective. For ζ complex, show that f fails to be injective. Determine the maximum number of inverse images of a point. With ζ real, show that the image of a neighborhood of 0 under f is not a smooth manifold.

EXERCISE 5.13. Define $f : \mathbf{C} \to \mathbf{C}$ by $f(z) = z^2 + z^4$. Find Γ_f. Then find a map g for which Γ_g is cyclic of order m.

EXERCISE 5.14. Assume the uniqueness property from Example 5.11. Show that the map $\gamma \to \psi = \Phi(\gamma)$ is a group homomorphism.

4. Maps between balls

In this section we assume that Ω_1 is the unit ball $\mathbb{B}_n$ in some $\mathbf{C}^n$ and that Ω_2 is the unit ball in $\mathbf{C}^N$. Recall from Chapter 4 the structure of the automorphism group $\mathrm{Aut}(\mathbb{B}_n)$. Each automorphism γ can be written $\gamma = U\phi_a$, where U is unitary and ϕ_a is the linear fractional transformation

$$\phi_a(z) = \frac{a - L_a z}{1 - \langle z, a \rangle}. \tag{10}$$

In (10), we have $||a|| < 1$. Put $s = \sqrt{1 - ||a||^2}$. We have

$$L_a(z) = \frac{\langle z, a \rangle a}{s + 1} + sz.$$

REMARK 5.5. The following facts are often used in the sequel.

- $L_a(a) = a$.
- $\langle L_a z, a \rangle = \langle z, a \rangle$.
- $\phi_a \circ \phi_a$ is the identity map.
- $||a - L_a z||^2 - |1 - \langle z, a \rangle|^2 = (||z||^2 - 1)(1 - ||a||^2)$.

We will often use homogeneous expansion. When $g : \mathbf{C}^n \to \mathbf{C}^N$ is a vector-valued polynomial, we write $g = \sum_{j=0}^d g_j$ to denote the expansion of g into homogeneous parts. Homogeneity nicely interacts with subgroups of $\mathrm{Aut}(\mathbb{B}_n)$. First consider the unit circle, which we can regard as a subgroup of $\mathrm{Aut}(\mathbb{B}_n)$ via the diagonal unitary map $z \mapsto e^{i\theta} z$. By homogeneity we have

$$g(e^{i\theta} z) = \sum_{j=0}^d e^{ij\theta} g_j(z). \tag{11}$$

Equation (11) is especially useful when $\frac{p}{q}$ is a rational proper mapping between balls. In this setting we will always assume that $\frac{p}{q}$ is reduced to lowest terms. Then q has no zeroes on the open ball, and by a result of Cima-Suffridge [CS], no zeroes on the closed ball as well. In particular $q(0) \neq 0$ and we will assume without loss of generality that $q(0) = 1$.

PROPOSITION 5.1. *Let* $f = \frac{p}{q} : \mathbb{B}_n \to \mathbb{B}_N$ *be a rational proper map. Then*

- *The degree of* q *is less than or equal to the degree of* p.
- *If* $p(0) = 0$, *then the degree of* q *is less than the degree of* p.

PROOF. Let d be the maximum of the two degrees. We write $p = \sum_{j=0}^d p_j$ and $q = \sum_{j=0}^d q_j$. For z on the sphere we have $||f(z)||^2 = 1$ and hence

$$||p(z)||^2 = \sum_{j,k} \langle p_j(z), p_k(z) \rangle = \sum_{j,k} q_j(z)\overline{q_k(z)} = |q(z)|^2. \tag{12}$$

Replace z by $e^{i\theta}$ and equate Fourier coefficients. We obtain various identities, the easiest of which is

$$\langle p_d, p_0 \rangle = q_d \overline{q_0} = q_d. \tag{13}$$

If $q_d \neq 0$, then $p_d \neq 0$ as well, and the first statement holds. If $p(0) = 0$, then $q_d = 0$ and the second statement holds. $\qquad\square$

Later on we will use the other identities arising from (12). We mention the other extreme case:

$$\sum_{j=0}^{d} ||p_j(z)||^2 = \sum_{j=0}^{d} |q_j(z)|^2.\tag{14}$$

Let $f : \mathbb{B}_n \to \mathbb{B}_N$ be a rational proper holomorphic mapping. There is a computational method from [DX1] to decide whether an automorphism $\gamma \in \mathrm{Aut}(\mathbb{B}_n)$ lies in Γ_f. This result depends upon the following approach in which one associates a Hermitian form to such a map.

DEFINITION 5.6. Let $f = \frac{p}{q}$ be a rational function with no singularities on the closed unit ball. Assume the fraction is reduced to lowest terms and $q(0) = 1$. Put

$$\mathcal{H}(f) = ||p||^2 - |q|^2.$$

EXAMPLE 5.14. For the identity map f, we have $\mathcal{H}(f) = ||z||^2 - 1$. Put $f(z) = \phi_a(z) = \frac{a - L_a z}{1 - \langle z, a \rangle}$. The formulas in Remark 5.5 yield

$$\mathcal{H}(f) = (1 - ||a||^2)(||z||^2 - 1).$$

In particular, the Hermitian norm gets multiplied by a constant.

Example 5.14 suggests the following crucial result. Because of this result, we call Γ_f the *Hermitian invariant group* of f.

THEOREM 5.3. *Let $f = \frac{p}{q}$ be a proper rational map from $\mathbb{B}_n$ to some $\mathbb{B}_N$. Let $\Gamma_f \leqslant \mathrm{Aut}(\mathbb{B}_n)$ be the Hermitian invariant group of f. Then $\gamma \in \Gamma_f$ if and only if there is a constant c_γ such that*

$$\mathcal{H}(f \circ \gamma) = c_\gamma \mathcal{H}(f).\tag{15}$$

PROOF. First assume (15) holds. Write $f = \frac{p}{q}$ and $f \circ \gamma = \frac{P}{Q}$. We assume $q(0) = Q(0) = 1$ and that the fractions are in lowest terms. After composing with automorphisms of the target we may also assume $p(0) = 0$ and $P(0) = 0$. Equation (15) yields

$$||P||^2 - |Q|^2 = C_\gamma(||p||^2 - |q|^2),\tag{16}$$

and thus the constant C_γ must equal 1. Write $Q = 1 + A$ and $q = 1 + a$ and plug in (16). Equating pure terms yields

$$2\mathrm{Re}(A) = 2\mathrm{Re}(a).$$

Since A, a are polynomials vanishing at 0 we obtain $A = a$. Equating mixed terms then gives

$$||P||^2 - ||A||^2 = ||p||^2 - |a|^2,$$

and hence $||P||^2 = ||p||^2$. By Theorem 4.3 of Chapter 4, there is a $U \in \mathbf{U}(N)$ such that $P = Up$. Thus $f \circ \gamma = g_\gamma \circ f$ for some automorphism g_γ.

The converse is easy. We are given an automorphism γ of the source ball for which there is an automorphism ψ_γ of the target ball with $f \circ \gamma = \psi_\gamma \circ f$. We need to prove (15). Consider an arbitrary automorphism φ of the target ball. We write $\varphi = U \circ \phi_a$. We may assume $f(0) = 0$. Hence

$$\mathcal{H}(\varphi \circ f) = (1 - ||a||^2)\mathcal{H}(f).\tag{17}$$

The equality $f \circ \gamma = \psi_\gamma \circ f$ and (17) guarantee that

$$\mathcal{H}(f \circ \gamma) = \mathcal{H}(\psi_\gamma \circ f) = c_\gamma \mathcal{H}(f)$$

for a non-zero constant c_γ. Hence (15) holds.   □

REMARK 5.6. Let us briefly reconsider the distinction between Γ_f and G_f. Suppose that f is a polynomial and $f(0) = 0$. To be in G_f, a unitary map γ must satisfy $f \circ \gamma = f$. To be in Γ_f, it must satisfy $||f \circ \gamma||^2 = ||f||^2$. This last equation is equivalent to the existence of a unitary V such that $f \circ \gamma = V \circ f$. The equality of squared norms is equivalent to γ being in Γ_f.

PROPOSITION 5.2. *Suppose f is a monomial proper map between balls. Then Γ_f contains an n-torus, namely the diagonal unitary mappings.*

PROOF. Put $f(z) = (..., c_\alpha z^\alpha, ...)$. Let γ be a diagonal unitary matrix with eigenvalues $e^{i\theta_j}$. Then, in multi-index notation,

$$(f \circ \gamma)(z) = (..., c_\alpha e^{i\alpha\theta} z^\alpha, ...).$$

Therefore $||f \circ \gamma||^2 = ||f||^2$ and thus $\mathcal{H}(f \circ \gamma) = \mathcal{H}(f) = ||f(z)||^2 - 1$. By Theorem 5.3, $\gamma \in \Gamma_f$.   □

REMARK 5.7. For a monomial map f, the group Γ_f can be larger than the torus. For the identity map, Γ_f is the full automorphism group. For the tensor product map $z \mapsto z^{\otimes m}$ for $m \geq 2$, the group Γ_f is the unitary group. Example 5.17 gives a proper monomial maps whose group is generated by the torus and the map that interchanges the variables. For a generic monomial map, however, Γ_f is the n-torus. When f is spherically equivalent to a monomial map, then (by Lemma 5.1) Γ_f contains a conjugate to the torus.

COROLLARY 5.2. *Suppose f is a rational proper map of degree 2. Then Γ_f contains a conjugate of the torus. In particular Γ_f is infinite.*

PROOF. By a theorem of Lebl [Le], a rational proper mapping of degree 2 is spherically equivalent to a monomial mapping.   □

Let f be a proper map between balls. Whether or not f is rational, it can be shown (see [DX2]) that Γ_f is non-compact if and only if T_f is noncompact. We will consider here the case when f is a rational proper map between balls. We next ask when is the group Γ_f noncompact. The answer is striking; f must be a linear fractional transformation, and hence spherically equivalent to the injection $z \mapsto (z, 0)$. For this map the group Γ_f is of course $\text{Aut}(\mathbb{B}_n)$. Therefore groups intermediate between the unitary group and the full automorphism group do not arise as Γ_f for any rational proper mapping f.

We will use an n-dimensional generalization of Schwarz's lemma. First we recall the classical result; we refer to [A] and [Kr] for geometric reinterpretations.

PROPOSITION 5.3. (Schwarz lemma) *Let $f : \mathbb{B}_1 \to \mathbf{C}$ be holomorphic. Suppose $f(0) = 0$ and $|f(z)| \leq 1$. Then the stronger inequality $|f(z)| \leq |z|$ holds on $\mathbb{B}_1$.*

PROOF. Since $f(0) = 0$, the function g defined by $g(z) = \frac{f(z)}{z}$ is also holomorphic. For any $r < 1$, its maximum absolute value on the disk $|z| \leq r$ is achieved on the circle $|z| = r$. Therefore

$$|g(z)| \leq \max_{|z| \leq r} \left(\frac{|f(z)|}{r} \right) \leq \frac{1}{r}. \tag{18}$$

Letting r tend to 1 in (18) shows that $|g(z)| \leq 1$ and hence $|f(z)| \leq |z|$. ☐

COROLLARY 5.3. *Let* $f : \mathbb{B}_n \to \mathbb{B}_N$ *be holomorphic and* $f(0) = 0$. *Then* $||f(z)|| \leq ||z||$ *holds for* $z \in \mathbb{B}_n$.

PROOF. Choose a nonzero $z \in \mathbb{B}_n$. Let l be a linear functional on $\mathbf{C}_N$ with $l(f(z)) = ||f(z)|| \leq 1$ and $||l|| \leq 1$. Define the linear map $L : \mathbf{C} \to \mathbf{C}^N$ by $\zeta \mapsto \frac{\zeta z}{||z||}$. Then $||L|| = 1$ as well. Put $g = l \circ f \circ L$. Then g satisfies the hypotheses of Schwarz's lemma in one dimension and therefore $|g(\zeta)| \leq |\zeta|$. Put $\zeta = ||z||$. We get

$$||f(z)|| = ||l(f(z))|| = ||l(f((L\zeta)))|| = ||g(\zeta)|| \leq |\zeta| = ||z||.$$

☐

THEOREM 5.4. *Let* $f : \mathbb{B}_n \to \mathbb{B}_N$ *be a rational proper map. Then* Γ_f *is non-compact if and only if* f *is a linear fractional transformation. If* Γ_f *is compact, then* Γ_f *lies in a maximal compact subgroup, that is, a conjugate of* $\mathbf{U}(n)$.

In order to prove Theorem 5.4, we first prove the following auxiliary result.

THEOREM 5.5. *Let* $f : \mathbb{B}_n \to \mathbb{B}_N$ *be a rational proper map of degree* d. *Put* $f = \frac{p}{q}$ *and assume that* $f(0) = 0$. *If* Γ_f *contains an automorphism* $\gamma = U\phi_a$ *that moves the origin, then*

$$(||p(a)||^2 - |q(a)|^2)(||p(Ua)||^2 - |q(Ua)|^2) = (1 - ||a||^2)^{2d}. \qquad (19)$$

PROOF. Since $p(0) = 0$, Proposition 5.1 implies that the degree of the denominator q is at most $d - 1$. We write

$$p(z) = \sum_{|\alpha|=1}^{d} A_\alpha z^\alpha$$

$$q(z) = \sum_{|\beta|=0}^{d-1} b_\beta z^\beta.$$

Without loss of generality we assume $b_0 = 1$. Assuming that $\gamma = U\phi_a$ exists as hypothesized, we will compute the coefficient c_γ in two ways. By definition we have

$$c_\gamma \left(||p||^2 - |q|^2 \right) = c_\gamma \mathcal{H}(f) = \mathcal{H}(f \circ (U\phi_a)).$$

Putting $f \circ (U\phi_a) = \frac{P}{Q}$, we have

$$\mathcal{H}(f \circ (U\phi_a)) = ||P||^2 - |Q|^2.$$

We have the following formulas for P and Q:

$$||P(z)||^2 = \frac{1}{|q(a)|^2} \left(|| \sum_{|\alpha|=1}^{d} A_\alpha (U(a - L_a(z))^\alpha (1 - \langle z, a \rangle)^{d-|\alpha|} ||^2 \right)$$

$$|Q(z)|^2 = \frac{1}{|q(a)|^2} \left(| \sum_{|\beta|=0}^{d-1} b_\beta (U(a - L_a(z))^\beta (1 - \langle z, a \rangle)^{d-|\beta|} |^2 \right).$$

The factor of $\frac{1}{|q(a)|^2}$ arises in order to make $Q(0) = 1$. We evaluate at 0 to get

$$||P(0)||^2 - |Q(0)|^2 = \frac{1}{|q(a)|^2} \left(||p(Ua)||^2 - |q(Ua)|^2 \right).$$

Then we evaluate at a, using $L_a(a) = a$, to get

$$||P(a)||^2 - |Q(a)|^2 = \frac{1}{|q(a)|^2}(-(1 - ||a||^2)^{2d}).$$

Evaluating $\mathcal{H}(f)$ at 0 and a, and using $c_\gamma \, \mathcal{H}(f) = \mathcal{H}(f \circ \gamma)$, yields both formulas

$$c_\gamma \, (||p(0)||^2 - 1) = -c_\gamma = \frac{1}{|q(a)|^2} \left(||p(Ua)||^2 - |q(Ua)|^2\right)$$

$$c_\gamma(||p(a)||^2 - |q(a)|^2) = -\frac{1}{|q(a)|^2}(1 - ||a||^2)^{2d}.$$

Formula (19) follows.   □

PROOF. We can now prove Theorem 5.4. Let $f = \frac{p}{q}$ be of degree d. After composition with an automorphism of the target, we may assume $f(0) = 0$. Assume $U\phi_a \in \Gamma_f$. In this case, Schwarz's lemma yields $||p(z)||^2 \leq |q(z)|^2 \, ||z||^2$ for z in the ball. Therefore

$$|q(z)|^2 - ||p(z)||^2 \geq |q(z)|^2(1 - ||z||^2).$$

Since $||Ua||^2 = ||a||^2$, we plug this inequality into (19) to get

$$(1 - ||a||^2)^{2d} \geq (1 - ||a||^2)^2 |q(a)|^2 |q(Ua)|^2$$

and therefore

$$(1 - ||a||^2)^{2d-2} \geq |q(a)|^2 |q(Ua)|^2. \tag{20}$$

If Γ_f is not compact, then we can find a sequence of automorphisms $U_k\phi_{a_k} \in \Gamma_f$ where $||a_k||$ tends to 1. Assume $d \geq 2$. By (20), a subsequence of $q(a_k)$ or of $q(U_k a_k)$ tends to 0. But the denominator q cannot vanish on the closed ball. This contradiction therefore implies $d = 1$. The degree of q is smaller than the degree of p. Thus, if the degree of p is 1, then q is constant. Hence f is a linear fractional transformation.

Suppose that f is not a linear fractional transformation. Since Γ_f is closed in $\mathrm{Aut}(\mathbb{B}_n)$, it is compact. By standard Lie group theory, Γ_f is contained in a maximal compact subgroup which must be a conjugate of $\mathbf{U}(n)$.   □

COROLLARY 5.4. *Let p be a proper polynomial map between balls with $p(0) = 0$. Unless p is of degree 1, we have $\Gamma_p \subseteq \mathbf{U}(n)$.*

PROOF. When $q = 1$, formula (20) forces $a = 0$ or $d = 1$. Thus, unless $d = 1$, there is no automorphism in Γ_p that moves the origin.   □

Corollary 5.4 gets used several times in this chapter. For polynomial maps f of degree at least 2 with $f(0) = 0$, only subgroups of $\mathbf{U}(n)$ are candidates for Γ_f.

EXERCISE 5.15. Verify the formula in Example 5.14 for the Hermitian norm.

5. Examples of unitary invariance

We pause to consider some examples of unitary invariance. These examples help illustrate the distinction between invariance for a map f and invariance for the map $||f||^2$.

EXAMPLE 5.15. Define a map $f : \mathbf{C}^2 \to \mathbf{C}$ by $f(z, w) = (1 + z)(1 + w)$. Notice that f is invariant under the group of two elements arising from permuting the variables. We ask whether there is a linear map L and an affine map T such that $f \circ L = T \circ f$. Supposing $L(z, w) = (az + bw, cz + dw)$ and $T(\zeta) = \alpha\zeta + \beta$, we obtain

$$(1 + az + bw)(1 + cz + dw) = \alpha(1 + z)(1 + w) + \beta.$$

Expanding and equating coefficients yields six equations:

$$ac = bd = 0$$

$$ad + bc = a + c = b + d = 1 - \beta = \alpha.$$

Note that $\alpha \neq 0$. The only solutions have $\alpha = 1$ and hence $\beta = 0$. We get $a = d = 1$ and $b = c = 0$ or $a = d = 0$ and $b = c = 1$. The set of such L is a permutation group of order 2.

EXAMPLE 5.16. Put $n = 3$ and consider the map f defined by

$$f(z) = (z_2 z_3, z_1 z_3, z_1 z_2).$$

Then f is **not** invariant under a non-trivial permutation of the coordinates. The Hermitian norm $||f(z)||^2 - 1$ is invariant under $L\sigma$, whenever L is a diagonal unitary matrix and σ is an arbitrary permutation of the coordinates. In fact the following result holds. See [DX1] for detailed discussion.

PROPOSITION 5.4. *For $n \geq 2$, define a quadratic monomial map f by*

$$f(z) = (..., z_j z_k, ...) \tag{21}$$

for all pairs j, k with $1 \leq j < k \leq n$. Assume that U is unitary and $||f \circ U||^2 = ||f||^2$. Then there are a diagonal L and a permutation σ such that $U = L\sigma$.

PROOF. Let the entries of U be denoted by u_{jk}. Computing both sides of $||f \circ U||^2 = ||f||^2$ and equating coefficients leads to the system

$$\sum_{1 \leq j < k \leq n} |u_{jl} u_{kl}|^2 = 0.$$

Hence, for each column of U, the product of any pair of distinct entries vanishes. Thus there can be at most one non-zero element in each column. Since U is invertible, each column contains exactly one non-zero element. Since U is unitary, each of these entries has modulus 1. Let L be the diagonal matrix with these non-zero entries and let σ denote the appropriate permutation. The result follows. $\square$

COROLLARY 5.5. *Let $f : \mathbb{B}_2 \to \mathbf{C}$ be defined by $f(z) = z_1 z_2$. Thus f is the map in (21) when $n = 2$. Then Γ_f is generated by the matrices A and σ, where*

$$A = \begin{pmatrix} e^{i\theta} & 0 \\ 0 & e^{i\phi} \end{pmatrix}$$

$$\sigma = \begin{pmatrix} 0 & 1 \\ 1 & 0 \end{pmatrix}.$$

We can use the proposition and corollary to construct proper maps f between balls for which Γ_f is the symmetric group. For example, to obtain a group of order 2, we need a way to avoid the matrix A as in Corollary 5.5. The first part of the following result is a special case of Corollary 5.1. We give a direct proof enabling us to prove the second conclusion.

THEOREM 5.6. *Let $f : \mathbf{C}^n \to \mathbf{C}^N$ be a polynomial map. Then there is $\epsilon > 0$, an integer K, and a polynomial map $g : \mathbf{C}^n \to \mathbf{C}^K$ such that $\epsilon f \oplus g$ is a proper map between balls. Furthermore we may choose g such that*

$$\epsilon^2 \|f(z)\|^2 + \|g(z)\|^2 = \sum_j |\lambda_j|^2 \|z\|^{2m_j} \tag{22}$$

for numbers λ_j satisfying $\sum |\lambda_j|^2 = 1$ and (distinct) positive integers m_j.

PROOF. Set $m_j = j$ for $0 \leq j \leq \deg(f)$. Choose λ_j with $\sum |\lambda_j|^2 = 1$. When each $\lambda_j \neq 0$, the right-hand side of (22) defines a positive definite Hermitian form $r(z, \overline{z})$ on a vector space of polynomials. Hence, for sufficiently small positive ϵ, the form $r(z, \overline{z}) - \epsilon^2 \|f(z)\|^2$ is also positive definite. Hence there exists a vector-valued polynomial $g(z)$ such that

$$r(z, \overline{z}) - \epsilon^2 \|f(z)\|^2 = \|g(z)\|^2.$$

Since $r(z, \overline{z}) = 1$ on the unit sphere, $\epsilon f \oplus g$ does the job. $\qquad \square$

REMARK 5.8. Consider a polynomial map p with $p(0) = 0$ such that

$$\|p(z)\|^2 = \sum_j |\lambda_j|^2 \|z\|^{2m_j}$$

and $\sum_j |\lambda_j|^2 = 1$. Assume $m_j \geq 1$ for each j. Then p is a proper polynomial map between balls and, by Corollary 5.4, Γ_p is a subgroup of $\mathbf{U}(n)$.

We next discuss one way we will use Theorem 5.6. Let G be a finite subgroup G of the unitary group $\mathbf{U}(n)$. We wish to find a polynomial proper map f for which $\Gamma_f = G$. It is easy to find a polynomial map h (not necessarily proper) for which $h \circ \gamma = h$ for all $\gamma \in G$, but doing so only shows that G is a subgroup of Γ_h. In other words, it is easier for $\|h\|^2$ to be invariant than it is for h to be invariant. We need a way to change h that makes the group smaller, and the result must map the sphere to a sphere. We illustrate the first idea with an example. Theorem 5.9 solves the problem by combining the two ideas.

EXAMPLE 5.17. The map $z \to f(z) = (z_1^3, \sqrt{3}z_1 z_2, z_2^3)$ is invariant under a cyclic group of order 3, as noted in Chapter 4, and G_f is cyclic of order 3. The squared norm $\|f(z)\|^2$ is invariant under both $\mathbf{U}(1) \oplus \mathbf{U}(1)$ and the group of order two obtained by interchanging the variables. In fact, Γ_f is generated by the diagonal unitary matrices and the matrix $\begin{pmatrix} 0 & 1 \\ 1 & 0 \end{pmatrix}$. This group is the semi-direct product of the torus $\mathbf{U}(1) \times \mathbf{U}(1)$ and the group of two elements. One can also find T_f. It is generated by the three-by-three unitary matrices with eigenvalues $e^{3i\theta}, e^{i(\theta+\phi)}, e^{3i\phi}$ and the permutation matrix that interchanges the first and third variables.

EXERCISE 5.16. Prove the statements in Example 5.17. Hint: To find T_f, for example, consider a $V \in \mathbf{U}(3)$ for which there is a $U \in \mathbf{U}(2)$ with $f \circ U = V \circ f$. Equate coefficients of the corresponding monomials in this equation and thereby determine the restrictions on V. Find the group homomorphism $\Phi : \Gamma_f \to T_f$ whose kernel is G_f.

EXERCISE 5.17. Put f as in Example 5.17. Define F by

$$F(z) = \mathbf{1} + f(z) = (1 + z_1^3, 1 + \sqrt{3}z_1 z_2, 1 + z_2^3).$$

Expanding the squared norm of F introduces the terms $z_1^3, z_1 z_2, z_2^3$. Verify their invariance under the cyclic subgroup of $\mathbf{U}(2)$ generated by the map $(z_1, z_2) \mapsto (\eta z_1, \eta^2 z_2)$; here η is a primitive cube root of 1.

EXERCISE 5.18. Put $f(z, w) = (z^{2k+1}, zw, w^{2k+1})$. Find all $U \in \mathbf{U}(2)$ for which there is a linear map V with $f \circ U = V \circ f$. Determine the group G_f.

EXERCISE 5.19. Formula (19) from Chapter 4 defines a polynomial $f_{5,2}$ for which $f_{5,2}(|z|^2, |w|^2)$ is the squared norm of a proper map p between balls. Find this map and determine the groups Γ_p and G_p.

6. Behavior of Γ_f under various constructions

The collection of proper holomorphic mappings between balls has considerable structure. For example, if $f : \mathbb{B}_n \to \mathbb{B}_N$ and $g : \mathbb{B}_n \to \mathbb{B}_K$ are proper holomorphic mappings, then the tensor product $f \otimes g$ also is, where the target ball can be regarded as $\mathbb{B}_L$ with $L = NK$. Another way to create a new map from f and g is *juxtaposition*, in which case the target dimensions add. These constructions strongly suggest that the question "Fix n, N. What are the proper holomorphic maps from $\mathbb{B}_n$ to $\mathbb{B}_N$" should be replaced by "Fix n. What are the proper holomorphic maps from $\mathbb{B}_n$ to $\mathbb{B}_N$ for some N."

We quickly recall these constructions (See [D1] for example) and then study how Hermitian invariant groups behave under them.

DEFINITION 5.7. Let f, g be holomorphic maps from $\mathbb{B}_n$ to possibly different dimensional complex Euclidean spaces. For $0 \le \theta \le \frac{\pi}{2}$ we consider the family of maps J_θ defined by

$$J_\theta(f, g)\ (z) = (\cos(\theta) f(z), \sin(\theta) g(z)) = \cos(\theta) f \oplus \sin(\theta) g.$$

We can iterate the juxtaposition operation. Given $\lambda = (\lambda_1, ..., \lambda_K) \in S^{2K-1}$ and maps $f_1, ..., f_K$ with the same source, we also consider the map F defined by

$$F = \lambda_1 f_1 \oplus \lambda_2 f_2 \oplus ... \oplus \lambda_K f_K.$$

We refer to F as a juxtaposition of the f_j. When each f_j is a proper map between balls, so is F.

DEFINITION 5.8. Let f, g be holomorphic maps from $\mathbb{B}_n$ to possibly different dimensional complex Euclidean spaces of dimensions N and K. We define a map $f \otimes g$ with source $\mathbb{B}_n$ and target $\mathbf{C}^{NK}$ by

$$f \otimes g = (f_1 g_1, f_1 g_2, ..., f_1 g_K, f_2 g_1, ..., f_2 g_K, ..., f_N g_K).$$

The constructions in Definitions 5.7 and 5.8 are particularly important when f, g are proper mappings between balls. Note that $||f \otimes g||^2 = ||f||^2 ||g||^2$. Therefore, if both f and g map the unit sphere to a unit sphere, then so does $f \otimes g$. Similarly, $J_\theta(f, g)$ also does. Note also that the map in Theorem 5.6 and Remark 5.8 is a juxtaposition of tensor powers.

THEOREM 5.7. *Let f, g be proper polynomial maps between balls with the same source. Let $\nu(g)$ denote the order of vanishing of g and $\deg(f)$ the degree of f. Assume that $2 \le \deg(f) < \nu(g)$. For $0 < \theta < 2\pi$, put $J = J_\theta = J_\theta(f, g)$. Then*

$$\Gamma_J = \Gamma_f \cap \Gamma_g.$$

COROLLARY 5.6. *Let f, g be proper polynomial maps between balls with the same source. Let m be larger than the degree of f. For $0 < \theta < 2\pi$, put*

$$j(f, g) = J_\theta(f, g \otimes z^{\otimes m}).$$

Then

$$\Gamma_{j(f,g)} \cap \mathbf{U}(n) = \Gamma_f \cap \Gamma_g \cap \mathbf{U}(n).$$

COROLLARY 5.7. *Suppose in Theorem 5.7 that f is a polynomial of degree at least 2 with $f(0) = 0$. Then $\Gamma_{j(f,g)} = \Gamma_f \cap \Gamma_g$.*

PROOF. Corollary 5.7 follows from Corollaries 5.4 and 5.6 because the hypothesis on f guarantees that the groups involved are subsets of $\mathbf{U}(n)$. Corollary 5.6 follows from Theorem 5.7 because the order of vanishing of $g \otimes z^{\otimes m}$ is at least m. We now prove the theorem. First let $c = \cos(\theta)$ and $s = \sin(\theta)$. Then

$$\mathcal{H}(J) = c^2 ||f||^2 + s^2 ||g||^2 - 1 \tag{23.1}$$

$$\mathcal{H}(J \circ \gamma) = c^2 ||f \circ \gamma||^2 + s^2 ||g \circ \gamma||^2 - 1. \tag{23.2}$$

The hypotheses imply that Γ_f and Γ_g are subgroups of the unitary group. If γ lies in both these groups, then the right-hand sides of (23.1) and (23.2) are equal, and hence the left-hand sides are also equal. Thus $\Gamma_f \cap \Gamma_g \subseteq \Gamma_J$. To prove the opposite inclusion, suppose that the left-hand sides are equal. Since γ is unitary (and hence linear), it preserves degrees. The hypotheses therefore prevent any interaction between the terms $||f \circ \gamma||^2$ and $||g \circ \gamma||^2$. We conclude that $||f \circ \gamma||^2 = ||f||^2$ and $||g \circ \gamma||^2 = ||g||^2$. $\square$

Next consider the mappings $z \mapsto z^{\otimes m}$, for m a positive integer. When $m = 1$, the map f is the identity map, and therefore its group Γ_f is the full automorphism group. For $m \geq 2$, the group Γ_f is the unitary group. The reason is clear; the Hermitian norm $||z||^{2m} - 1$ is obviously invariant under the unitary group, and it is not invariant under any automorphism that moves the origin. Corollary 5.7 has therefore the following additional corollary.

COROLLARY 5.8. *Assume that $K \geq 2$, that $\sum_{j=1}^{K} |\lambda_j|^2 = 1$, and that $m_1, ..., m_K$ are distinct positive integers. Define F, a juxtaposition of tensor powers, by*

$$F(z) = \lambda_1 z^{\otimes m_1} \oplus ... \oplus \lambda_K z^{\otimes m_K}.$$

Then $\Gamma_F = \mathbf{U}(n)$.

If we take partial tensor products as in Chapter 4, then we might make the group smaller. For example, if we begin with the identity map (z, w) the group is the full automorphism group. Put $f(z, w) = (z^2, zw, w)$, a partial tensor product of the identity map. Then Γ_f is a torus. But, when we take a partial tensor product of f, the group gets larger. See Exercise 5.20.

Next we want to show how to find maps whose Hermitian group is a given finite group. We start with a simple result in one dimension.

THEOREM 5.8. *Let m be a positive integer. Let $f_m : \mathbf{C} \to \mathbf{C}^2$ be defined by $f_m(z) = \frac{1}{2}(z^m + z^{2m}, z^{2m} - z^{3m})$. Then f_m is a proper mapping from the disk to the ball. Furthermore, Γ_{f_m} is the cyclic subgroup of the unit circle generated by $e^{\frac{2\pi i}{m}}$. In particular, when $m = 1$, the Hermitian invariant group Γ_{f_1} is the trivial group.*

PROOF. A simple computation shows that $||f(z)||^2 < 1$ on the disk and that $||f(z)||^2 = 1$ on the circle. Therefore f is a proper mapping from $\mathbb{B}_1$ to $\mathbb{B}_2$. Corollary 5.4 implies that Γ_f is a subgroup of the unitary group $\mathbf{U}(1)$, which is the unit circle. We apply Theorem 5.3. Suppose $0 \le \theta < 2\pi$ and $e^{i\theta} \in \Gamma_{f_m}$. Assume that

$$||f_m(e^{i\theta}z)||^2 - 1 = c \,(||f_m(z)||^2 - 1). \tag{24}$$

Putting $z = 0$ shows that $c = 1$ and hence

$$||f_m(e^{i\theta}z)||^2 = ||f_m(z)||^2.$$

Using the formula for f_m yields

$$|z^m + z^{2m}|^2 + |z^{2m} - z^{3m}|^2 = |z^m e^{im\theta} + z^{2m}e^{2im\theta}|^2 + |z^{2m}e^{2im\theta} - z^{3m}e^{3im\theta}|^2.$$

Hence, for every z and each such θ we have

$$2\mathrm{Re}(z^m \bar{z}^{2m} - z^{2m}\bar{z}^{3m}) = 2\mathrm{Re}(e^{im\theta}(z^m\bar{z}^{2m} - z^{2m}\bar{z}^{3m})).$$

Equating coefficients of the $z^m\bar{z}^{2m}$ term gives $e^{im\theta} = 1$, and the other coefficients are then equal as well. Hence $m\theta = 2\pi$ and the result follows. □

We have the following decisive general result from [DX1].

THEOREM 5.9. *Let G be a finite subgroup of the unitary group $\mathbf{U}(n)$. Then there is an N and a proper polynomial map $f : \mathbb{B}_n \to \mathbb{B}_N$ with $\Gamma_f = G$.*

PROOF. Given a finite subgroup G of the unitary group, one can find a finite collection of holomorphic polynomials $h(z) = (h_0(z), ..., h_K(z))$ that generate the algebra of polynomials invariant under G. We may assume that h_0 is the constant map 1, and that $h_j(0) = 0$ for $j \ge 1$. We then consider the map

$$H(z) = \left((1 + h_1(z)) \otimes z^{\otimes m_1}, ..., (1 + h_K(z)) \otimes z^{\otimes m_K}\right).$$

We choose the integers m_j in such a way that none of the components have any terms of the same degree as those of any other component. Theorem 5.6 guarantees that we can find a map g such that, for ϵ small enough, $\epsilon H \oplus g$ maps the unit sphere to some unit sphere. If we replace g by $g \otimes z^{\otimes m}$ for some large m, then $v = \epsilon H \oplus (g \otimes z^{\otimes m})$ also maps the sphere to a sphere, and (25) holds. If we choose m large, then the order of vanishing of $g \otimes z^{\otimes m}$ exceeds the degree of H. If each $m_j \ge 1$, then Γ_v is a subgroup of $\mathbf{U}(n)$.

We show both containments $G \subseteq \Gamma_v$ and $\Gamma_v \subseteq G$. Note that

$$\epsilon^2||H||^2 + ||g \otimes z^{\otimes m}||^2 = ||v||^2 = \sum |\lambda_j|^2 ||z||^{2k_j}. \tag{25}$$

Suppose first that $L \in G$. Then $h_j \circ L = h_j$ for each j. Since also $||z \circ L||^2 = ||z||^2$, we obtain $||H \circ L||^2 = ||H||^2$. Since L is unitary, the right-hand side of (25) shows that $||v \circ L||^2 = ||v||^2$. Since the terms in H and $g \otimes z^{\otimes m}$ are of different degrees, both $||H \circ L||^2 = ||H||^2$ and $||g \circ L||^2 = ||g||^2$. Hence $L \in \Gamma_v$ and $G \subseteq \Gamma_v$.

Next suppose $L \in \Gamma_v$. Again because the terms of H and those of $g \otimes z^{\otimes m}$ do not interact, we must have $||H \circ L||^2 = ||H||^2$. Furthermore the terms $|1 + h_j|^2 ||z||^{2m_j}$ are of different degrees, and hence each is invariant under composition with L. Since $||z \circ L||^2 = ||z||^2$, we conclude for each j that

$$|1 + h_j|^2 = |1 + h_j \circ L|^2.$$

Expanding the squared norm then shows that $h_j \circ L = h_j$ for each j. But the h_j are precisely G invariant. Thus $L \in G$. Hence $\Gamma_v \subseteq G$. □

REMARK 5.9. One crucial point in the proof is worth elaborating. Let t be a holomorphic polynomial with $t(0) = 0$. The equation $|t \circ L|^2 = |t|^2$ tells us much less than the equation $|1 + t \circ L|^2 = |1 + t|^2$. The second equation implies the first but also implies $t \circ L = t$.

EXERCISE 5.20. For the map f given by $f(z, w) = (z^2, zw, w)$ show directly that Γ_f is $S^1 \times S^1$. Let $g(z, w) = (z^2, zw, zw, w^2)$. Show that g is obtained from f by a partial tensor product operation. Show that Γ_g is the unitary group $\mathbf{U}(2)$.

EXERCISE 5.21. Consider the cyclic group G of order 3 from Example 5.17. Find a proper polynomial map f for which $\Gamma_f = G$.

EXERCISE 5.22. Put $n = 2$ and write (z, w) for the variables. Let η be a primitive odd p-th root of unity with $p \geq 5$. Consider the cyclic subgroup G of the unitary group generated by the map $(z, w) \mapsto (\eta z, \eta^2 w)$. Find a basis for the algebra of polynomials invariant under G. When G has order 5, construct a proper polynomial map f for which $\Gamma_f = G$.

EXERCISE 5.23. Consider the dihedral group D_4 with 8 elements. Represent it as a subgroup of the unitary group $\mathbf{U}(2)$. Suggestion: you can do so with matrices whose entries are $0, 1, -1$. If you are ambitious, find a proper polynomial map f for which $\Gamma_f = D_4$. (The author does not know the smallest possible degree of such a map.)

We end this section with a short summary of some of the main results about the group Γ_f. Assume that $f : \mathbb{B}_n \to \mathbb{B}_N$ is a rational proper map.

- $\Gamma_f = \mathrm{Aut}(\mathbb{B}_n)$ if and only if f is a linear fractional transformation.
- Γ_f is noncompact if and only if $\Gamma_f = \mathrm{Aut}(\mathbb{B}_n)$. Otherwise Γ_f is contained in a conjugate of $\mathbf{U}(n)$.
- Γ_f is a conjugate of $\mathbf{U}(n)$ if and only if f is spherically equivalent to a juxtaposition of tensor powers.
- $\Gamma_f = \mathbf{U}(n)$ if and only if f is a juxtaposition of tensor powers.
- Γ_f contains an n-torus if and only if f is spherically equivalent to a monomial map.
- Put $f(z, w) = (z^3, \sqrt{3}zw, w^3)$. Then Γ_f is generated by the diagonal unitary matrices and the unitary matrix that permutes the two variables. This group is the semi-direct product of the torus (diagonal unitaries) and the group of two elements. In particular, it is non-Abelian.
- Let G be an arbitrary finite subgroup of $\mathrm{Aut}(\mathbb{B}_n)$. Then there is an N and a proper rational map $f : \mathbb{B}_n \to \mathbb{B}_N$ such that $\Gamma_f = G$.
- Let G be an arbitrary finite subgroup of $\mathbf{U}(n)$. Then there is an N and a proper polynomial map $f : \mathbb{B}_n \to \mathbb{B}_N$ such that $\Gamma_f = G$.

7. A criterion for a power series being a polynomial

In this section we use circular symmetries to give a criterion for when a power series must be a polynomial.

Before doing so we recall a basic standard fact about Lie groups and their Lie algebras. See [Wa] for a nice account of the basics. A *Lie group* G is a smooth manifold that is also a group, and the group operations (composition and taking inverses) are smooth functions. As a manifold, G has a tangent space at each point. The Lie algebra of G, usually written $\mathfrak{g}$, consists of the left-invariant vector fields on G. The Lie algebra can be identified with the tangent space at the identity. We illustrate with a simple example we need.

EXAMPLE 5.18. Consider the Lie group $\mathbf{U}(n)$. Its defining equation is $U^*U = I$. Imagine that U depends smoothly on a real parameter t and assume that $U(0) = I$. Differentiating the defining relation and using $U(0) = I$ give

$$0 = (U^*)'U + U^*U' = (U^*)'(0) + U'(0).$$

Therefore the Lie algebra of $\mathbf{U}(n)$ consists of skew-Hermitian matrics. Conversely, the exponential of a skew-Hermitian matrix is unitary.

In this section we work with diagonal unitary matrices. In Definition 5.9 below, we regard the torus $\mathbf{U}(1) \oplus ... \oplus \mathbf{U}(1)$ as a subgroup of $\mathbf{U}(n)$.

DEFINITION 5.9. Let $\mathcal{C} = \mathcal{C}(n)$ denote the collection of continuously differentiable maps $\gamma : (-\pi, \pi) \to \mathbf{U}(1) \oplus ... \oplus \mathbf{U}(1)$ such that $\gamma(0) = \mathbf{I}_n$.

For $\gamma \in \mathcal{C}$, Example 5.18 shows that $\gamma'(0) = iL$, where $i^2 = -1$ and $L = L^*$ is real and diagonal. The eigenvalues of L are real. When they all have the same sign, we obtain a criterion guaranteeing that a holomorphic map is in fact a polynomial.

PROPOSITION 5.5. Let Ω be a ball about 0 in $\mathbf{C}^n$ and suppose $f : \Omega \to \mathbf{C}^N$ is holomorphic. For some $\gamma \in \mathcal{C}$, assume that $||f \circ \gamma||^2 = ||f||^2$. Put $L = -i\gamma'(0)$. If all the eigenvalues of L have the same sign, then f is a polynomial. More generally, if L has k eigenvalues of the same sign, then there is a k-dimensional vector subspace V such that the restriction of f to V is a polynomial.

PROOF. Put $\theta(t) = (\theta_1(t), ..., \theta_n(t))$, where $\gamma(t)$ is diagonal with eigenvalues $e^{i\theta_j(t)}$. We expand f in a (vector-valued) power series convergent in Ω:

$$f(z) = \sum_\alpha c_\alpha z^\alpha.$$

We are given that $||f(z)||^2 = ||f(\gamma(t)z)||^2$ for all $t \in (-\pi, \pi)$ and all $z \in \Omega$. Equating Taylor coefficients yields, for each pair α, β of multi-indices, that

$$\langle c_\alpha, c_\beta \rangle = e^{i\theta(t)\cdot(\alpha-\beta)}\langle c_\alpha, c_\beta \rangle$$

and hence we have

$$0 = \langle c_\alpha, c_\beta \rangle \left(1 - e^{i\theta(t)\cdot(\alpha-\beta)}\right). \tag{26}$$

Write $\mathbf{m} = (m_1, ..., m_n)$, where the m_j are the eigenvalues of L. Differentiate (26) and evaluate at $t = 0$ to obtain

$$0 = \langle c_\alpha, c_\beta \rangle (\mathbf{m} \cdot (\alpha - \beta)). \tag{27}$$

When all the m_j have the same sign we will show that $c_\alpha = 0$ for all but finitely many α.

Assume that the eigenvalues are L have the same sign. After replacing t by $-t$ we may assume they are all positive. Let $K_1 = \min(m_j)$ and $K_2 = \max(m_j)$. Thus $K_2 \geq K_1 > 0$. Let α and β be multi-indices for which $\mathbf{m} \cdot (\alpha - \beta) = 0$; then

$$K_1|\alpha| = K_1\sum \alpha_j \leq \sum m_j\alpha_j = \sum m_j\beta_j \leq K_2\sum \beta_j = K_2|\beta|. \tag{28}$$

Let W be the (finite-dimensional) span of the coefficients c_α. Choose $\alpha_1, ..., \alpha_\nu$ such that the c_{α_j} span W. Choose a multi-index η with $|\eta| > \frac{K_2}{K_1}|\alpha_j|$ for $1 \leq j \leq \nu$. By (28), $\mathbf{m} \cdot (\eta - \alpha_j) \neq 0$ for all j. Since (27) holds for all α, β, we conclude that

$\langle c_\eta, c_{\alpha_j} \rangle = 0$ for all j. Therefore $c_\eta = 0$ and hence there are only finitely many non-vanishing coefficient vectors. Thus f is a polynomial.

Next suppose that L has k eigenvalues of the same sign. After renumbering the coordinates and replacing t by $-t$ if necessary, we may assume that these eigenvalues are positive and correspond to the first k coordinates. Setting the rest of the variables equal to 0 puts us in the situation above. The conclusion follows. $\square$

REMARK 5.10. One can draw stronger conclusions. For example, when $n = 1$, equation (27) implies that the vectors c_α are mutually orthogonal and hence that f is an orthogonal sum of monomials. Therefore, if f maps to a one-dimensional space, and $||f \circ \gamma||^2 = ||f||^2$ for a non-trivial γ, then f must be a monomial!

8. A criterion for a formal power series to be a rational function

How does one decide whether a formal power series represents a rational function? The question evokes something from elementary mathematics. How does one decide when a decimal expansion defines a rational number? The well-known answer of course is that the expansion is eventually periodic. An analogous result holds for formal power series. We begin with a one-variable result dating back to Kronecker. We formulate it in a way allowing for a generalization to higher dimensions.

Consider a formal power series in one complex variable z:

$$\sum_{n=0}^{\infty} a_n z^n.$$

We denote it by $f(z)$. We let $j_k f(z) = \sum_{n=0}^{k} a_n z^n$. Thus $j_k f$ is a polynomial of degree k; if f were smooth, then $j_k f$ would be the k-th order Taylor polynomial of f at 0. We use the letter j because such Taylor polynomials are called *jets*. We write $R_k(z)$ for the formal remainder term. Note that R_k is divisible by z^{k+1}. Thus, by definition, for each k there is a formal series $g_k(z)$ such that

$$f(z) = j_k f(z) + R_k(z) = j_k f(z) + z^{k+1} g_k(z). \tag{29}$$

THEOREM 5.10. *Let $f(z) = \sum_{n=0}^{\infty} a_n z^n$ be a formal power series in one variable z. Consider the infinite collection of formal series g_k defined in (29). The following statements are equivalent:*

- *f is the series of a rational function.*
- *There are an integer d and constants $q_0, ..., q_m$ (not all 0) such that, for $l > d$,*

$$\sum_{k=0}^{m} a_{l-k} q_k = 0.$$

- *There is an integer K such that g_K is a linear combination of the g_j for $0 \le j < K$.*

PROOF. First suppose g_K is such a linear combination. Then there are constants c_n such that

$$\frac{f(z) - j_K f(z)}{z^{K+1}} = g_K = \sum_{n=0}^{K-1} c_n g_n = \sum_{n=0}^{K-1} c_n \frac{f(z) - j_n f(z)}{z^{n+1}}. \tag{30}$$

Clearing denominators in (30) yields an equation for the unknown series $f(z)$:

$$f(z) - j_K f(z) = \sum_{n=0}^{K-1} c_n \left(z^{K-n}(f(z) - j_n f(z)) \right). \tag{31}$$

Equation (31) is an affine equation for f. Gathering terms in f yields

$$f(z) \left(1 - \sum_{n=0}^{K-1} c_n z^{K-n} \right) = j_K f(z) - \sum_{n=0}^{K-1} c_n j_n f(z) z^{K-n}. \tag{32}$$

Equation (32) gives polynomials p and q such that $qf = p$ and shows that f is rational. More explicitly,

$$f(z) = \frac{p(z)}{q(z)} = \frac{j_K f(z) - \sum_{n=0}^{K-1} c_n j_n f(z) z^{K-n}}{1 - \sum_{n=0}^{K-1} c_n z^{K-n}}. \tag{33}$$

Note that q is of degree at most K, and has degree K precisely when $c_0 \neq 0$. The numerator in (33) is also of degree at most K.

Next suppose that $f = \frac{p}{q}$ is a rational function where the degree of q is m and the degree of p is d. Put $q(z) = \sum_{k=0}^{m} q_k z^k$. Then $fq = p$ is a polynomial of degree d. We have

$$f(z)q(z) = \sum_{n=0}^{\infty} a_n z^n \sum_{k=0}^{m} q_k z^k = \sum_{n=0}^{\infty} \sum_{k=0}^{m} a_n q_k z^{k+n} = \sum_{l=0}^{d} p_l z^l = p(z). \tag{34.1}$$

Since p is of degree d, the coefficient of z^{k+n} is 0 when $k + n > d$. Equation (34.1) thus implies

$$0 = \sum_{k=0}^{m} a_{l-k} q_k = \sum_{n=l}^{m+l} a_n q_{l-n}, \tag{34.2}$$

and the second statement holds.

Finally the second statement implies the third. The second statement implies that there is a non-zero column vector with entries $q_m, \ldots q_0$ in the null space of the $(m+1)$ by $(m+1)$ matrix A formed from the coefficients a_j as follows:

$$\begin{pmatrix} a_K & a_{K+1} & \cdots & a_{K+m} \\ a_{K+1} & a_{K+2} & \cdots & a_{K+m+1} \\ \cdots & & & \\ a_{K+m} & a_{K+m+1} & \cdots & a_{K+2m} \end{pmatrix}. \tag{34.3}$$

The second statement holds for all large enough l. Hence the equations repeat, and this vector is in the null space of a matrix with $m + 1$ columns and infinitely many rows. The column entries are of the form

$$\begin{pmatrix} a_{t+1} & a_{t+2} & \cdots \end{pmatrix},$$

and each precisely corresponds to the formal series g_t. Since the same non-zero vector $q_m, \ldots, q_0$ is annihilated by each of these columns, we conclude that a non-trivial linear combination of these columns vanishes. In other words, there is a K such that g_K is a linear combination of m previous g_n. $\square$

We provide some intuition for the proof that f rational implies the third statement. Suppose first that f is a polynomial of degree d. Then we have $g_d = 0$. Hence $g_d = \sum_{l=0}^{d-1} c_l g_l$, where each coefficient c_l equals 0. Thus (30) holds with $K = d$. Furthermore, for each positive n, we also have

$$g_{d+n} = 0 = \sum_{l=0}^{d-1} c_l g_l.$$

Next suppose that f is the reciprocal of a polynomial of degree m and $f(0) = 1$. Then, using the geometric series, we can write

$$f(z) = \frac{1}{q(z)} = \frac{1}{1 - \sum_{l=1}^{m} b_l z^l} = \frac{1}{1 - r(z)} = \sum_{n=0}^{\infty} r(z)^n. \qquad (35)$$

Thus $fq = \sum_{n=0}^{\infty} r^n$ and hence

$$g_m = \sum_{l=1}^{m} b_l g_{m-l} = \sum_{k=0}^{m-1} b_{m-k} g_k.$$

Again (30) holds.

Now suppose that $f(z) = \frac{p(z)}{q(z)} = \frac{p(z)}{1-r(z)}$, where p has degree d and $r(z)$ is as in (35). We claim that

$$g_{d+m} = \sum_{l=1}^{d+m} c_l g_{d+m-l}. \qquad (36)$$

Here we have set $c_l = b_l$ for $1 \leq l \leq m$ and $c_l = 0$ for $m + 1 \leq l \leq m + d$. Using (36) yields the needed linear combination:

$$g_{d+m} - \sum_{l=0}^{d+m-1} c_{d+m-l} g_l.$$

Hence (30) holds with $K = d + m$. It remains to prove the claim. By definition, $g_k = \frac{f - j_k f}{z^{k+1}}$. By (32), the claim is equivalent to

$$f(z) \left(1 - \sum_{l=1}^{m} b_l z^l \right) = j_{m+d} f(z) - \sum_{l=1}^{m+d} b_l z^l j_{m+d-l} f(z).$$

The left-hand side is $f(z)q(z)$ and one can show that the right side is $p(z)$.

EXAMPLE 5.19. Put $f(z) = \frac{p(z)}{1-cz}$ for a polynomial p of degree m. When $m = 0$ we see that

$$f(z) = f(0) \left(\sum_{n=0}^{\infty} c^n z^n \right).$$

Hence $g_1 = c g_0$ and in fact $g_k = c^k g_0$. Consider next $\frac{z^m}{1-cz}$. This time we obtain $g_{m+1} = c g_m$ and in fact $g_{m+k} = c^k g_m$. Thus we need to take the first several constants equal to 0. The idea is analogous to the decimal expansion of a rational number; the expansion *eventually repeats* but it need not repeat from the outset.

EXAMPLE 5.20. Put $q(z) = 1 - (z + z^2 + z^3)$. Then $\frac{1}{q(z)}$ has expansion

$$\sum_{n=0}^{\infty} (z + z^2 + z^3)^n.$$

We take $K = 3$ and we obtain the relation $g_3 = g_0 + g_1 + g_2$. If we consider $\frac{p(z)}{q(z)}$ where the degree of p is large, then we obtain an analogous third-order recurrence only after setting enough coefficients equal to 0.

EXERCISE 5.24. Verify (36).

EXERCISE 5.25. Suppose that f is a formal power series whose coefficients repeat the pattern $1, 2, 3$. Find an explicit formula for f.

EXERCISE 5.26. Suppose that $f(z) = \sum_{n=0}^{\infty} nz^n$. Find the lowest order recurrence satisfied by the g_k in the proof of Theorem 5.10. Use the proof of the theorem to find f explicitly.

EXERCISE 5.27. Suppose that $f(z) = \sum_{n=0}^{\infty} n^2 z^n$. Find the lowest order recurrence satisfied by the g_k in the proof of Theorem 5.10. Use the proof of the theorem to find f explicitly.

EXERCISE 5.28. Let $p(n)$ be a polynomial of degree d. Show, by any valid method, that $\sum_{n=0}^{\infty} p(n)z^n$ is a polynomial in the expression $\frac{1}{1-z}$ of degree $d + 1$.

We now extend Theorem 5.10 to higher dimensions. Let $z \in \mathbf{C}^t$ for $t \geq 1$. We begin by writing a formal power series $f(z)$ in terms of its homogeneous parts,

$$f(z) = \sum_{k=0}^{\infty} f_k(z), \tag{37}$$

where f_k is homogeneous of degree k. We again let $j_K f$ denote the K-jet of f. We then have

$$f(z) - j_K f(z) = \sum_{n=K+1}^{\infty} f_n(z).$$

THEOREM 5.11. *The formal power series in (37) represents a rational function if and only if there is an integer K with the following property. For each l with $1 \leq l \leq K$, there is a polynomial b_l, either 0 or homogeneous of degree l, such that*

$$f - j_K f = \sum_{l=1}^{K} b_l(f - j_{K-l}f). \tag{38}$$

PROOF. First suppose (38) holds. This formula corresponds to (31) from the one variable case, which was obtained by clearing denominators. It is an affine equation in the unknown f, and we solve it as before to obtain

$$f = \frac{j_K f - \sum_{l=1}^{K} b_l j_{K-l} f}{1 - \sum_{l=1}^{K} b_l}. \tag{39}$$

Thus $f = \frac{p}{q}$, where each of p, q is of degree at most K.

Consider the converse. As in the one variable case, we require the analogues of (34.1), (34.2), and (34.3). We note when f is a polynomial of degree d that (38) holds for $K > d$ by setting all the b_l equal to 0. Assume $f = \frac{p}{q}$ is rational. In terms of homogeneous parts we have

$$p(z) = \sum_{i=0}^{m} q_i \sum_{k=0}^{\infty} f_k = \sum_{n=0}^{\infty} \sum_{k=n}^{n+m} q_{n-k} f_k.$$

For n greater than the degree of p, we obtain the analogue of (34.2):

$$\sum_{k-n}^{n+m} q_{n-k} f_k = 0.$$

As before this system of equations shows that the analogue of the null space (34.3) holds. The rows of this matrix correspond to the formal series $f - j_K f$.

We note the following. When f is the reciprocal of the polynomial $1 - \sum_{l=1}^{K} q_l$, then (38) holds by setting $b_n = q_n$. When p is of degree d and $q = 1 - \sum_{l=1}^{N} q_l$, then (38) holds by choosing $K = d + N$ and then setting $b_l = 0$ for $N+1 \le l \le N+d$.   □

Formula (38) follows from knowing q and the degree of p. We put $K = d + N$. Then we set $b_l = q_l$ for $1 \le l \le N$ and $b_l = 0$ when $N + 1 \le l \le N + d$. The coefficients of p do not matter. In formula (33), the expression $c_n z^{K-n}$ corresponds to the homogeneous polynomial b_{K-n} in formula (39).

We close with two simple comments. First we note an analogy with showing that an eventually periodic decimal expansion represents a rational number. The first few terms in the expansion do not matter; they correspond to the numerator of f. The proof in the periodic case is analogous to the proof of (38) when $f = \frac{1}{q}$. Second we mention that (38) is a statement about infinitely many terms in the series. It is obvious that no finite jet of f can answer whether f is rational. The expressions $f - j_k f$ depend on all the terms of order more than k. Nonetheless (38) is an elegant formula. If one wishes, after putting $b_0 = -1$, one can rewrite (38) as

$$0 = \sum_{l=0}^{K} b_l (f - j_{K-l} f). \tag{40}$$

Formula (40) is a linear dependence condition, where the coefficents are homogeneous polynomials rather than constants.

CHAPTER 6

Appendix

1. The real and complex number systems

In this appendix we organize some of the mathematical prerequisites for reading this book. The reader must be thoroughly informed about basic real analysis (see [Ro] and [F1]) and should know a bit of complex variable theory (see [A] and [D2]).

The real number system $\mathbf{R}$ is characterized by being a *complete ordered field*. The field axioms enable the usual operations of addition, subtraction, multiplication, and division (except by 0). These operations satisfy familiar laws. The order axioms allow us to manipulate inequalities as usual. The completeness axiom is more subtle; this crucial property distinguishes $\mathbf{R}$ from the rational number system $\mathbf{Q}$. One standard way to state the completeness axiom uses the least upper bound property:

DEFINITION 6.1. If S is a non-empty subset of $\mathbf{R}$ and S is bounded above, then S has a least upper bound α, written $\sup(S)$, and called the *supremum* of S.

Recall that a sequence of real numbers is a function $n \mapsto x_n$ from the natural numbers to $\mathbf{R}$. (Sometimes we also allow the indexing to begin with 0.) The sequence $\{x_n\}$ *converges* to the real number L if, for all $\epsilon > 0$, there is an integer N_ϵ such that $n \geq N_\epsilon$ implies $|x_n - L| < \epsilon$.

The least upper bound property enables us to prove that a bounded monotone nondecreasing sequence $\{x_n\}$ of real numbers converges to the supremum of the values of the sequence. It also enables a proof of the fundamental result of basic real analysis: a sequence of real numbers converges if and only if it is a Cauchy sequence. Recall that a sequence is *Cauchy* if, for every $\epsilon > 0$, there is an N_ϵ such that $n, m \geq N_\epsilon$ implies $|x_n - x_m| < \epsilon$. Thus a sequence has a limit L if the terms are eventually as close to L as we wish, and a sequence is Cauchy if the terms are eventually all as close to each other as we wish. The equivalence of the concepts suggests that the real number system has no gaps.

For clarity we highlight these fundamental results as a theorem. The ability to prove Theorem 6.1 should be regarded as a prerequisite for reading this book.

THEOREM 6.1. *If a sequence $\{x_n\}$ of real numbers is bounded and monotone, then $\{x_n\}$ converges. A sequence $\{x_n\}$ converges to a real number L if and only if $\{x_n\}$ is Cauchy.*

COROLLARY 6.1. *A monotone sequence converges if and only if it is bounded.*

REMARK 6.1. The first statement in Theorem 6.1 is considerably easier than the second. It is possible to prove the difficult (*if*) part of the second statement by extracting a monotone subsequence and using the first part. It is also possible to prove the second statement by using the Bolzano-Weierstrass property from Theorem 6.2 below.

© Springer Nature Switzerland AG 2019
J. P. D'Angelo, *Hermitian Analysis*, Cornerstones,
https://doi.org/10.1007/978-3-030-16514-7_6

The complex number system $\mathbf{C}$ is a field, but it has no ordering. As a set $\mathbf{C}$ is simply the Euclidean plane $\mathbf{R}^2$. We make this set into a field by defining addition and multiplication:

$$(x, y) + (a, b) = (x + a, y + b)$$
$$(x, y) * (a, b) = (xa - yb, xb + ya).$$

The additive identity 0 is then the ordered pair $(0, 0)$ and the multiplicative identity 1 is the pair $(1, 0)$. Note that $(0, 1) * (0, 1) = (-1, 0) = -(1, 0)$. As usual we denote $(0, 1)$ by i and then write $x + iy$ instead of (x, y). We then drop the $*$ from the notation for multiplication, and the law becomes obvious. Namely, we expand $(x + iy)(a + ib)$ by the distributive law and set $i^2 = -1$. These operations make $\mathbf{R}^2$ into a field called $\mathbf{C}$.

Given $z = x + iy$ we write $\overline{z} = x - iy$ and call $\overline{z}$ the complex conjugate of z. We define $|z|$ to be the Euclidean distance of z to 0; thus $|z| = \sqrt{x^2 + y^2}$ and $|z|^2 = z\overline{z}$.

The non-negative real number $|z - w|$ equals the Euclidean distance between complex numbers z and w. The following properties of distance make $\mathbf{C}$ into a complete metric space. (See the next section.)

- $|z - w| = 0$ if and only if $z = w$.
- $|z - w| \geq 0$ for all z and w.
- $|z - w| = |w - z|$ for all z and w.
- $|z - w| \leq |z - \zeta| + |\zeta - w|$ for all z, w, ζ. (the triangle inequality)

Once we know that $|z - w|$ defines a distance, we can repeat the definition of convergence.

DEFINITION 6.2. Let $\{z_n\}$ be a sequence of complex numbers, and suppose $L \in \mathbf{C}$. We say that z_n converges to L if, for all $\epsilon > 0$, there is an N_ϵ such that $n \geq N_\epsilon$ implies $|z_n - L| < \epsilon$.

Let $\{a_n\}$ be a sequence of complex numbers. We say that $\sum_{n=1}^{\infty} a_n$ converges to L, if

$$\lim_{N \to \infty} \sum_{n=1}^{N} a_n = L.$$

We say that $\sum_{n=1}^{\infty} a_n$ converges absolutely if $\sum_{n=1}^{\infty} |a_n|$ converges. It is often easy to establish absolute convergence; a series of non-negative numbers converges if and only if the sequence of partial sums is bounded. The reason is simple: if the terms of a series are non-negative, then the partial sums form a monotone sequence, and hence the sequence of partial sums converges if and only if it is bounded. See Corollary 6.1 above. We also use the following standard comparison test; we include the proof because it beautifully illustrates the Cauchy convergence criterion.

PROPOSITION 6.1. Let $\{z_n\}$ be a sequence of complex numbers. Assume for all n that $|z_n| \leq c_n$, and that $\sum_{n=1}^{\infty} c_n$ converges. Then $\sum_{n=1}^{\infty} z_n$ converges.

PROOF. Let S_N denote the N-th partial sum of the series $\sum z_n$, and let T_N denote the N-th partial sum of the series $\sum c_n$. For $M > N$ we have

$$|S_M - S_N| = |\sum_{N+1}^{M} z_n| \leq \sum_{N+1}^{M} |z_n| \leq \sum_{N+1}^{M} c_n = T_M - T_N. \tag{1}$$

Since $\sum c_n$ is convergent, $\{T_N\}$ is a Cauchy sequence of real numbers. By (1), $\{S_N\}$ is also Cauchy, and hence $\sum_{n=1}^{\infty} z_n$ converges by Theorem 6.1.   □

We pause to recall and discuss the notion of equivalence class, which we presume is familiar to the reader. Let S be a set. An *equivalence relation* on S is a relation $\sim$ such that, for all $a, b, c \in S$,

> **Reflexive property:** $a \sim a$
> **Symmetric property:** $a \sim b$ if and only if $b \sim a$
> **Transitive property:** $a \sim b$ and $b \sim c$ implies $a \sim c$.

Given an equivalence relation on a set S, we can form a new set, sometimes written $S/\sim$, as follows. We say that a and b are *equivalent*, or lie in the same *equivalence class*, if $a \sim b$ holds. The elements of $S/\sim$ are the equivalence classes; the set $S/\sim$ is called the quotient space.

We mention three examples. The first is trivial, the second is easy but fundamental, and the third is profound.

EXAMPLE 6.1. Let S be the set of ordered pairs (a, b) of integers. We say that $(a, b) \sim (c, d)$ if $100a + b = 100c + d$. If we regard the first element of the ordered pair as the number of dollars, and the second element as the number of cents, then two pairs are equivalent if they represent the same amount of money. (Note that we allow negative money here.)

EXAMPLE 6.2. Let S be the set of ordered pairs (a, b) of integers, with $b \neq 0$. We say that $(a, b) \sim (A, B)$ if $aB = Ab$. The equivalence relation restates, without mentioning division, the condition that $\frac{a}{b}$ and $\frac{A}{B}$ define the same rational number. Then $S/\sim$ is the set of rational numbers. It becomes the system $\mathbf{Q}$ after we define addition and multiplication of equivalence classes and verify the required properties.

EXAMPLE 6.3. The real number system $\mathbf{R}$ is sometimes defined to be the *completion* of the rational number system $\mathbf{Q}$. In this definition, a real number is an equivalence class of Cauchy sequences of rational numbers. Here we define a sequence of rational numbers $\{q_n\}$ to be Cauchy if, for each positive integer K, we can find a positive integer N such that $m, n \geq N$ implies $|q_m - q_n| < \frac{1}{K}$. (The number $\frac{1}{K}$ plays the role of ϵ; we cannot use ϵ because real numbers have not yet been defined!) Two Cauchy sequences are equivalent if their difference converges to 0. Thus Cauchy sequences $\{p_n\}$ and $\{q_n\}$ of rational numbers are equivalent if, for every $M \in \mathbf{N}$, there is an $N \in \mathbf{N}$ such that $|p_n - q_n| < \frac{1}{M}$ whenever $n \geq N$. Intuitively, we can regard a real number to be the collection of all sequences of rational numbers which appear to have the same *limit*. We use the language of the next section; as a set, $\mathbf{R}$ is the metric space *completion* of $\mathbf{Q}$. As in Example 6.2, we need to define addition, multiplication, and order and establish their properties before we get the real number system $\mathbf{R}$.

We are also interested in convergence issues in higher dimensions. Let $\mathbf{R}^n$ denote real Euclidean space of dimension n and $\mathbf{C}^n$ denote complex Euclidean space of dimension n. In the next paragraph, we let $\mathbf{F}$ denote either $\mathbf{R}$ or $\mathbf{C}$.

As a set, $\mathbf{F}^n$ consists of all n-tuples of elements of the field $\mathbf{F}$. We write $z = (z_1, \ldots, z_n)$ for a point in $\mathbf{F}^n$. This set has the structure of a real or complex vector space with the usual operations of vector addition and scalar multiplication:

$$(z_1, z_2, \ldots, z_n) + (w_1, w_2, \ldots, w_n) = (z_1 + w_1, z_2 + w_2, \ldots, z_n + w_n).$$

$$c(z_1, z_2, \ldots, z_n) = (cz_1, cz_2, \ldots, cz_n)$$

DEFINITION 6.3 (norm). A *norm* on a real or complex vector space V is a function $v \mapsto ||v||$ satisfying the following three properties:

(1) $||v|| > 0$ for all nonzero v.
(2) $||cv|| = |c|\,||v||$ for all $c \in \mathbf{C}$ and all $v \in V$.
(3) (The triangle inequality) $||v + w|| \leq ||v|| + ||w||$ for all $v, w \in V$.

We naturally say *normed vector space* for a vector space equipped with a norm. We can make a normed vector space into a metric space by defining $d(u, v) = ||u - v||$.

For us the notations $\mathbf{R}^n$ and $\mathbf{C}^n$ include the vector space structure and the *Euclidean squared norm* defined by (2):

$$||z||^2 = \langle z, z \rangle. \tag{2}$$

These norms come from the *Euclidean inner product*. In the real case we have

$$\langle x, y \rangle = \sum_{j=1}^{n} x_j y_j \tag{3.1}$$

and in the complex case we have

$$\langle z, w \rangle = \sum_{j=1}^{n} z_j \overline{w}_j. \tag{3.2}$$

In both cases $||z||^2 = \langle z, z \rangle$.

2. Metric spaces

The definitions of convergent sequence in various settings are so similar that it is natural to put these settings into one abstract framework. One such setting is *metric spaces*.

We assume that the reader is somewhat familiar with metric spaces. We recall the definition and some basic facts. Let $\mathbf{R}_+$ denote the non-negative real numbers.

DEFINITION 6.4. Let X be a set. A *distance* function on X is a function $d : X \times X \to \mathbf{R}_+$ satisfying the following properties:

(1) $d(x, y) = 0$ if and only if $x = y$.
(2) $d(x, y) = d(y, x)$ for all x, y.
(3) $d(x, z) \leq d(x, y) + d(y, z)$ for all x, y, z.

If d is a distance function on X, then the pair (X, d) is called a *metric space* and d is called the *metric*.

The real numbers, the complex numbers, real Euclidean space, and complex Euclidean space are all metric spaces under the usual Euclidean distance function. One can define other metrics, with very different properties, on these sets. For example, on any set X, the function $d : X \times X \to \mathbf{R}_+$, defined by $d(x, y) = 1$ if $x \neq y$ and $d(x, x) = 0$, is a metric. In general sets admit many different useful distance functions. When the metric is understood, one often says "Let X be a metric space". This statement is convenient but a bit imprecise.

Metric spaces provide a nice conceptual framework for convergence.

DEFINITION 6.5. Let $\{x_n\}$ be a sequence in a metric space (X, d). We say that x_n *converges* to x if, for all $\epsilon > 0$, there is an N such that $n \geq N$ implies $d(x_n, x) < \epsilon$. We say that $\{x_n\}$ is *Cauchy* if, for all $\epsilon > 0$, there is an N such that $m, n \geq N$ implies $d(x_m, x_n) < \epsilon$.

DEFINITION 6.6. A metric space (M, d) is *complete* if every Cauchy sequence converges.

If a metric space (M, d) is not complete, then we can form a new metric space called its *completion*. The idea precisely parallels the construction of **R** given **Q**. The completion consists of equivalence classes of Cauchy sequences of elements of (M, d). The distance function extends to the larger set by taking limits.

Here are several additional examples of metric spaces. We omit the needed verifications of the properties of the distance function, but we mention that in some instances proving the triangle inequality requires effort.

EXAMPLE 6.4. Let X be the space of continuous functions on $[0, 1]$. Define $d(f, g) = \int_0^1 |f(x) - g(x)| dx$. Then (X, d) is a metric space. More generally, for $1 \leq p < \infty$, we define $d_p(f, g)$ by

$$d_p(f, g) = \left(\int_0^1 |f(x) - g(x)|^p dx \right)^{\frac{1}{p}}.$$

We define $d_\infty(f, g)$ by $d_\infty(f, g) = \sup |f - g|$.

Of these examples, only (X, d_∞) is complete. Completeness in this case follows because the uniform limit of a sequence of continuous functions is itself continuous.

A subset Ω of a metric space is called *open* if, whenever $p \in \Omega$, there is a positive ϵ such that $x \in \Omega$ whenever $d(p, x) < \epsilon$. In particular the empty set is open and the whole space X is open. A subset K is called *closed* if its complement is open.

PROPOSITION 6.2. *Let (X, d) be a metric space. Let $K \subseteq X$. Then K is closed if and only if, whenever $\{x_n\}$ is a sequence in K, and x_n converges to x, then $x \in K$.*

PROOF. Left to the reader.   □

Let (M, d) and (M', d') be metric spaces. The natural collection of maps between them is the set of continuous functions.

DEFINITION 6.7 (Continuity). $f : (M, d) \to (M', d')$ is *continuous* if, whenever U is open in M', then $f^{-1}(U)$ is open in M.

PROPOSITION 6.3. *Suppose $f : (M, d) \to (M', d')$ is a map between metric spaces. The following are equivalent:*

(1) *f is continuous*
(2) *Whenever x_n converges to x in M, then $f(x_n)$ converges to $f(x)$ in M'.*
(3) *For all $\epsilon > 0$, there is a $\delta > 0$ such that*

$$d(x, y) < \delta \implies d'(f(x), f(y)) < \epsilon.$$

EXERCISE 6.1. Prove Propositions 6.1 and 6.2.

We next mention several standard and intuitive geometric terms. The *interior* of a set S in a metric space is the union of all open sets contained in S. The *closure* of a set S is the intersection of all closed sets containing S. Thus a set is open if and only if it equals its interior, and a set is closed if and only if it equals its closure. The *boundary* $b\Omega$ of a set Ω consists of all points in the closure of Ω but not in the interior of Ω. Another way to define boundary is to note that $x \in b\Omega$ if and only if, for every $\epsilon > 0$, the ball of radius ϵ about x has a non-empty intersection with both Ω and its complement.

Continuity often gets used together with the notion of a *dense* subset of a metric space M. A subset S is *dense* if each $x \in M$ is the limit of a sequence of points in S. In other words, M is the closure of S. For example, the rational numbers are dense in the real numbers. If f is continuous on M, then $f(x) = \lim_n f(x_n)$, and hence f is determined by its values on a dense set.

One of the most important examples of a metric space is the collection $C(M)$ of continuous complex-valued functions on a metric space M. Several times in the book we use compactness properties in $C(M)$. We define compactness in the standard *open cover* fashion, called the Heine-Borel property. What matters most for us is the Bolzano-Weierstrass property.

We quickly review some of the most beautiful results in basic analysis.

DEFINITION 6.8. Let M be a metric space and let $K \subseteq M$. K is *compact* if, whenever K is contained in an arbitrary union $\cup A_\alpha$ of open sets, then K is contained in a finite union $\cup_{k=1}^N A_{\alpha_k}$ of these open sets. This condition is often called the *Heine-Borel property*.

This definition of compact is often stated informally "every open cover has a finite subcover", but these words are a bit imprecise.

DEFINITION 6.9. Let (M, d) be a metric space. A subset $K \subseteq M$ satisfies the *Bolzano-Weierstrass property* if, whenever $\{x_n\}$ is a sequence in K, then there is a subsequence $\{x_{n_k}\}$ converging to a limit in K.

THEOREM 6.2. *Let (M, d) be a metric space and let $K \subseteq M$. Then K is compact if and only if K satisfies the Bolzano-Weierstrass property.*

THEOREM 6.3. *A subset of Euclidean space is compact if and only if it is closed and bounded.*

EXERCISE 6.2. Prove Theorems 6.2 and 6.3.

DEFINITION 6.10 (Equicontinuity). A collection $\mathcal{K}$ of complex-valued functions on a metric space (M, d) is called *equicontinuous* if, for all x and for all $\epsilon > 0$, there is a $\delta > 0$ such that

$$d(x, y) < \delta \implies |f(x) - f(y)| < \epsilon$$

for all $f \in \mathcal{K}$.

DEFINITION 6.11 (Uniformly bounded). A collection $\mathcal{K}$ of complex-valued functions on a metric space (M, d) is called *uniformly bounded* if there is a C such that $|f(x)| \leq C$ for all $x \in M$ and for all $f \in \mathcal{K}$.

We refer to [F1] for a proof of the following major result in analysis. The statement and proof in [F1] apply in the more general context of locally compact Hausdorff topological spaces. In this book we use Theorem 6.4 to show that certain integral operators are compact. See Sections 10 and 11 of Chapter 2.

THEOREM 6.4 (Arzela-Ascoli theorem). *Let M be a compact metric space. Let $C(M)$ denote the continuous functions on M with $d(f, g) = \sup_M |f(x) - g(x)|$. Let $\mathcal{K}$ be a subset of $C(M)$. Then $\mathcal{K}$ is compact if and only if the following three items are true:*

(1) *$\mathcal{K}$ is equicontinuous.*
(2) *$\mathcal{K}$ is uniformly bounded.*
(3) *$\mathcal{K}$ is closed.*

COROLLARY 6.2. *Let $\mathcal{K}$ be a closed, uniformly bounded, and equicontinuous subset of $C(M)$. Let $\{f_n\}$ be a sequence in $\mathcal{K}$. Then $\{f_n\}$ has a convergent subsequence. That is, $\{f_{n_k}\}$ converges uniformly to an element of $\mathcal{K}$.*

PROOF. By the theorem $\mathcal{K}$ is compact; the result then follows from the Bolzano-Weierstrass characterization of compactness. □

EXERCISE 6.3. Let M be a compact subset of Euclidean space. Fix $\alpha > 0$. Let H_α denote the subset of $C(M)$ satisfying the following properties:

(1) $\|f\|_\infty \le 1$.
(2) $\|f\|_{H_\alpha} \le 1$. Here

$$\|f\|_{H_\alpha} = \sup_{x \ne y} \frac{|f(x) - f(y)|}{|x - y|^\alpha}.$$

Show that H_α is compact.

A function f for which $\|f\|_{H_\alpha}$ is finite is said to satisfy a Hölder condition of order α. See Definition 2.13.

3. Integrals

This book presumes that the reader knows the basic theory of the Riemann-Darboux integral, which we summarize. See [Ro] among many possible texts.

Let $[a, b]$ be a closed bounded interval on $\mathbf{R}$, and suppose $f : [a, b] \to \mathbf{R}$ is a bounded function. We define $\int_a^b f(t)dt$ by a standard but somewhat complicated procedure. A partition P of $[a, b]$ is a finite collection of points p_j such that $a = p_0 < \cdots < p_j < \cdots < p_N = b$. Given f and a partition P, we define the lower and upper sums corresponding to the partition:

$$L(f, P) = \sum_{j=1}^{N} (p_j - p_{j-1}) \inf_{[p_{j-1}, p_j]} (f(x))$$

$$U(f, P) = \sum_{j=1}^{N} (p_j - p_{j-1}) \sup_{[p_{j-1}, p_j]} (f(x)).$$

DEFINITION 6.12. A bounded function $f : [a, b] \to \mathbf{R}$ is *Riemann integrable* if $\sup_P L(f, P) = \inf_P U(f, P)$. If so, we denote the common value by $\int_a^b f(t)dt$ or simply by $\int_a^b f$.

An equivalent way to state Definition 6.12 is that f is integrable if, for each $\epsilon > 0$, there is a partition P_ϵ such that $U(f, P_\epsilon) - L(f, P_\epsilon) < \epsilon$.

In case f is complex-valued, we define it to be integrable if its real and imaginary parts are integrable, and we put

$$\int_a^b f = \int_a^b u + iv = \int_a^b u + i \int_a^b v.$$

The integral satisfies the usual properties:

(1) If f, g are Riemann integrable on $[a, b]$, and c is a constant, then $f + g$ and cf are Riemann integrable and

$$\int_a^b f + g = \int_a^b f + \int_a^b g,$$

$$\int_a^b cf = c \int_a^b f.$$

(2) If f is Riemann integrable and $f(x) \geq 0$ for $x \in [a, b]$, then $\int_a^b f \geq 0$.
(3) If f is continuous on $[a, b]$, then f is Riemann integrable.
(4) If f is monotone on $[a, b]$, then f is Riemann integrable.

We assume various other basic results, such as the change of variables formula, without further mention.

The collection of complex-valued integrable functions on $[a, b]$ is a complex vector space. We would like to define the distance $\delta(f, g)$ between integrable functions f and g by

$$\delta(f, g) = ||f - g||_{L^1} = \int_a^b |f(x) - g(x)| dx,$$

but a slight problem arises. If f and g agree for example everywhere except at a single point, and each is integrable, then $\delta(f, g) = 0$ but f and g are not the same function. This point is resolved by working with equivalence classes of functions. Two functions are called equivalent if they agree except on what is called a set of measure zero. See Section 7 of Chapter 1. Even after working with equivalence classes, this vector space is not complete (in the metric space sense). One needs to use the Lebesgue integral to identify its completion.

Often one requires so-called *improper* integrals. Two possible situations arise; one is when f is unbounded on $[a, b]$, the other is when the interval is unbounded. Both situations can happen in the same example. The definitions are clear, and we state them informally. If f is unbounded at a, for example, but Riemann integrable on $[a + \epsilon, b]$ for all positive ϵ, then we define

$$\int_a^b f = \lim_{\epsilon \to 0} \int_{a+\epsilon}^b f$$

if the limit exists. If f is Riemann integrable on $[a, b]$ for all b, then we put

$$\int_a^\infty f = \lim_{b \to \infty} \int_a^b f.$$

The other possibilities are handled in a similar fashion. Here are two simple examples of improper integrals:

(1) $\int_0^1 x^\alpha dx = \frac{1}{\alpha+1}$ if $\alpha > -1$.
(2) $\int_0^\infty e^{-x} dx = 1$.

At several points in this book, whether an improper integral converges will be significant. We mention specifically Section 8 of Chapter 3, where one shows that a function has k continuous derivatives by showing that an improper integral is convergent.

The following theorem is *fundamental* to all that we do in this book.

THEOREM 6.5 (Fundamental theorem of calculus). *Assume f is continuous on $[a, b]$. For $x \in (a, b)$ put $F(x) = \int_a^x f(t)dt$. Then F is differentiable and $F'(x) = f(x)$.*

The final theorem in this section is somewhat more advanced. We state this result in Section 7 of Chapter 1, but we never use it. It is important partly because its statement is so definitive, and partly because it suggests connections between the Riemann and Lebesgue theories of integration.

THEOREM 6.6. *A function on a closed interval $[a, b]$ is Riemann integrable if and only if the set of its discontinuities has measure zero.*

EXERCISE 6.4. Establish the above properties of the Riemann integral.

EXERCISE 6.5. Verify that $\int_a^b cf = c \int_a^b f$ when c is complex and f is complex-valued. Check that $\text{Re}(\int_a^b f) = \int_a^b \text{Re}(f)$ and similarly with the imaginary part.

EXERCISE 6.6. Verify the improper integrals above.

The next three exercises involve finding sums. Doing so is generally much harder than finding integrals.

EXERCISE 6.7. Show that $\sum_{j=0}^n \binom{j}{k} = \binom{n+1}{k+1}$. Suggestion. Count the same thing in two ways.

EXERCISE 6.8. For p a nonnegative integer, consider $\sum_{j=1}^n j^p$ as a function of n. Show that it is a polynomial in n of degree $p + 1$ with leading term $\frac{n^{p+1}}{p+1}$. If you want to work harder, show that the next term is $\frac{n^p}{2}$. Comment: The previous exercise is useful in both cases.

EXERCISE 6.9. For p a positive integer, prove that $\int_0^1 t^p dt = \frac{1}{p+1}$ by using the definition of the Riemann integral. (Find upper and lower sums and use the previous exercise.)

EXERCISE 6.10. Prove the fundamental theorem of calculus. The idea of its proof recurs throughout this book.

EXERCISE 6.11. Put $f(0) = 0$ and $f(x) = x \, \text{sgn}(\sin(\frac{1}{x}))$. Here $\text{sgn}(t) = \frac{t}{|t|}$ for $t \neq 0$ and $\text{sgn}(0) = 0$.

- Sketch the graph of f.
- Determine the points where f fails to be continuous.
- Show that f is Riemann integrable on $[-1, 1]$.

4. Exponentials and trig functions

The unit circle is the set of complex numbers of unit Euclidean distance from 0, that is, the set of z with $|z| = 1$.

The complex exponential function is defined by

$$e^z = \sum_{n=0}^{\infty} \frac{z^n}{n!}.$$

The series converges absolutely for all complex z. Furthermore the resulting function satisfies $e^0 = 1$ and $e^{z+w} = e^z e^w$ for all z and w.

We define the complex trig functions by

$$\cos(z) = \frac{e^{iz} + e^{-iz}}{2}$$

$$\sin(z) = \frac{e^{iz} - e^{-iz}}{2i}.$$

When z is real these functions agree with the usual trig functions. The reader who needs convincing can express both sides as power series.

Note, by continuity of complex conjugation, we have $e^{\bar{z}} = \overline{e^z}$. Combining this property with the addition law gives (assuming t is real)

$$1 = e^0 = e^{it} e^{-it} = |e^{it}|^2.$$

Thus $z = e^{it}$ lies on the unit circle. Its real part x is given by $x = \frac{z+\bar{z}}{2}$ and its imaginary part y is given by $y = \frac{z-\bar{z}}{2i}$. Comparing with our definitions of cosine and sine, we obtain the famous Euler identity (which holds even when t is complex):

$$e^{it} = \cos(t) + i\sin(t).$$

Complex logarithms are quite subtle. For a positive real number t we define $\log(t)$, sometimes written $\ln(t)$, by the usual formula

$$\log(t) = \int_1^t \frac{du}{u}.$$

For a nonzero complex number z, written in the form $z = |z|e^{i\theta}$, we provisionally define its logarithm by

$$\log(z) = \log(|z|) + i\theta. \tag{4}$$

The problem with this formula is that θ is defined only up to multiples of 2π. We must therefore restrict θ to an interval of length 2π. In order to define the logarithm precisely, we must choose a *branch cut*. Thus we first choose an open interval of length 2π, and then we define the logarithm only for θ in that open interval. Doing so yields a *branch* of the logarithm. For example, we often write (4) for $0 \neq z = |z|e^{i\theta}$ and $-\pi < \theta < \pi$. Combining the identity $e^{\alpha+\beta} = e^\alpha e^\beta$ with (4), we obtain $e^{\log(z)} = |z|e^{i\theta} = z$. For a second example, suppose our branch cut is the non-negative real axis; then $0 < \theta < 2\pi$. Then $\log(-1) = i\pi$, but logs of positive real numbers are not defined! To correct this difficulty, we could assume $0 \leq \theta < 2\pi$ and obtain the usual logarithm of a positive number. The logarithm, as a function on the complement of the origin in $\mathbf{C}$, is then discontinuous at points on the positive real axis.

5. Complex analytic functions

The geometric series arises throughout mathematics. Suppose that z is a complex number not equal to 1. Then we have the *finite geometric series*

$$\sum_{j=0}^{n-1} z^j = \frac{1 - z^n}{1 - z}.$$

When $|z| < 1$, we let $n \to \infty$ and obtain the *geometric series*

$$\sum_{j=0}^{\infty} z^j = \frac{1}{1 - z}.$$

The geometric series and the exponential series lie at the foundation of complex analysis. We have seen how the exponential function informs trigonometry. The geometric series enables the proof of Theorem 6.7 below; the famous Cauchy integral formula (Theorem 6.8) combines with the geometric series to show that an arbitrary complex analytic function has a local power series expansion.

A subset Ω of $\mathbf{C}$ is called *open* if, for all $p \in \Omega$, there is an open ball about p contained in Ω. In other words, there is a positive ϵ such that $|z - p| < \epsilon$ implies $z \in \Omega$. Suppose that Ω is open and $f : \Omega \to \mathbf{C}$ is a function. We say that f is *complex analytic* on Ω if, for each $z \in \Omega$, f is complex differentiable at z. (in other words, if the limit in (5) exists).

$$\lim_{h \to 0} \frac{f(z + h) - f(z)}{h} = f'(z) \tag{5}$$

A continuously differentiable function $f : \Omega \to \mathbf{C}$ satisfies the *Cauchy-Riemann equations* if $\frac{\partial f}{\partial \bar{z}} = 0$ at all points of Ω. The complex partial derivative is defined by

$$\frac{\partial}{\partial \bar{z}} = \frac{1}{2}\left(\frac{\partial}{\partial x} + i\frac{\partial}{\partial y}\right).$$

In most elementary books on complex variables, one writes $f = u + iv$ in terms of its real and imaginary parts, and writes the Cauchy-Riemann equations as the pair of equations

$$\frac{\partial u}{\partial x} = \frac{\partial v}{\partial y}$$

$$\frac{\partial u}{\partial y} = -\frac{\partial v}{\partial x}.$$

Perhaps the most fundamental theorem in basic complex analysis relates complex analytic functions, convergent power series, and the Cauchy-Riemann equations. Here is the precise statement:

THEOREM 6.7. *Assume that Ω is open and $f : \Omega \to \mathbf{C}$ is a function. The following are equivalent:*

(1) *f is complex analytic on Ω.*
(2) *For all p in Ω, there is a ball about p on which f is given by a convergent power series:*

$$f(z) = \sum_{n=0}^{\infty} a_n(z - p)^n.$$

(3) *f is continuously differentiable and $\frac{\partial f}{\partial \bar{z}} = 0$ on Ω.*

The key step used in establishing Theorem 6.7 is the Cauchy integral formula. Readers unfamiliar with complex line integrals should consult [A] or [D2], and should read about Green's theorem in Section 1 of Chapter 4 in this book. We mention that, in the research literature on several complex variables, the word *holomorphic* is commonly used instead of *complex analytic*.

THEOREM 6.8 (Cauchy integral theorem and Cauchy integral formula). *Let f be complex analytic on and inside a positively oriented, simple closed curve γ. Then*

$$\int_\gamma f(z)dz = 0.$$

For z in the interior of γ, we have

$$f(z) = \frac{1}{2\pi i} \int_\gamma \frac{f(\zeta)}{\zeta - z} d\zeta.$$

We close this review of complex variable theory by recalling the Fundamental Theorem of Algebra. Many proofs are known, but all of them require the methods of analysis. No purely algebraic proof can exist, because the completeness axiom for the real numbers must be used in the proof.

THEOREM 6.9 (Fundamental theorem of algebra). *Let $p(z)$ be a non-constant polynomial with complex coefficients and of degree d. Then p factors into a product of d linear factors:*

$$p(z) = c \prod_{j=1}^{d} (z - z_j),$$

where the z_j need not be distinct.

6. Probability

Many of the ideas in this book are closely connected with probability theory. We barely glimpse these connections.

We begin by briefly discussing probability densities, and we restrict our consideration to continuous densities. See a good text such as [HPS] for more information and the relationship with Fourier transforms.

Let J be a closed interval on $\mathbf{R}$; we allow the possibility of infinite endpoints. Assume that $f : J \to [0, \infty)$ is continuous. Then f is called a *continuous probability density* on J if $\int_J f = 1$. Let a denote the left-hand endpoint of J. We define the *cumulative distribution* function F by

$$F(x) = \int_a^x f(t)dt.$$

For $y < x$, we interpret $F(x) - F(y) = \int_y^x f(t)dt$ as the probability that a *random variable* lies in the interval $[x, y]$.

We do not attempt to say precisely what the phrase "Let X be a random variable" means. In our setting, we are given the continuous density function f, and we say "X is a random variable with continuous density f" to indicate the situation we have described. The intuition for the term random variable X is the following. Suppose X is a real-valued function defined on some set, and for each $x \in \mathbf{R}$, the probability that X takes on a value at most x is well-defined. We write $F(x)$ for this probability. Thus $F(x) - F(y)$ denotes the probability that F takes on a value in the interval $(y, x]$. In the case of continuous densities, the probability that X takes on any specific value is 0. This property is sometimes taken as the *definition* of continuous random variable. Hence $F(x) - F(y)$ denotes the probability that X takes on a value in the interval $[y, x]$.

Let X denote a random variable on an interval J, with continuous density f. We say that X has *finite expectation* if

$$\int_J |t| f(t) dt < \infty.$$

We say that X has *finite variance* if

$$\int_J (t - \mu)^2 f(t) dt < \infty.$$

When these integrals are finite, we define the *mean* μ and *variance* σ^2 of X by

$$\mu = \int_J t f(t) dt$$

$$\sigma^2 = \int_J (t - \mu)^2 f(t) dt.$$

The mean is also known as the expected value. More generally, if g is any function we call $\int_J g(t) f(t) dt$ the expected value of g. Thus the variance is the expected value of $(t - \mu)^2$ and hence measures the deviation from the mean.

PROPOSITION 6.4. *The variance satisfies* $\sigma^2 = \int_J t^2 f(t) dt - \mu^2$.

PROOF. Expanding the square in the definition of the variance gives:

$$\int_J (t - \mu)^2 f(t) dt = \int_J t^2 f(t) dt - 2\mu \int_J t f(t) dt + \mu^2 \int_J f(t) dt.$$

Since $\mu = \int_J t f(t) dt$ and $1 = \int_J f(t) dt$, the last two terms combine to give $-\mu^2$. □

The computation in Proposition 6.4 arises in many contexts. It appears, for example, in the proof of the parallel axis theorem for moments of inertia. The same idea occurs in verifying the equivalence of two ways of stating Poincaré inequalities in Chapter 4. Compare also with the proof of Bessel's inequality, Proposition 2.2.

EXAMPLE 6.5 (The normal, or Gaussian, random variable). For $0 < \sigma^2 < \infty$ and $x \in \mathbf{R}$, put $g(x) = \frac{1}{\sqrt{2\pi}\sigma} e^{\frac{-x^2}{2\sigma^2}}$. See Example 1.7. Then the mean of the random variable with density g is 0 and the variance is σ^2.

EXAMPLE 6.6 (The uniform random variable). Let $f(x) = \frac{1}{b-a}$ for $a \leq x \leq b$. Then f is a probability density. Its cumulative distribution function F is given on $\mathbf{R}$ by $F(x) = 0$ if $x < a$, by $F(x) = 1$ if $x > b$, and by $F(x) = \frac{x-a}{b-a}$ for $x \in [a, b]$.

EXERCISE 6.12. Show that the mean of the uniform random variable on $[a, b]$ is $\frac{a+b}{2}$. Compute its variance.

Let X be a random variable with continuous density function f. The probability that $X \leq x$ is by definition the integral $\int_{-\infty}^x f(t) dt$. We write:

$$\mathrm{Prob}(X \leq x) = \int_{-\infty}^x f(t) dt.$$

Let ϕ be a strictly monotone differentiable function of one real variable. We can use the fundamental theorem of calculus to compute the density of $\phi(X)$. Assuming that ϕ is increasing, we have

$$\mathrm{Prob}(\phi(X) \leq x) = \mathrm{Prob}(X \leq \phi^{-1}(x)) = \int_{-\infty}^{\phi^{-1}(x)} f(t)dt.$$

Differentiating and using the fundamental theorem of calculus, we see that the density of $\phi(X)$ is given by $f \circ \phi^{-1}(\phi^{-1})'$. An example of this situation gets briefly mentioned in Exercise 4.68, where X is the Gaussian and $\phi(x) = x^2$ for $x \geq 0$. In case ϕ is decreasing a similar calculation gives the answer $-f \circ \phi^{-1}(\phi^{-1})'$. Hence the answer in general is $f \circ \phi^{-1}|(\phi^{-1})'|$.

We end this appendix by glimpsing the connection between the Fourier transform and probability. Given a continuous random variable on $\mathbf{R}$ with density f, we defined above the expected value of a function g by $\int_{-\infty}^{\infty} g(t)f(t)dt$. Take $g(t) = \frac{1}{\sqrt{2\pi}}e^{-it\xi}$. Then the expected value of g is the Fourier transform of f. The terminology used in probability theory often differs from that in other branches of mathematics; for example, the expected value of e^{itX}, where X is a random variable, equals $\int_{-\infty}^{\infty} e^{itx}f(x)dx$. This function is called the *characteristic function* of the random variable X rather than (a constant times the inverse of) the Fourier transform of f.

The *central limit theorem* is one of the major results in probability theory and statistics. Most readers should have heard of the result, at least in an imprecise fashion ("everything is asymptotically normal"), and we do not state it here. See [F1] or [HPS] for precise statements of the central limit theorem. Its proof relies on several things discussed in this book: the Fourier transform is injective on an appropriate space, the Fourier transform of a Gaussian of mean zero and variance one is itself, and the Gaussian defines an approximate identity as the variance tends to 0.

EXERCISE 6.13. Show that there is a continuous probability density f on $\mathbf{R}$, with finite expectation, such that $f(n) = n$ for all positive integers n.

Notation used in this book

Nearly all of the notation used in this book is standard. Occasionally a symbol can be used to denote different things; we mention some of these ambiguities below. In all cases the context should make the meaning clear. This summary is organized roughly by topic.

Basic notation

$\mathbf{R}$ is the real number system, $\mathbf{C}$ is the complex number system, and i denotes the imaginary unit with $i^2 = -1$. Usually x, y denote real variables, and z, w, ζ denote complex variables. $\bar{z}$ denotes the complex conjugate of z, and $\text{Re}(z)$ denotes the real part of z.

$\{a_n\}$ denotes a sequence; the objects could be real numbers, complex numbers, elements in a Hilbert space, etc.

When $\sum_{n=1}^{\infty} a_n$ is an infinite series, we let A_N denote the N-th partial sum. (The small letter denotes the terms and the capital letter denotes the partial sums, analogous to f denoting a function and F denoting its integral.)

$\mathbf{R}^n$ denotes n-dimensional real Euclidean space, and $\mathbf{C}^n$ denotes n-dimensional complex Euclidean space. $\mathcal{H}$ denotes a Hilbert space.

$||v||$ denotes the norm of a vector v in any of these spaces. We also use $||L||$ to denote the operator norm of a bounded linear mapping L.

We write $||f||_{L^p}$ to denote the L^p norm of a function, and $||f||_{L^\infty}$ to denote the sup norm. We write $||z||_2$ to denote the l^2 norm of a sequence $\{z_n\}$.

$L^2([a, b])$ denotes the space (of equivalence classes of) square-integrable functions on $[a, b]$.

$\langle z, w \rangle$ denotes the Hermitian inner product of elements z, w in a Hilbert space. In Chapter 3, Section 3, the same notation denotes the pairing of a distribution and a function. In Chapter 4, Section 10 it denotes the pairing of a vector field and a 1-form. In these pairings, there is no complex conjugation involved.

If u is a differentiable function of several variables, we sometimes denote the partial derivative $\frac{\partial u}{\partial x_j}$ by u_{x_j} or u_j.

The letter δ sometimes denotes the Dirac delta distribution, defined by $\delta(f) = f(0)$. Sometimes it denotes a positive real number.

S^1 denotes the unit circle and S^{k-1} the unit sphere in $\mathbf{R}^k$. Very often we use S^{2n-1} for the unit sphere in $\mathbf{C}^n$.

$\mathbb{B}_n$ denotes the unit ball in $\mathbf{C}^n$. Ω denotes an open, connected set in $\mathbf{C}^n$. $\mathcal{A}^2(\Omega)$ denotes the Hilbert space of square-integrable holomorphic functions on Ω.

ω often denotes an m-th root of unity. Sometimes ω or ω_k is a differential form.

© Springer Nature Switzerland AG 2019
J. P. D'Angelo, *Hermitian Analysis*, Cornerstones,
https://doi.org/10.1007/978-3-030-16514-7_0

Δ denotes the Laplacian, or Laplace operator. On $\mathbf{R}^n$,

$$\Delta(f) = \sum_{j=1}^{n} \frac{\partial^2 f}{\partial x_j^2} = \sum_{j=1}^{n} f_{jj}.$$

Several times the Laplacian is expressed in terms of complex derivatives. See Section 11 of Chapter 1 and Section 10 of Chapter 4.

$\binom{n}{k}$ denotes the binomial coefficient $\frac{n!}{k!(n-k)!}$.

$\binom{m}{\alpha}$ denotes the multinomial coefficient $\frac{m!}{\alpha_1! \dots \alpha_n!}$. See below for more information on multi-indices.

Notations with similar or several uses

The letter d often means the exterior derivative. We use it in other ways that arise in calculus; for example, ds represents the arc-length element along a curve, dV denotes the volume form, etc. In the appendix, $d(x, y)$ denotes the distance between points x and y in a metric space.

The letter D typically denotes some kind of derivative. It often means the linear operator sending f to its derivative Df.

In Chapter 2, $\mathbf{H}_m$ denotes the space of harmonic homogeneous polynomials of degree m on $\mathbf{R}^n$.

In Chapter 2, Section 12 and in Chapter 3, Section 9, $H_n(x)$ denotes the n-th Hermite polynomial.

In Chapter 4, H_m is used often to denote a certain polynomial mapping from $\mathbf{C}^n \to \mathbf{C}^N$, defined and discussed in detail in Section 4 of Chapter 4.

In Chapter 1, σ_N denotes Cesàro means. In Chapter 4, σ_{N-1} denotes the volume form on S^{N-1}. Also note that σ^2 denotes variance.

The symbol α also has many uses; it can be a real number, a complex number, a multi-index, or a differential form.

In Chapter 3, Λ^s denotes the standard pseudo-differential operator of order s.

In Chapter 4, Section 6, $\Lambda^k(V^*)$ denotes the space of k-linear alternating forms on a vector space V.

The letter Γ is used to denote the Gamma function. Thus $\Gamma(x)$ denotes its value at x. We also write Γ for certain groups; in particular, Γ_f denotes the Hermitian-invariant group of a proper holomorphic map. See Chapter 5.

The letter γ has several uses. In Chapter 5 it often denotes an automorphism of the unit ball. In earlier chapters it sometimes denotes a curve.

Fourier series and Fourier transforms

$f * g$ denotes the convolution of f and g, either on the circle or on the real line.

$\hat{f}(n)$ denotes the n-th Fourier coefficient of a function on the circle; $\hat{f}(\xi)$ denotes the Fourier transform of a function on $\mathbf{R}$. Sometimes we write $\mathcal{F}(f)$ instead of $\hat{f}$.

$S_N(f)$ denotes the symmetric partial sums $\sum_{-N}^{N} \hat{f}(n) e^{inx}$ of a Fourier series. Approximate identities:

- F_N denotes the Fejér kernel.
- P_r denotes the Poisson kernel.
- In Chapter 1, $\mathcal{G}_t$ denotes the Gaussian kernel.
- We later use G_σ to denote the Gaussian with mean 0 and variance σ^2.
- The relationship between the parameter t in $\mathcal{G}_t$ and σ is given by $t = \frac{1}{2\sigma^2}$.

σ_N denotes the N-th Cesaro mean of a sequence $\{A_M\}$. Often A_M will itself denote the M-th partial sum of $\sum a_n$.

$\mathcal{S}$ denotes the Schwartz space of smooth functions of rapid decrease and $\mathcal{S}'$ denotes its dual space.

$\mathbf{S}^m$ denotes the collection of symbols of order m.

W^s denotes the L^2 Sobolev space of index s.

Hilbert spaces and linear algebra

$\mathcal{L}(\mathcal{H})$ denotes the bounded linear transformations on $\mathcal{H}$, and $\mathcal{L}(\mathcal{H}, \mathcal{H}')$ the bounded linear transformations from $\mathcal{H}$ to $\mathcal{H}'$.

I usually denotes the identity transformation.

L^* denotes the adjoint of L.

$\mathcal{N}(L)$ denotes the nullspace of L and $\mathcal{R}(L)$ denotes the range of L.

$V^\perp$ denotes the orthogonal complement of a subspace V.

$V \oplus W$ denotes the orthogonal sum of the subspaces V and W; if v and w are orthogonal vectors, the notation $v \oplus w$ denotes their sum $v + w$, but emphasizes the orthogonality.

det denotes the determinant.

λ usually means an eigenvalue.

Functions of several variables

$T_x(\mathbf{R}^n)$ denotes the tangent space of $\mathbf{R}^n$ at x. $T_x^*(\mathbf{R}^n)$ denotes the cotangent space.

$\otimes$ denotes the tensor product; we often write $z^{\otimes m}$ with a particular meaning. See Section 4 of Chapter 4.

$\mathbf{U}(n)$ denotes the group of unitary transformations on $\mathbf{C}^n$; often we write Γ for a finite subgroup of $\mathbf{U}(n)$.

Multi-index notation:

- A multi-index α is an n-tuple $\alpha = (\alpha_1, \ldots, \alpha_n)$ of non-negative integers.
- z^α is multi-index notation for $\prod_{j=1}^n (z_j)^{\alpha_j}$.
- When $\sum \alpha_j = m$, we write $\binom{m}{\alpha}$ for the multinomial coefficient $\frac{m!}{\alpha_1! \, \ldots \, \alpha_n!}$.
- We write $|z|^{2\alpha}$ for the product

$$\prod_j |z_j|^{2\alpha_j}.$$

- $\beta(\alpha)$ is the Beta function (of n indices) defined in Section 8 of Chapter 4.

Some special functions:

- $P_n(x)$ is the n-th Legendre polynomial.
- $T_n(x)$ is the n-th Chebyshev polynomial.
- $L_n(x)$ is the n-th Laguerre polynomial.
- $H_n(x)$ is the n-th Hermite polynomial. (Chapters 2 and 3)
- $H_m(z)$ denotes the polynomial map $z \mapsto z^{\otimes m}$. (Chapter 4)
- $\Gamma(x)$ denotes the Gamma function of x.

As noted above, we also use Γ to denote a finite subgroup of $\mathbf{U}(n)$ and we use Γ_f to denote the Hermitian-invariant group of a proper holomorphic mapping.

$f_{p,q}$ denotes a certain group-invariant polynomial, defined in formula (18) from Chapter 4.

If η is a differential form, then $d\eta$ denotes its exterior derivative.

$\alpha \wedge \beta$ denotes the wedge product of differential forms α and β.

Ψ denotes the differential form $\sum dz_j \wedge d\bar{z}_j$ or $\sum d\zeta_j \wedge d\bar{\zeta}_j$.

$\frac{\partial}{\partial \bar{z}_j}$ denotes the complex partial derivative defined in Section 4.6. The notations ∂ and $\bar{\partial}$ are also defined there. The one-dimensional versions are defined in Section 11 of Chapter 1.

$[L, K]$ denotes the commutator $LK - KL$ of operators L and K (their Lie bracket when they are vector fields).

References

Let us comment on the references.

Both [SS] and [Y] are intended for advanced undergraduate audiences and each has considerable overlap with this book. I feel that [SS] is a tremendous book; when I first began planning the course that led to this book, I imagined simply going through as much of [SS] as possible. Gradually I realized I wanted to include enough additional material (primarily parts of Chapter 2 and all of Chapter 4) to justify writing my own text. Chapter 4 in this book is primarily motivated by my own research, and hence does not appear in other texts.

The books [A] and [F1] are masterful treatments of complex and real analysis. The book [K] is considerably more advanced than this book, but can be consulted for more information on the material in Chapters 1 and 3. We barely touch probability, about which there are many texts. I particularly like [HPS], which does quite a bit with probability densities at a level similar to this book.

The books [E], [F2], [G], and [GS] are intended for different audiences than for those of this one, but they all provide tremendous physical insight and each overlaps with this book. I thank Mike Stone for giving me a copy of [GS]; I am amazed how good physicists are at mathematics! The book [E] covers a lot of territory, in part because the subject of Medical Imaging uses so much mathematics. It includes considerable discussion of Fourier transforms and distributions.

The book [HLP] is a classic. Its style contrasts with more abstract and more modern books, but it is packed with hard information and compelling examples. I also adore the book [S]. The Cauchy-Schwarz inequality is the most used result in my book, and the more one sees it used, the more one appreciates lots of mathematics.

The book [HH] is one of the best textbooks in multi-variable calculus; it includes differential forms, a proof of Stokes' theorem, and the usual large number of computational problems. My two books [D1] and [D2] are written at the undergrad level; [D2] contains all the complex analysis needed in this book, and [D1] has a large intersection with Chapter 2 of this book. I also refer to my papers [D3], [D4] and [D5] for some of the results in Chapter 4. The bibliographies in these papers also provide additional entries to the literature.

For the second edition I have included additional research papers mentioned in the new Chapter 5. The reader should also consult the bibliographies of these papers for a more complete picture.

[A] Ahlfors, Lars V., Complex Analysis: an introduction to the theory of analytic functions of one complex variable, Third edition, International Series in Pure and Applied Mathematics, McGraw-Hill Book Co., New York, 1978.

© Springer Nature Switzerland AG 2019
J. P. D'Angelo, *Hermitian Analysis*, Cornerstones,
https://doi.org/10.1007/978-3-030-16514-7_0

[AL] Andersen, E. and Lempert, L., On the group of holomorphic automorphisms of $\mathbf{C}^n$, Invent. Math. 110 (1992), no. 2, 371–388.

[B] Borwein, J. M., Hilbert's inequality and Witten's zeta-function, Amer. Math. Monthly 115 (2008), no. 2, 125–137.

[BER] Baouendi, M. Salah, Ebenfelt, Peter, and Rothschild, Linda Preiss, Real submanifolds in complex space and their mappings, Princeton Mathematical Series 47, Princeton University Press, Princeton, NJ, 1999.

[CD] Catlin, D. W and D'Angelo, J. P., A stabilization theorem for Hermitian forms and applications to holomorphic mappings, Math. Res. Lett. 3 (1996), no. 2, 149–166.

[CS] Cima, J. A. and Suffridge, T. J., Boundary behavior of rational proper maps, Duke Math. J. 60 (1990), no. 1, 135–138.

[CeS] Celik, M. and Sahutoglu, S., Compactness of the $\overline{\partial}$-Neumann operator and commutators of the Bergman projection with continuous functions, J. Math. Anal. Appl. 409 (2014), no. 1, 393–398.

[D1] D'Angelo, J., Inequalities from complex analysis, Carus Mathematical Monograph No. 28, Mathematics Association of America, Washington, DC, 2002.

[D2] D'Angelo, J. An introduction to complex analysis and geometry, Pure and Applied Mathematics Texts, American Math Society, Providence, 2011.

[D3] D'Angelo, J., Invariant CR mappings, pages 95–107 in Complex Analysis: Several complex variables and connections with PDEs and geometry (Fribourg 2008), Trends in Math, Birkhäuser-Verlag.

[D4] D'Angelo, J., A monotonicity result for volumes of holomorphic images, Michigan Math J. 54 (2006), 623–646.

[D5] D'Angelo, J., The combinatorics of certain group invariant maps, Complex Variables and Elliptic Equations, Volume 58, Issue 5 (2013), 621–634.

[DT] D'Angelo, J. and Tyson, J., An invitation to Cauchy-Riemann and sub-Riemannian geometries, Notices Amer. Math. Soc. 57 (2010), no. 2, 208–219.

[DX1] D'Angelo, J. and Xiao, M., Symmetries in CR complexity theory, Adv. Math. 313 (2017), 590–627.

[DX2] D'Angelo, J. and Xiao, M., Symmetries and regularity for holomorphic maps between balls, Math. Res. Lett., Vol. 25, no. 5 (2018), 1389–1404.

[Dar] Darling, R. W. R., Differential forms and connections, Cambridge University Press, Cambridge, 1994.

[E] Epstein, C., Introduction to the mathematics of medical Imaging, Prentice Hall, Upper Saddle River, New Jersey, 2003.

[F] Forstnerič, F., Extending proper holomorphic mappings of positive codimension, Invent. Math. 95 (1989), no. 1, 31–61.

[F1] Folland, G., Real analysis: Modern techniques and their applications, Second Edition, John Wiley and Sons, New York, 1984.

[F2] Folland, G., Fourier analysis and its applications, Pure and Applied Mathematics Texts, American Math Society, Providence, 2008.

[F3] Folland, G., Harmonic analysis in Phase Space, Princeton University Press, Princeton, 1989.

[G] Greenberg, M., Advanced engineering mathematics, Prentice Hall, Upper Saddle River, New Jersey, 1998.

[Gr] Grundmeier, D., Signature pairs for group-invariant Hermitian polynomials, Internat. J. Math. 22 (2011), no. 3, 311–343.

[GS] Goldbart, P. and Stone, M., Mathematics for physics, Cambridge Univ. Press, Cambridge, 2009.

[H] Hunt, Bruce J., Oliver Heaviside: A first-rate oddity, Phys. Today 65(11), 48(2012), 48–54.

[Ha] Haslinger, F., The $\overline{\partial}$-Neumann problem and Schrödinger operators, De Gruyter Expositions in Mathematics, 59, De Gruyter, Berlin, 2014.

[HT] Hill, C. D. and Taylor, M., Integrability of rough almost complex structures, J. Geom. Anal. 13 (2003), no. 1, 163–172.

[HLP] Hardy, G. H., Littlewood, J. E., and Polya, G, Inequalities, Second Edition, Cambridge Univ. Press, Cambridge, 1952.

[HH] Hubbard, J. H, and Hubbard, B., Vector calculus, linear algebra, and differential Forms, Prentice-Hall, Upper Saddle River, New Jersey, 2002.

[HPS] Hoel P., Port S., and Stone C., Introduction to probability theory, Houghton Mifflin, Boston, 1971.

[J] Jacobowitz, H., An introduction to CR structures, Mathematical Surveys and Monographs, 32, American Mathematical Society, Providence, RI, 1990.

[K] Katznelson, Y., An introduction to harmonic analysis, Dover Publications, New York, 1976.

[Kr] Krantz, S., Complex Analysis: the geometric viewpoint. Carus Mathematical Monographs No. 23, Mathematical Association of America, Washington, DC, 1990.

[Le] Lebl, J., Normal forms, Hermitian operators, and CR maps of spheres and hyperquadrics, Michigan Math. J. 60 (2011), no. 3, 603–628.

[Q] Quillen, Daniel G., On the representation of hermitian forms as sums of squares, Invent. Math. 5 (1968) 237–242.

[Ra] Rauch, J., Partial differential equations, Springer Verlag, New York, 1991.

[Ro] Rosenlicht, M., Introduction to analysis, Dover Publications, New York, 1968.

[S] Steele, J. Michael, The Cauchy-Schwarz master class, MAA Problem Book Series, Cambridge Univ. Press, Cambridge, 2004.

[SR] Saint Raymond, X., Elementary introduction to the theory of pseudodifferential operators, CRC Press, Boca Raton, 1991.

[SS] Stein, E. and Shakarchi, R., Fourier analysis: An introduction, Princeton Univ. Press, Princeton, New Jersey, 2003.

[SZ] Stroethoff, Karel and Zheng, De Chao, Toeplitz and Hankel operators on Bergman spaces, Trans. Amer. Math. Soc. 329 (1992), no. 2, 773–794.

[Wa] Warner, F., Foundations of differentiable manifolds and Lie groups, Springer-Verlag, New York, 1983.

[Y] Young, N., An introduction to Hilbert space, Cambridge Univ. Press, Cambridge, 1988.

Index

© Springer Nature Switzerland AG 2019
J. P. D'Angelo, *Hermitian Analysis*, Cornerstones,
https://doi.org/10.1007/978-3-030-16514-7_